MODIFIED ATMOSPHERE PACKAGING OF FOOD

ELLIS HORWOOD SERIES IN FOOD SCIENCE AND TECHNOLOGY

Editor-in-Chief: I. D. MORTON, Professor and formerly Head of Department of Food and Nutritional Science, King's College, London.
Series Editors: D. H. WATSON, Ministry of Agriculture, Fisheries and Food; and
M. J. LEWIS, Department of Food Science and Technology, University of Reading

MODIFIED ATMOSPHERE PACKAGING OF FOOD

Editors

B. OORAIKUL Dip.Agr., B.Food Tech., M.Tech., Ph.D.

M. E. STILES B.Sc., M.Sc., Ph.D.

both of Department of Food Science
Faculty of Agriculture and Forestry
University of Alabama, Edmonton, Canada

ELLIS HORWOOD

NEW YORK   LONDON   TORONTO   SYDNEY   TOKYO   SINGAPORE

First published in 1991 by
ELLIS HORWOOD LIMITED
Market Cross House, Cooper Street,
Chichester, West Sussex, PO19 1EB, England

A division of
Simon & Schuster International Group
A Paramount Communications Company

Printed and bound in Great Britain
by Bookcraft Ltd, Midsomer Norton, Avon

**Exclusive distribution by Van Nostrand Reinhold (International),
an imprint of Chapman & Hall, 2–6 Boundary Row, London SE1 8HN**

Chapman & Hall, 2–6 Boundary Row, London SE1 8HN, England

Van Nostrand Reinhold Inc., 115 5th Avenue, New York, NY10003, USA

Nelson Canada, 1120 Birchmont Road, Scarborough, Ontario M1K 5G4, Canada

Chapman & Hall Japan, Thomson Publishing Japan, Hirakawacho Nemoto Building, 7F,
1-7-11 Hirakawa-cho, Chiyoda-ku, Tokyo 102, Japan

Chapman & Hall Australia, Thomas Nelson Australia, 102 Dodds Street, South Melbourne,
Victoria 3205, Australia

Chapman & Hall India, R. Seshadri, 32 Second Main Road, CIT East,
Madras 600 035, India

Rest of the world:
Thomson International Publishing, 10 Davis Drive, Belmont, California 94002, USA

British Library Cataloguing in Publication Data

Modified atmosphere packaging of food
1. Food. Packaging
I. Ooraikul, B. II. Stiles, M. E.
664.092
ISBN 0–7476–0064–3

Library of Congress Cataloging-in-Publication Data

Modified atmosphere packaging of food / editors, B. Ooraikul, M. E. Stiles
p. cm. — (Ellis Horwood series in food science and technology)
Includes biblographical references and index.
ISBN 0–7476–0064–3
1. Food — Packaging. 2. Protective atmospheres. I. Ooraikul, B. II. Stiles, M. E. III. Series.
TP374.M63   1990
664'.092–dc20   90–5063
                                                                                          CIP

CONTRIBUTORS

MURRAY G. FIERHELLER, M.Sc., Food Scientist, Food Processing Development Centre, Alberta Agriculture, P.O. Box 3476, Leduc, AB, Canada T9E 6M2

BUNCHA OORAIKUL, Ph.D., Professor, Department of Food Science, University of Alberta, Edmonton, AB, Canada T6G 2P5

WILLIAM D. POWRIE, Ph.D., Professor, Department of Food Science, University of British Columbia, Vancouver, B.C., Canada V6T 1W5

BRENT J. SEKURA, Ph.D., Associate Professor, Department of Food Science, University of British Columbia, Vancouver, B.C., Canada V6T 1W5

MICHAEL E. STILES, Ph.D., Professor, Department of Food Science, University of Alberta, Edmonton, AB, Canada T6G 2P5

TABLE OF CONTENTS

PREFACE

At the 50th Anniversary Meeting of the Institute of Food Technologists the ten most significant innovations in food science developed during the past 50 years were named (Food Technology, September 1989). Among the "Top 10" innovations, controlled atmosphere packaging (CAP) for fruits and vegetables was listed 5th in order of importance. Of course, CAP is a forerunner of MAP (modified atmosphere packaging) in which a variety of food products are packaged under selective mixtures of atmospheric gases, but without the on-going maintenance (control) of the gas mixture. Development of packaging systems and films that are selectively permeable to specific gases has been the key element in the commercialization of controlled and modified atmosphere packaging of foods.

It may not be far from the truth to say that since then there has been an explosion of activities around MAP/CAP, especially in research and development into various aspects of this technology. The application of MAP to some bakery products, fresh fruits and salads and fresh meats and meat products has reached a significant level both in Europe and North America. The increasing consumer demand for fresh or near-fresh products and convenient, microwavable foods has added impctus to the growth of MAP/CAP technology. It is, therefore, timely that a comprehensive book that provides scientific background and practical applications of the technology should be written.

In writing this book a group of scientists and technologists who have had firsthand experience in various areas of MAP technology was assembled to contribute chapters in their areas of expertise. The book is a collection of up-to-date reviews of scientific and technological developments as well as some research data which have not yet been published. It is hoped that the text will stimulate the interests of both researchers and users of the technology. The dynamism of this field of technological development is such that it is impossible to keep current with all of the developments taking place. This book is an attempt to provide a reference base, but it will likely require updating with subsequent editions on a regular basis for the foreseeable future.

B. Ooraikul
M.E. Stiles
September 23, 1991

Chapter 1

INTRODUCTION: REVIEW OF THE DEVELOPMENT OF MODIFIED ATMOSPHERE PACKAGING

B. Ooraikul and M. E. Stiles
Department of Food Science, University of Alberta

1.1 Historical perspective

Man has always been interested in the preservation of food to extend its storage life. Preservation methods may have begun with sun drying in the tropics; or in the temperate zones, with sun drying during the summer, and atmospheric freezing during the winter. Natural fermentation is another means of food preservation used since ancient times, especially by the Asians. Presumably, most of the food so preserved was for household consumption as the scale of operation would have been too small to have commercial significance.

The modern food processing industry began in 1810 when Nicholas Appert invented the canning process. This was followed by the invention of a mechanical ammonia refrigeration system in 1875. In the 1960s Louis Pasteur had discovered the connection between microorganisms and food spoilage, and this put the development of food processing and preservation techniques on a firm scientific foundation. The two world wars and the Korean war during the first half of the 20th century accelerated these developments. This, together with the recognition of Food Science and Technology as an important discipline of study in its own right, did much for the progress of the food industry, to the point where the consumer interest is now focused more on quality rather than quantity of our food supply.

In recent years, with increasing concern about individual health and well-being as affected by food and the environment, greater attention has been given to how food is produced, processed, packaged, stored, distributed and consumed. Part of this has been exemplified by the revival of the "back-to-nature" sentiment advocated by a large number of people. Words such as "organically grown" or "natural" have increasing use in advertising of food products. Environmental pollution resulting from packaging materials such as metal cans, bottles and plastic materials is also a subject for study and debate. The frozen food industry has grown at a very rapid rate not only because of the pollution concerns about used cans and bottles, but mainly because freezing offers foods with quality closest to "fresh", compared with other processed foods. The advent of microwave ovens has even further accelerated the growth of frozen foods.

The current growing demand for "near-fresh" quality and shelf-stable products has spurred the development of many innovative processing and preservation techniques. Foremost among these are retortable pouches, aseptic packaging, controlled-atmosphere storage, and vacuum and modified-atmosphere packaging, the last being one of the most promising and most extensively studied at the present time.

Modified-atmosphere packaging (MAP) has been erroneously described as synonymous with controlled-atmosphere storage (CAS) or packaging (CAP). Sometimes vacuum packaging is considered to be a simple variation of MAP. In this book, the three will be regarded as different techniques. MAP is defined by Hintlian and Hotchkiss [25] as "the packaging of a perishable product in an atmosphere which has been modified so that its composition is other than that of air". This is in contrast with CAS, which involves maintaining a precisely defined atmosphere in the storage chamber, and vacuum packaging, which is the packaging of a product in a high-barrier package from which the air is removed. CAP may be regarded as a misnomer for MAP, since it is technically impossible or impractical to maintain the original atmosphere around the product once it is sealed inside a package. This is particularly true with fresh and nonsterile products, owing to their dynamic chemical and microbial nature, and the physical characteristics of the package and packaging material.

MAP is not a new technology. Its origins date back to 1922 when Brown [8], under the auspices of the Food Investigation Board of the Department of Scientific and Industrial Research in the United Kingdom, investigated the effect of different concentrations of O_2 and CO_2 at various temperatures on the germination and growth of fruit-rotting fungi. The investigation was done at the Imperial College of Science and Technology, London, England. The study revealed the effectiveness of CO_2, at 10% concentration or higher, in retarding the germination and growth of these fungi, especially at 10°C or less. He recommended the use of gas storage in combination with cold storage to extend the shelf life of fruits.

Inspired by these experiments, Killefer in 1930 [29] showed that pork and lamb kept fresh twice as long in 100% CO_2 at 4-7°C compared with storage in air at the same temperature. Subsequent experiments with beef [37], pork and bacon [9], and fish [16] produced similar improvements in keeping quality. Tomkins [64] and Moran *et al.* [37] reported that mould growth on meat could be retarded with as low as 4% CO_2 in the storage atmosphere. The higher the CO_2 concentration, and the lower the storage temperature, the more effective the inhibition. Haines [23] found that it took twice as long for some common meat bacteria to multiply to the same number if stored in 10% CO_2 at 0°C, rather than in air at the same temperature. Nitrogen was found to be ineffective compared with CO_2 in the extension of storage life of pork at 0°C [9].

The potential for use of gases in the storage atmosphere to improve keeping quality of food was recognized more than half a century ago, though how it worked was not fully understood. Initial commercial application of the technique, which began much later, was in the form of CAS. In this approach, the selected concentration of gases, chiefly CO_2, in the storage atmosphere is carefully maintained at all times. CAS has been used, for example, to delay physiological changes such as ripening in fruits and vegetables [65], and to transport chilled beef carcases from New Zealand and Australia [51].

However, the science and technology of gas preservation of food evolved slowly, especially in North America, where refrigeration and highly developed transportation and distribution systems are available. The previously "lightly restricted use" of food preservatives also did little to encourage the development of gas preservation of foods. In the last two decades, with the rising costs of raw food products, labour and energy, and the tightening controls on some preservatives and additives, interest in the use of gases for food preservation has been renewed.

The availability of packaging films with a wide range of physical characteristics and versatile packaging equipment has enabled attention to be turned to MAP of food in small, convenient retail or distribution units. In the MAP system, the cumbersome continuous control of the atmosphere surrounding the product is eliminated. This makes it much cheaper for a large-scale operation, and the process became practical for application at the retail level.

By the 1960s, vacuum packaging had become a common practice for many dry products as well as fresh meat [14]. Experimental MAP products appeared in Europe, first with baked goods such as bread and cakes packaged with CO_2 [52, 53]. Research and development activities on MAP were accelerated during the 1970s and 1980s, resulting in successful commercial applications. A large number of studies on MAP were conducted during this period. Examples of these studies include beef, pork and poultry [12, 13, 15, 22, 59], fish and other seafoods [2, 4, 7, 10, 11, 35, 45, 50, 61, 66, 67], cereals and nuts [26, 43], fruits and vegetables [28, 30, 31, 48], and bakery products [1, 40, 41, 42, 52, 53, 55, 56, 57, 58]. Gases studied include: CO_2, N_2, ethylene oxide, propylene oxide, sulphur dioxide (SO_2), carbon monoxide (CO) and ozone [14, 31].

Whether or not the food industry is adequately prepared for MAP it has emerged as an important storage technology. Its influence on food quality and cost, especially on fresh and near-fresh products, will be extremely important in the years to come. How well it will serve, and whether its full potential will be realized, depends on how thoroughly food scientists understand various aspects of this promising technology, and how it is applied. MAP has still to be fully developed. It is hoped that this book will serve as a catalyst for this development.

1.2 Perishability of food

Biological materials, from which most food is derived, will deteriorate or spoil, given adequate time, especially when kept under unfavourable conditions. The deterioration is due essentially to microbial degradation and/or chemical reactions, producing changes in the product which may lower product quality, causing a potential health hazard and, ultimately, a substantial economic loss.

1.2.1 Microbial problems

The microbial problems of foods can be predicted from a knowledge of their a_w, pH, oxidation-reduction potential, inherent or extrinsic inhibitory substances, processing parameters, storage temperature and competitive microflora. The effect of these factors on microbial spoilage and safety of foods is reviewed in many texts on food microbiology, for example Banwart [3], and the ICMSF [27] book on *Microbial Ecology of Foods*, Volume I, *Factors Affecting the Life and Death of Microorganisms*. Some foods have natural protective barriers to microbial contamination, such as the shell and membranes of eggs and the protective outer surfaces of many fruits, vegetables, grains and nuts. These foods become far more susceptible to microbial spoilage when these protective layers are damaged or removed.

Fluid milk left at room temperature has a storage life of less than one day. With pasteurization and refrigeration the manufacturer's storage life on fluid milk is generally 10 d. In contrast, fresh carcase beef has a refrigerated storage life of 28 d or more. Most contamination is on the exterior of the carcase, which is dry and the surface area to weight ratio is low. As the carcase is broken into wholesale (primal) or retail cuts, the surface area to weight ratio is increased. Under aerobic, refrigerated conditions of storage, retail cuts of meat generally spoil within 4 d. Wholesale cuts of beef, vacuum packaged in gas-impermeable plastic film, have a refrigerated storage life of 28 d or more. Lamb carcases packaged under carefully controlled anaerobic conditions have a storage life at -1°C of up to 12 weeks [21]. In comminuted meats the microorganisms are distributed throughout the meat, and refrigerated storage life is even more dramatically reduced. However, extended storage life can be achieved even with refrigerated storage of vacuum-packaged product in a gas-impermeable plastic film.

Temperature is the key factor influencing storage life and safety of fresh and processed foods. As the temperature approaches 60°C the maximum temperature at which microorganisms grow in foods is achieved and exceeded. At 60 to 70°C, vegetative cells are killed. The efficiency of cell death depends on the physical environment of the food, microbial load and duration of heat treatment. The only likely survivors of efficient heat treatment are bacterial spores (endospores) which require drastic heating, up to 121°C, for their destruction. As the temperature is decreased to points above freezing, the growth of microorganisms is slowed or stopped. For example, bacteria which have a generation time of 15-20 min at optimum growth temperature will only double once every 6-8 h or even every 100 h at -1 to 7°C [27].

Many public health agencies have developed charts similar to that shown in Figure 1.1, designating 4 to 60°C as the *danger zone* for microbial growth. This is predicated on the death of vegetative cells above 60°C and the fact that few pathogenic microorganisms grow at 4°C. Psychrotrophs are those microorganisms that grow at about 0°C, but their growth optima and maxima do not fit the definition of psychrophiles [18]. Psychrotrophs capable of growth in foods include many species in a wide range of genera of bacteria, yeasts and moulds [27]. Under aerobic conditions the principal psychrotrophic spoilage microflora is composed of gram negative, nonsporeforming rod-shaped bacteria, including *Flavobacterium, Moraxella* and *Pseudomonas* spp. [17]. Under anaerobic or elevated CO_2 storage conditions the principal psychrotrophic spoilage microflora includes lactic acid bacteria such as *Lactobacillus* and *Leuconostoc* spp., and/or *Enterobacter* and *Serratia* spp. [6].

The realization that *Clostridium botulinum* type E and nonproteolytic types B and F grow at 3.3°C [24] destroyed the myth that pathogenic bacteria were not capable of growing in refrigerated foods. Psychrotrophic enteric pathogens have also been reported, including *Aeromonas hydrophila* [62] and *Yersinia enterocolitica* [49]. Of even greater concern is the high incidence of *Listeria monocytogenes* in foods, and its ability to grow at 0°C over a wide pH range, and to survive in many processed foods, especially cheeses [34]. Although human listeriosis is not a new disease, the severity of its symptoms and its confirmation as a foodborne bacterial pathogen has made this one of the most important foodborne pathogens of the 1980s and 1990.

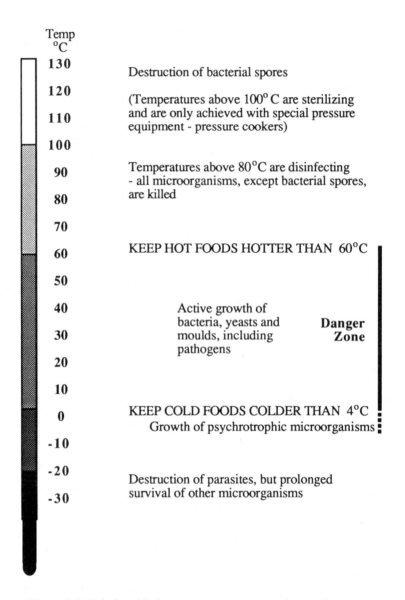

Figure 1.1. Relationship between temperature and growth
and survival of microorganisms.

MAP foods have great promise for marketing of prepared and semi-prepared foods to meet consumer demands for minimally processed, "fresh" foods. Predictive microbiology of these foods suggests potential hazards, notably the hazard of *C. botulinum* growth. Public health authorities express concern about *C. botulinum* and other psychrotrophic pathogens and the fact that the safety of minimally processed refrigerated foods relies on the fine thread of adequacy of refrigeration. However, MAP foods are not new to the North

American food market. Vacuum-packaged sliced luncheon meats have been marketed over the past 25 years without apparent incident with *C. botulinum* or other pathogens. However, the presence of inhibitory agents used in the manufacture of these meats and the growth of the competitive lactic acid-producing microflora acts as a protective hurdle for safety [60]. In cooked, unpreserved foods part of this protective hurdle is removed, and in some foods refrigeration could be the only hurdle for microbial safety [36]. Clearly, the microbiology of this new and emerging generation of foods requires careful analysis and research to assure safety and extended storage life of the foods.

1.2.2 Physicochemical problems

Physicochemical changes in food during storage can result in deterioration of its quality and, in more severe cases, may render the product unacceptable or even pose health hazards to consumers. The types of changes and the rate at which they occur depend on the physical and chemical nature of the food, how it has been handled and the environment in which it is stored. Fresh foods may have added problems caused by the activity of naturally occurring enzymes. These activities are normally eliminated or minimized in processed food.

1.2.2.1   Fruits and vegetables

After harvest most of the physiological changes in fruits and vegetables are related to oxidative metabolism. Fruits and vegetables vary greatly in their chemical composition but, commonly, they are high in water (70-98%) and carbohydrate, which is mainly in the form of cellulose, starch and sugars, and low in lipids. Respiration causes changes in these components which result in alteration of appearance, odour, flavour, and texture. Starch is hydrolysed to sugars such as sucrose, glucose and fructose, increasing the sweetness and appealing aroma of the fruit while its texture becomes softer. Softening of fruit tissue is also associated with a decrease in cellulose and insoluble pectins through enzymatic activities during storage. Parallel to this is a drop in acid content and a slight increase in protein and some amino acids and aromatic compounds [5, 38, 44].

In most fruits, the colour changes after harvest through loss of chlorophyll and synthesis of carotenoid pigments, resulting in a progressive change from green to yellow, red or orange as the fruit ripens. These changes are influenced by storage temperature, maturity and variety of fruits. Generally, the changes are more dramatic in climacteric varieties; thus the fruits reach senescence sooner than the nonclimacteric varieties.

In vegetables, similar changes occur. For example, potatoes stored at low temperature accumulate reducing sugars through metabolism of starch. Ascorbic acid and some amino acids are lost during storage, while protein may increase slightly. Part of chlorophyll is converted to pheophytin, which is olive green or brown; while the amount of carotenoids increases, turning the colour yellow or orange, making the vegetable less attractive and visually unacceptable. There is also a toughening of vegetable tissue with increasing length of storage. This is due to the loss of turgor as well as increases in hemicellulose and lignin.

Another problem that can develop during storage is enzymatic browning. This occurs if fruit or vegetable tissue is damaged. The damage may be caused by disease, insects or other pests, by freezing, or through mechanical handling during or after harvest. When the injured fruit or vegetable is exposed to air the injured part undergoes rapid darkening due to the action of polyphenolase and peroxidase enzymes. Without preventive measures these changes render produce unacceptable within a relatively short period of time. Therefore, to extend the storage life of produce, control measures must be used, with primary aims to reduce the respiration rate and minimize activity of spoilage microorganisms.

The most common method to prolong storage life of fresh produce is cool storage at high humidity. Another common method is CAS, where the environment in which the produce is stored is continually controlled with respect to temperature, relative humidity and type of gas atmosphere. CAS has been used quite extensively, especially with fruits. Hypobaric storage, where gases in the storage environment are removed to lower pressure has been introduced, but the technique is uneconomical for use with most produce [54].

Most of these techniques are suitable for bulk storage, but usually not for retail units. However, once the produce leaves storage, normal rates of respiration and microbial activities resume to limit retail shelf life. Modified-atmosphere (MA) packaging and storage, where the produce can be packaged under an MA in barrier plastics in sizes suitable for shipping or retail, appears to be the most suitable method when the produce is to remain on retail shelves for an extended period, or when it is to be shipped to distant markets.

MAP of fresh cut vegetables by Liquid Air Corp. has found commercial success in Europe [47]. By washing, cutting or shredding and packaging in oxygen-permeable film in an atmosphere of N_2 or an appropriate mixture of CO_2/N_2, potatoes, lettuce or cauliflower can be stored for 15-35 d, compared with 6-8 d without MAP.

1.2.2.2   Meat

The major factor influencing storage life of fresh meat is microbial growth. However, during storage physicochemical changes take place which may affect the organoleptic quality of meat. A series of chemical reactions occur after slaughter which ultimately determine meat quality. Of prime importance is the onset of *rigor mortis* during which skeletal muscles stiffen and meat is tough. The time for onset of *rigor mortis* varies with the size of the animals and pre-slaughter conditions from 2 to 24 h, beef being 10-24 h, pork 4-18 h and chicken 2-4 h [5]. The toughness of meat during rigor is the result of the stiffening of the protein matrix of the muscles. As the rigor passes the muscles become soft and pliable again.

Meat is not normally marketed immediately after slaughter. *Rigor mortis* must have passed before it is suitable for consumption. During ageing the tenderness and aroma of meat improve. Beef requires as long as 14 d to age at 0°C or 4 d at 18°C; other red meats and poultry require shorter times. Proteolytic enzymes play an important role in ageing, which results in slight increases in pH and water-holding capacity of meat, with concomitant improvement in tenderness and juiciness [5, 33, 38, 46].

Storage-life extension of retail cuts of fresh meat is not only economically desirable, but for centrally prepared retail cuts of meat it may also be necessary for improvement of eating quality. However, problems of microbial spoilage and fat oxidation must be overcome. These problems are intensified at retail level as the meat has to be cut, chopped or minced into portion sizes, exposing greater surface areas to microbial contamination and oxidative reactions.

The most common preservation methods for fresh meat are cooling or freezing. At 0°C the storage life of carcase meat is 3-6 weeks, during which microbial growth and lipid oxidation are retarded. Weight loss is an economic problem which can be reduced if high relative humidity is maintained, but under such conditions microbial growth is enhanced. Freezing and storage of meat at -18° to -20°C and 90% RH will extend its storage life to 9-15 months. The quality of frozen meat depends on when it is frozen and the rate at which it is frozen and thawed. However, freezing and freezer storage of meat is an expensive method of preservation.

Controlled-atmosphere (CA) and hypobaric storage of meat have been used with some success. These methods are costly and they are not readily applicable to portion packs for retail markets. MAP has a demonstrated potential for storage of wholesale or retail cuts of meat. The principal limiting factor for its use with retail meat cuts, especially beef, is the maintenance of myoglobin in the reduced form, so that the meat is a dark (purple) red colour. During MAP storage the ageing process proceeds naturally, and weight loss is eliminated except as moisture purge in the package.

1.2.2.3   Fish and marine products

Fish is generally far more perishable than meat. With a final pH $\geq$ 6.0 and readily available nonprotein and protein nitrogen it is an excellent substrate for microbial growth. Fish lipids are largely unsaturated and highly susceptible to oxidative deterioration. Fish do not have an anal sphincter muscle so soft faecal matter is readily released after harvest; as a result fish readily become contaminated after harvest. Furthermore, the outer mucous layer of the skin absorbs water and harbours bacteria. Spoilage bacteria are almost entirely on the surface of fish, so the surface to volume ratio markedly affects the rate of spoilage. Spoilage rate increases as fish progress from whole to gutted, to fillets or steaks, to minced fish. The gills of fish are generally highly contaminated and susceptible to microbial growth [27].

The situation with shellfish is very different. Although shrimp die when they are harvested and spoilage conditions are similar to those outlined for fish, crustaceans such as lobster and crab are marketed live, as are molluscs such as oysters, clams and mussels. Therefore, the quality of the water source from which they were harvested and the captive care have a direct influence on the safety of these foods for consumption. Extended storage quality *per se* is not an issue.

The natural spoilage microflora of fish and other marine animals is adapted to growth at low temperatures and includes psychrotrophic and psychrophilic bacteria. The predominating aerobic spoilage microflora is generally *Pseudomonas* spp., but species of *Acinetobacter, Alteromonas, Flavobacterium* and *Moraxella* are also involved. In addition to the spoilage microflora, special consideration must be given to the survival and growth of *C. botulinum* types B and E, and other pathogens including *Vibrio parahaemolyticus* and other vibrios associated with human infection [39, 63, 66]. The solution to the problems of microbial spoilage of fish has been to can or freeze the product. Fish and other marine foods may be preserved by other means, and MAP can be expected to be of value to control the natural aerobic, gram negative rod-shaped microflora that generally causes spoilage of fish. Unfortunately for extended fish storage, but fortunately for public health, researchers discovered that the hurdle to botulinal safety of MAP fresh fish is good refrigeration and that the margin of safety is low [19, 20].

1.2.2.4   Other foods

MAP of foods has great potential for application in many other areas of food processing. Notable developments have already occurred, as outlined earlier in this chapter. In most circumstances, the efficiency of MAP storage depends on reliable refrigeration of the product throughout the marketing chain. However, in the case of bakery products, shelf-stable products have been developed [1, 40, 41, 42, 52, 53, 55, 56, 57, 58].

Extensive markets have been established for MAP sandwiches with distribution systems requiring 28-35 d storage life for successful marketing. In a study of MAP sandwiches [36], it was shown that untrained panellists considered processed meat sandwiches acceptable up to 35 d storage at 4°C; roast beef sandwiches for 28-35 d; and hamburgers for 14-35 d depending on the presence or absence of oxygen in the gas atmosphere, respectively. The production and marketing of such products requires re-evaluation of microbial criteria by producers and public health authorities. Furthermore, producers must exercise exceptional care in selecting high-quality food ingredients, in sanitation and hygiene, and in refrigeration from time of production to point of sale.

The potential for MAP foods is best illustrated by the French development of *sous vide* (under vacuum) foods. Unlike many other foods, development of *sous vide* originated in the food service industry rather than the retail food trade.

1.3   Conclusions

Although our knowledge of the profound effects of modified gas atmospheres on the storage life of foods dates back some 100 years, the possibility of applying this knowledge to our modern food technology has only become possible with the advances in polymer science that have made available a wide range of plastic films suitable for storage of foods. Initially, the CA or MA storage of foods was only applicable to bulk foods, for example transoceanic shipments of meat [67] and controlled ripening of fruits and vegetables [32]. CA and/or MA storage is not a panacea for food storage, and many learned that errors could be economically disastrous. Nonetheless, this technology, in concert with careful temperature control, has contributed significantly to the efficiency of our food system.

An efficient food system implies a food supply that is readily available, economically attainable, convenient to use, nutritious, acceptable, and safe. CA storage has had a profound influence on the post-harvest storage and the extended availability of many fruits and vegetables. The use of gas impermeable plastic film has allowed MA (though not CA) storage to be applied to smaller units of food. This has been most widely used for fresh wholesale meats, especially beef, and for processed, sliced luncheon meats. This has significantly affected the convenience of beef retailing and the convenience and availability of processed meats for the consumer.

When vacuum packaging was introduced for fresh wholesale beef and sliced luncheon meats it was not without expressions of concern about their safety. After all, harsh lessons had been learned during the development of cooked sausage meats and low-acid canned foods [24]. Vacuum-packaged meats have been marketed in excess of 20 years. Now, as consumer demands increase for fresh (implying *never frozen*) foods, opportunities for MAP foods have increased. The 35-day storage life "fresh" sandwiches, shelf-stable English-style crumpet, *sous vide* preservation of fine dining meals, salads and fresh fruits have all made their debut, in general without the care and precaution desired by the public health officials. However, this has occurred at a time when botulism has been reported in hitherto unsuspected foods, such as garlic in oil, fermented bean curds and caviar. Furthermore, previously unknown bacteria and some known but unsuspected bacteria have been shown to be agents of foodborne illness. What degree of hazard does this new generation of foods really represent?

The diversity of plastic films, ingenious combinations of gases, double packaging and, doubtless, many new developments create untold advantages and opportunities for MAP foods. At the same time, if refrigeration is the ultimate hurdle for safety assurance of MAP foods, do these foods represent a sufficient advantage that refrigeration during transportation and retail will be improved and temperature abuse indicators will be included on MAP packages? Is the day at hand that the manufacturer and/or the retailer will accept responsibility for the food up to (and beyond?) sale of the food to the consumer? MAP foods could represent a "revolution in the making" in food processing and marketing.

References

1. Aboagye, N.Y., Ooraikul, B., Lawrence, R. and Jackson, E.D. 1986. Energy costs in modified atmosphere packaging and freezing processes as applied to a baked product. Proceedings of the Fourth International Congress of Engineering and Food. Edmonton, AB. Food Engineering and Process Applications Vol. 2 Unit Operations. Le Maguer, M and Jelen, P. (eds.). Elsevier Appl. Sci. Publishers, London. pp. 417-425.
2. Banks, H., Ranzell, N. and Finne, G. 1980. Shelf-life studies on carbon dioxide packaged finfish from the Gulf of Mexico. J. Food Sci. 45: 157-162.
3. Banwart, G.J. 1989   Basic Food Microbiology. 2nd edition. Van Nostrand Reinhold, New York.
4. Barnett, H.J., Stone, F.E., Roberts, G.C., Hunter, P.J., Nelson, R.W. and Kwok, J. 1982. A study in the use of a high concentration of CO_2 in a modified atmosphere to preserve fresh salmon. Mar. Fish. Rev. 44(3): 7-11.
5. Belitz, H.-D. and Grosch, W. 1986. Food Chemistry. Translated by D. Hadziyev. Springer Verlag, Berlin.
6. Blickstad, E., Enfors, S.-O. and Molin, G. 1981. Effect of high concentrations of CO_2 on the microbial flora of pork stored at 4°C and 14°C, pp. 345-357. *In* Psychrotrophic Microorganisms in Spoilage and Pathogenicity. Roberts, T.A., Hobbs, G., Christian, J.H.B. and Skovgaard, N. (eds.). Academic Press, New York.
7. Bremner, H.A. and Statham, J.A. 1987. Packaging in CO_2 extends shelf-life of scallops. Food Technol. Aust. 39: 177-179.
8. Brown, W. 1922. On the germination and growth of fungi at various temperatures and in various concentrations of oxygen and carbon dioxide. Ann. Bot. 36: 257-283.
9. Callow, E.H. 1932. Gas storage of pork and bacon. 1. Preliminary experiments. J. Soc. Chem. Ind. 51: 116T-119T.
10. Cann, D.C., Smith, G.L. and Houston, N.G. 1983. Further studies on marine fish stored under modified atmosphere packaging. Torry Research Station, Aberdeen, U.K.
11. Chen, H.C., Myers, S.P., Hardy, R.W. and Biede, S.L. 1984. Color stability of astaxanthin pigmented rainbow trout under various packaging conditions. J. Food Sci. 49: 1337-1340.
12. Christopher, F.M., Vanderzant, C., Carpenter, Z.L. and Smith, G.C. 1979. Microbiology of pork packaged in various gas atmospheres. J. Food Prot. 42: 323-327.
13. Clark, D.S. and Lentz, C.P. 1969. The effect of carbon dioxide on the growth of slime producing bacteria on fresh beef. Can. Inst. Food Sci. Technol. J. 2: 72-75.
14. Clark, D.S. and Takács, J. 1980. Gases as preservatives. *In:* Microbial Ecology of Foods. Academic Press, N.Y.
15. Clark, D.S. Lentz, C. and Roth, L.A. 1976. Use of carbon monoxide for extending shelf-life of prepackaged fresh beef. Can. Inst. Food Sci. Technol. J. 9: 114-117.
16. Coyne, F.P. 1932. The effect of carbon dioxide on bacterial growth with special reference to the preservation of fish. J. Soc. Chem. Ind. 51: 119T-121T.

17. Dainty, R.H., Shaw, B.G. and Roberts, T.A. 1983. Microbial and chemical changes in chill-stored red meats, p.151. *In* Food Microbiology - Advances and Prospects. Roberts, T.A. and Skinner, F.A. (eds.). Academic Press, Toronto.
18. Eddy, B.P. 1960. The use and meaning of the term "Psychrophilic". J. Appl. Bacteriol. 23: 189-190.
19. Eklund, M.W. 1982. Significance of *Clostridium botulinum* in fishery products preserved short of sterilization. Food Technol. 36(12): 107-112, 115.
20. Genigeorgis, C.A. 1985. Microbial and safety implications of the use of modified atmospheres to extend the storage life of fresh meat and fish. Int. J. Food Microbiol. 1: 237-251.
21. Gill, C.O. and Penney, N. 1985. Modification of in-pack conditions to extend the storage life of vacuum packaged lamb. Meat Sci. 14: 43-60.
22. Gray, R.J.H., Elliott, P.H. and Tomlins, R.I. 1984. Control of two major pathogens on fresh poultry using a combination potassium sorbate/carbon dioxide packaging treatment. J. Food Sci. 49: 142-145, 179.
23. Haines, R.B. 1933. The influence of carbon dioxide preservation on the rate of multiplication of certain bacteria as judged by viable counts. J. Soc. Chem. Ind. 52: 13T-17T.
24. Hauschild, A.H.W. 1989. *Clostridium botulinum,* pp. 111-189. *In* Foodborne bacterial pathogens, M.P. Doyle (ed.) Marcel Dekker, Inc., New York.
25. Hintlian, C.B. and Hotchkiss, J.H. 1986. The safety of modified atmosphere packaging: a review. Food Technol. 40(12): 70-76.
26. Holaday, C.E., Pearson, J.L. and Slay, W.O. 1979. A new packaging method for peanuts and pecans. J. Food Sci. 44: 1530-1533.
27. ICMSF. 1980 Microbial Ecology of Foods, Volume I. Factors Affecting Life and Death of Microorganisms. International Commission on Microbiological Specifications for Foods. Academic Press, New York.
28. Kader, A.A. 1986. Biochemical and physiological basis for effects of controlled and modified atmospheres on fruits and vegetables. Food Technol. 40(5): 99-104.
29. Killefer, D.H. 1930. Carbon dioxide preservation of meat and fish. Ind. Eng. Chem. 22: 140-143.
30. Kohno, S. 1985. Research on quality presentation technique for cut vegetables. National Food Research Institute of Ministry of Agriculture, Japan, as cited by R.W. Gates in the Proceedings of the Third International Conference on Controlled/Modified Atmosphere/Vacuum Packaging, Sept. 16-18, 1987, Itasca, IL, published by Schotland Business Research, Inc., Princeton, NJ., pp. 266-291.
31. Kramer, A., Solomos, T., Wheaton, F., Puri, A., Sirivichaya, S., Lotem, Y., Fowke, M. and Ehrman, L. 1980. A gas-exchange process for extending the shelf life of raw foods. Food Technol. 34(7): 65-74.
32. Labuza, T.P. and Breene, W.M. 1989. Applications of "active packaging" for improvement of shelf-life and nutritional quality of fresh and extended shelf-life foods. J. Food Proc. Preserv. 13: 1-69.

33. Lawrie, R.A. 1974. Meat Science. 2nd. ed. Pergamon Press, Oxford.
34. Lovett, J. 1989. *Listeria monocytogenes*, pp. 283-310. *In* Foodborne bacterial pathogens, Doyle, M.P. (ed.). Marcel Dekker, Inc., New York.
35. Matches, J.R. and Leyrisse, M.E. 1985. Controlled atmosphere storage of spotted shrimp (*Pandalus platyceros*). J. Food Prot. 48: 709-711.
36. McMullen, L. and Stiles, M.E. 1989. Storage life of selected meat sandwiches at 4°C in modified gas atmospheres. J. Food Prot. 52: 792-798.
37. Moran, T., Smith, E.C. and Tomkins, R.G. 1932. The inhibition of mould growth on meat by carbon dioxide. J. Soc. Chem. Ind. 51: 114T-116T.
38. Myer, L.H. 1960. Food Chemistry. Reinhold Publ. Corp., New York.
39. Oliver, J.D. 1989. *Vibrio vulnificus*, pp. 569-600. *In* Foodborne Bacterial Pathogens. Doyle, M.P. (ed.). Marcel Dekker, Inc., New York.
40. Ooraikul, B. 1982. Gas packaging for a bakery product. Can. Inst. Food Sci. Technol. J. 15(4): 313-315.
41. Ooraikul, B., Smith, J.P. and Jackson, E.D. 1983. Evaluation of factors controlling spoilage microbiota from gas packaged crumpets with response surface methodology. Research in Food Science and Nutrition. Vol. 2. Basic Studies in Food Science. McLoughlin, J.V. and McKenna, B.M. (eds.). Proceedings of the Sixth International Congress of Food Science and Technology, Dublin, September 18-23, 1983, pp. 92-93.
42. Ooraikul, B., Koersen, W.J., Smith, J.P., Jackson, E.D. and Lawrence, R. 1988. Modified atmosphere packaging of selected bakery products. Proceedings of the 7th World Congress of Food Science and Technology 1987. Singapore.
43. Ory, R.L., Delucca, A.J., St. Angelo, A.J. and Dupuy, H.P. 1980. Storage quality of brown rice as affected by packaging with and without carbon dioxide. J. Food Prot. 43: 929-932.
44. Pantastico, Er. B. 1975. Postharvest Physiology, Handling and Utilization of Tropical and Subtropical Fruits and Vegetables. The AVI Publ. Co., Inc., Westport, CT.
45. Passy, N., Mannheim, C.H. and Cohen, D. 1983. Effect of a modified atmosphere and pretreatments on quality of chilled fresh water prawns (*Macrobachium rosenbergii*). Lebensm. Wiss. u. Technol. 16: 224-229.
46. Price, J.F. and Schweigert, B.S. 1987. The Science of Meat and Meat Products, 3rd ed. Food and Nutrition Press, Inc. Westport, CT.
47. Rice, J. 1987. Gas-pack technology requires close attention to tricky variables. Food Proc. 48(6): 64-65.
48. Rij, R.E. and Stanley, R.R. 1987. Quality retention of fresh broccoli packaged in plastic films of defined CO_2 transmission rates. Packaging Technol. May/June, 1987: 22.
49. Schiemann, D.A. 1989. *Yersinia enterocolotica* and *Yersinia pseudotuberculosis*, pp. 601-672. *In* Foodborne bacterial pathogens, Doyle, M.P. (ed.). Marcel Dekker, Inc., New York.

50. Scott, D.N., Fletcher, G.C. and Hogg, M.G. 1986. Storage of snapper fillets in modified atmospheres at -1°C. Food Technol. Aust. 38: 234-238.

51. Scott, W.J. 1936. The growth of organisms on ox muscle. III. The influence of 10 per cent carbon dioxide on rates of growth at -1°C. J. Council Sci. Ind. Res. 11: 266.

52. Seiler, D.A.L. 1965. Factors influencing the mould free shelf life of cake with particular reference to the use of antimould agents. Brit. Baking Ind. Res. Assoc. Report No. 81.

53. Seiler, D.A.L. 1978. The microbiology of cake and its ingredients. Food Trade Rev. 48: 339-344.

54. Shewfelt, R.L. 1986. Postharvest treatment for extending the shelf life of fruits and vegetables. Food Technol. 40(5): 70-80, 89.

55. Smith, J.P., Jackson, E.D. and Ooraikul, B. 1983. Storage study of a gas-packaged bakery product. J. Food Sci. 48: 1370-1371, 1375.

56. Smith, J.P., Ooraikul, B. and Jackson, E.D. 1984. Linear programming: a tool in reformulation studies to extend the shelf life of English-style crumpets. Food Technol. Aust. 36: 454-459.

57. Smith, J.P., Ooraikul, B., Koersen, W.J., Jackson, E.D. and Lawrence, R.A. 1986. Novel approach to oxygen control in modified atmosphere packaging of bakery products. Food Microbiol. 3: 315-320.

58. Smith, J.P., Ooraikul, B., Koersen, W.J., Jackson, E.D. and Lawrence, R.A. 1987. Shelf life extension of a bakery product using ethanol vapor. Food Microbiol. 4: 329-337.

59. Spahl, A., Reineccius, G. and Tatini, S. 1980. Storage life of pork chops in CO_2-containing atmospheres. J. Food Prot. 44: 670-673.

60. Steele, J.E., and Stiles, M.E. 1981. Food poisoning potential of artificially contaminated vacuum packaged sliced ham in sandwiches. J. Food Prot. 44: 430-434.

61. Steir, R.F., Bell, L., Ito, K.A., Shafer, B.D., Brown, L.A., Seeger, M.L., Allen, B.H., Porcuna, M.N. and Lerke, P.A. 1981. Effect of modified atmosphere storage on C. botulinum toxigenesis and the spoilage microflora of salmon fillets. J. Food Sci. 46: 1639-1642.

62. Stelma, G.N. Jr. 1989. Aeromonas hydrophila, pp. 1-19. In Foodborne Bacterial Pathogens. Doyle, M.P. (ed.). Marcel Dekker, Inc., New York.

63. Stiles, M.E. 1989. Less recognized or presumptive foodborne pathogenic bacteria, pp. 673-733. In Foodborne Bacterial Pathogens. Doyle, M.P. (ed.). Marcel Dekker, Inc., New York.

64. Tomkins, R.G. 1932. The inhibition of the growth of meat attacking fungi by carbon dioxide. J. Soc. Chem. Ind. 51: 261T-264T.

65. Tomkins, R.G. 1962. The conditions produced in film packages by fresh fruits and vegetables and the effect of those conditions on storage life. J. Appl. Bacteriol. 25: 290-307.

66. Twedt, R.M. 1989. Vibrio parahaemolyticus, pp. 543-568. In Foodborne bacterial pathogens, Doyle, M.P. (ed.). Marcel Dekker, Inc., New York.

67. Villemure, G., Simard, R.E. and Picard, G. 1986. Bulk storage of cod fillets and gutted cod (Gadus morhua) under carbon dioxide atmosphere. J. Food Sci. 51: 317-320.

68. Wolfe, S.K. 1980. Use of CO- and CO_2-enriched atmospheres for meats, fish and produce. Food Technol. 34(3): 55-58, 63.

Chapter 2

SCIENTIFIC PRINCIPLES OF CONTROLLED/MODIFIED ATMOSPHERE PACKAGING

M.E. Stiles
Department of Food Science, University of Alberta

2.1 Introduction

Control of the spoilage of food has challenged the ingenuity of humans since the earliest of times. Concentrating and drying foods by natural (and artificial) means was the mainstay of food preservation. Fermentation also became a popular means of food preservation, relying on the anaerobic environment and the growth of microorganisms to produce inhibitory substances that ultimately preserve the food. In the developed world, refrigeration and freezing have become major factors in the extension of the storage life of foods. Under refrigeration the growth rate of microorganisms and the respiration rate of the food itself is markedly reduced. However, refrigeration has a selective effect on microorganisms. At all storage temperatures, down to the minimum at which water remains in the liquid state, there are microorganisms that will grow. Linear decreases in temperature result in geometric decreases in reaction or growth rate. Frozen storage eliminates microbial growth, reduces the rate of chemical oxidation, and has variable effects on texture depending on the food type, but it has the disadvantage of being energy expensive. To many, the advent of modified-atmosphere

(MA) or controlled-atmosphere (CA) storage provides foods that are "fresh", meaning "never frozen." This is viewed as advantageous to food producers because it reduces costs compared with frozen storage, it avoids problems of texture change and it extends the storage life or protects foods from spoilage.

The key factor in CA and MA storage is the gas atmosphere, and primarily the concentration of carbon dioxide. When plant and animal tissues respire they take up oxygen and release carbon dioxide. Some microorganisms also use up oxygen and release carbon dioxide. By the chemical laws of mass action, increased carbon dioxide or decreased oxygen causes a reduced rate of tissue respiration. Decreased respiration rate reduces the energy available for the biochemical changes that occur in fruits and vegetables. This results in slower rates of ripening and prolonged preripening storage of produce. Bacteria are more adaptable than plants. As CO_2 concentration increases many aerobic bacteria (organisms which require oxygen for growth) are actively inhibited; however, anaerobic or facultative (adaptable) anaerobic bacteria can grow and dominate the microbial population. The technology involves the containment of the packaging gas, achieved originally in bulk "controlled" atmosphere containers. Development of plastics has resulted in "active packaging", a term used by Labuza and Breene [10] to describe the relationship between gas atmosphere and packaging material. The review by Labuza and Breene is especially valuable because it brings together a range of information from widely dispersed publications.

CA generally refers to a circumstance in which the desired atmosphere is maintained throughout storage. The atmosphere can subsequently be altered to induce changes, such as the ripening of produce. MAP generally refers to a package in which air is removed, i.e. vacuum packaging, or air is removed and replaced with desired gas(es). An MA may also be achieved by gas flushing, in which air is replaced dependent upon the degree of gas flushing. Some films are gas permeable, others are impermeable to gases; this dictates the stability of the gas in the package. However, in MAP atmosphere changes not only as a function of gas transmission but also as a result of respiration of food and bacteria in the package. Key factors in food preservation by CAP/MAP include gas atmosphere, temperature, moisture and pressure. Of these, the gas atmosphere is paramount.

Temperature is also a key factor in CA/MA storage of foods. The lower the temperature, the slower the respiration rate and the slower the rate of spoilage. This principle is readily applied to storage of meat, dairy and bakery products, but for produce, there are critical temperatures at which tissue damage or so-called "chill injury" occurs. For example, many subtropical and tropical fruits are susceptible to chill injury below 10°C [8]. Tomatoes are also subject to chill injury when stored under refrigeration. Mature, unripe tomatoes will never ripen if they are stored below 12°C; however, partially ripened tomatoes can be stored between 8 and 12°C. In contrast, depending on varietal type, apples can be stored between 0 and 5°C with excellent preservation of eating quality for up to 8 months.

Pressure is also another factor in CA/MA storage of foods. Low pressure (hypobaric) storage is especially important for produce because it allows the removal of the ripening agent, ethylene [8]. Removal of ethylene can be important because in addition to acting as a ripening agent it promotes senescence (ageing) of produce.

2.2   Effect of food type

The effect of the gas atmosphere differs with different foods, and in the case of produce, it differs between different items of produce.

2.2.1   Fruit and vegetable produce

Fresh fruit and vegetables have different needs for MA or CA storage. In the case of ripening fruits and vegetables, MA/CA is used to control ripening; in the case of plant products that are marketed in an immature form, MA/CA is used to retard senescence, for example, yellowing in broccoli; and in cut vegetables, such as lettuce or cabbage, CA/MA may be used to prevent browning of the cut surfaces. In fruits and vegetables, not only the concentration of CO_2 but also that of O_2 is of paramount importance. Total absence of O_2 (anaerobiosis) results in off-flavours developing in the products. Unfortunately, each type of produce has a specific MA/CA that best suits or optimizes its storage life. Levels of O_2 above 3% but below 20% or levels of CO_2 above 20% influence the respiratory (Krebs) cycle in plant tissues. The critical oxygen concentration that initiates anaerobic respiration depends on the respiration rate. Each produce type has its own tolerance of oxygen and response to lowering of O_2 level.

The effect of reduced pO_2 on delay of ripening can be reversed by ethylene, the natural "hormone" of ripening. Reduced pO_2 decreases the biosynthesis of ethylene, while elevated pCO_2 does not affect ethylene production but acts as a competitive inhibitor of ethylene [8]. Much the same applies to elevation of CO_2 level. At very high CO_2 levels, off-flavours develop.

Use of MA/CA for storage of produce is not a panacea because lack of attention to detail of differences between produce types can lead to extensive spoilage. Conversely, careful application of the principles of MA/CA storage can markedly extend the storage life of unripe, semi-prepared or prepared fruits and vegetables.

2.2.2   Meats and fish

The principles of MA/CA storage of meats and fish are less complex than those for plant produce. In meats the presence of oxygen permits the growth of an aerobic putrefactive microflora, which spoils the meat as a result of off-odours (produced by the time that the microbial load reaches about 10^7 colony forming units per gram or square centimetre). The aerobic microflora break down proteins to produce volatile compounds, such as free amines [3]. Increased concentration of CO_2 in the atmosphere of refrigerated meats retards the growth of the aerobic spoilage bacteria and allows a fermentative microflora to develop, which consists primarily of lactic acid bacteria (LAB). At or above 7 to 10°C a spoilage microflora develops in the meat that causes off-odours, but these tend to be extrinsic, not intrinsic to the meat.

The mechanism of CO_2 preservation of meats in not clear. The minimum concentration of CO_2 for an inhibitory effect is between 20 and 30%. This enables vacuum packaging to serve as a good chilled, storage environment for meats. In a gas impermeable vacuum package, the concentration of CO_2 rapidly increases to the required levels as a result of the respiration of meat and microorganisms.

The use of vacuum or MA packaging with elevated CO_2 can be readily applied to poultry and processed meats. In the case of red meats, the colour change of meat myoglobin is an important consideration. Anaerobic packaging causes myoglobin to develop the dark, purplish red colour of its deoxygenated state. Claims that high levels of CO_2 cause surface darkening due to metmyoglobin formation has lead to the suggestion that carbon monoxide should also be added to produce the stable, bright red carboxymyoglobin pigment in meat [13]. The use of nitrogen has the same effect on colour as CO_2, but it does not have the desirable antimicrobial effects. Because CO_2 is absorbed by the moisture in the meat, some meats are packaged with a mixture of CO_2 and N_2. The N_2 gives a pillowing effect to the package. Because CO_2 acts as an inhibitor of certain spoilage bacteria, it is possible to package red meats with mixtures of CO_2 and O_2, to retard spoilage bacteria on the one hand, while promoting the oxygenated, bright red oxymyoglobin on the other. However, the extension of storage life in oxygenated packages is limited compared with an anaerobic environment. The lactic acid bacteria do not cause oxidation of the myoglobin to form metmyoglobin, hence, despite high counts of lactic acid bacteria, the vacuum- or MA-packaged red meats "bloom" to produce bright red coloration on exposure to air.

The safety of MAP of fish is questionable as a result of studies by Genigeorgis [5]. Concern is expressed that *Clostridium botulinum* type E which grows at temperatures as low as 3.3°C, may grow on fish and produce toxin before the fish spoils. This has generally limited the application of MAP to fish; however, storage of fish in CO_2-enriched atmospheres does retard microbial growth and the rate of amine production is also reduced. The mechanism of amine inhibition is not clear [2].

2.2.3   Baked goods

These foods decline in quality by loss of moisture, staling and mould growth. Packaging retards moisture loss but it promotes mould growth. Elevated CO_2 retards mould growth under such conditions. A 10% increase in CO_2 concentration and a 5.5°C (10°F) reduction in storage temperature doubled the mould-free storage life of bread and cake (see Chapter 4). CO_2 in the headspace of crumpets enabled these products to be stored at ambient temperatures for 3 weeks or more [12].

2.2.4   Dairy foods

The use of MA/CA in dairy foods has only limited application compared with produce and meats. Increased pCO_2 in refrigerated milk does inhibit growth of psychrotrophic bacteria. In the storage of cut cheeses, an anaerobic environment is used to prolong storage life through inhibition of mould growth.

2.3   Gas atmosphere

The effect of gas atmosphere on foods is not universal, but in all cases it has a profound effect on storage life and keeping quality. In the storage of grains, airtight packaging has valuable applications to prevent insect and mould spoilage. The MA storage of grains is extensively treated by Calderon and Barkai-Golan [1] in their text on *Food Preservation by Modified Atmospheres*. They emphasize the application of MA storage of grains in developing countries; however, anaerobic storage of grains could eliminate or reduce the need for fumigants in cereal storage and it would also eliminate or reduce mould growth. The inhibition of mould growth requires exceptionally low levels of dissolved oxygen in the food. Other environmental factors such as relative humidity, temperature and the competitive microflora also affect mould growth.

In fruits and vegetables the principle form of storage is CA. CA effectively maintains the quality and extends the storage life of a variety of fruits and vegetables, but each fruit and vegetable type has its own specific requirement and tolerance for gas atmosphere and temperature of storage. Decreased respiration rate reduces the rate at which fruit ripens and thereby extends its storage life. This is particularly important in climacteric produce, in which the ripening process proceeds at an increasing rate, for example in apples, bananas, pears etc. Associated with respiration is heat generation. If the respiration rate is reduced, then less heat is evolved. Furthermore, in the absence of oxygen or with elevated concentrations of CO_2, ethylene biosynthesis is stopped [9]. Reduced oxygen and increased CO_2 also inhibit mould growth in produce.

By comparison, the storage of other foods in MA or CA is uncomplicated. The gas atmosphere (and temperature) generally create an environment in which storage life is extended by selection of a relatively innocuous microflora in the foods, for example the lactic acid bacteria that predominate in bakery and meat products under anaerobic conditions of storage. In general the important factors are increased CO_2, decreased or absent O_2 and reduced temperature. Vacuum packaging produces these conditions in meats because CO_2 is generated by natural respiration in fresh meat or by growth of heterofermentative (mixed end products of fermentation, including CO_2) lactic acid bacteria in fresh and processed meats.

The inhibitory effect of CO_2 on growth and metabolism of microorganisms was reviewed by Dixon and Kell [2]. CO_2 is inhibitory to the growth of many microorganisms. This can be exploited in the preservation of refrigerated foods. CO_2 hydrates and dissociates in water. In most food systems this involves the following equilibrium because they are at pH values <8.

$$CO_2 + H_2O \rightleftharpoons H_2CO_3 \rightleftharpoons HCO_3^- + H^+$$

The amount of CO_2 in solution depends on the partial pressure of CO_2 (pCO_2) in the gas phase, temperature and pH. As the temperature decreases, the solubility of CO_2 increases. As pH increases above 8.0, carbonate ions are formed, so that the equilibrium shown in the equation above moves further to the right [2].

The antimicrobial properties of CO_2 are well-known and have been used to preserve different foods, including meats, fish and plant produce. In meats and fish there is a direct antimicrobial effect; whereas in plant produce, other factors in addition to reduced mould growth are involved. Despite the practical knowledge of the inhibitory effects of CO_2 on microbial growth, the mechanism of inhibition is not clear. However, CO_2 has several mechanisms whereby it acts against microorganisms: (i) functioning of the cell membrane; (ii) inhibition of metabolic processes; both of which may be related to (iii) disruption of enzyme activity. CO_2 reacts with proteins, affecting the rate and extent of solution in water. During the storage of proteinaceous foods in atmospheres containing CO_2, the antimicrobial action is achieved through solubilization and absorption of the CO_2, its penetration of the microbial membrane and reduction of the intracellular pH. The pH changes induced by CO_2 at medium to high partial pressures are sufficient to disrupt enzymatic function [13]. However, organisms that are CO_2-sensitive when grown aerobically are CO_2-insensitive when grown anaerobically [6]. This suggests changes in the sites of inhibition between aerobic and anaerobic growth.

In contrast, other gases except ethylene are inert. Nitrogen has been tested for an antimicrobial anoxic effect on microorganisms similar to that of CO_2. N_2 is an inert gas and has a similar effect to vacuum packaging. An N_2 flush removes residual O_2 but it also reduces (dilutes) the CO_2. In general, N_2 has limited value in fresh meat packaging, except as a package filler and to reduce liquid purge from meat [11].

Carbon monoxide has a limited effect on microorganisms.   A study by Gee and Brown [4] showed that CO has no effect on growth of *Pseudomonas aeruginosa*, slows the growth rate of *Escherichia coli* in proportion to the CO concentration, and increases the lag phase of growth of *Achromobacter* and *Pseudomonas fluorescens*.   Clearly there is no universal effect of CO on common microorganisms associated with food spoilage, and at 1% CO no effect on bacterial growth was detected.   CO does affect the colour of meat by the formation of carboxymyoglobin.   This compound is spectrally similar to oxymyoglobin (the bright red pigment of red meats) and far more stable to oxidation.   In produce, CO reduces the ageing process, presumably by its interaction with some cytochromes which blocks the oxidative breakdown processes.   CO also inhibits some yeasts and moulds, presumably by the same mechanism [13].

2.4   Hypobaric   storage

Hypobaric or low pressure storage is another form of MA or CA storage in which pressure, temperature and humidity are accurately controlled.   Jamieson [7] in his review noted that in addition to these three factors in hypobaric storage, the rate at which air is changed in the storage environment is closely regulated.   These systems can operate with air, but if a hypobaric CA or MA is required, other gases are added.   Because hypobaric storage systems rely to a large extent on evaporation for cooling, the need for convection systems for cooling is removed.   In hypobaric storage there is "out-gassing" of the product [7].   The CO_2 and other volatiles produced during respiration, including ethylene, are removed and the controlled level of O_2 is maintained.   Systems for hypobaric storage are commercially available, and the application of this form of storage to pork loins, strawberries, papaya and other produce has been demonstrated.

2.5   Conclusions

Despite widespread use of CA and MA storage of foods, there is still room for improved understanding of the mechanisms whereby CO_2 inhibits microbial growth.   Too often these systems have been promoted by nontechnical sales representatives, so that incorrect claims have been made and inadequate information has been distributed.   There is an urgent need to integrate gas and packaging technology with a sound knowledge of the foods to be packaged and basic scientific principles.   This is most apparent in the concerns expressed by Public Health Officials regarding the "safety" of many MA or CA foods.

References

1. Calderon, M. and Barkai-Golan, R.   1990.   Food Preservation by Modified Atmospheres.   CRC Press, Boca Raton, Florida.
2. Dixon, N.M. and Kell, D.B.   1989.   The inhibition by CO_2 of the growth and metabolism of micro-organisms. J. Appl. Bacteriol. 67: 109-136.
3. Edwards, R.A. and Dainty, R.H.   1987.   Volatile compounds associated with the spoilage of normal and high pH vacuum-packed pork. J. Sci. Food Agric. 38: 57-66.
4. Gee, D.L. and Brown, W.D.   1980/81.   The effect of carbon monoxide on bacterial growth.   Meat Sci. 5: 215-222.
5. Genigeorgis, C.A.   1985.   Microbial and safety implications of the use of modified atmospheres to extend the storage life of fresh meat and fish.   Int. J. Food Microbiol. 1: 237-251.
6. Gill, C.O. and Tan, K.H.   1980.   Effect of carbon dioxide on growth of meat spoilage bacteria.   Appl. Environ. Microbiol. 39: 317-319.
7. Jamieson, W.   1980.   Use of hypobaric conditions for refrigerated storage of meats, fruits and vegetables.   Food Technol. 34(3): 63-71.
8. Kader, A.A.   1980.   Prevention of ripening in fruits by use of controlled atmospheres.   Food Technol. 34(3): 51-54.
9. Knee, M. 1990.   Ethylene effects in controlled atmosphere storage of horticultural crops, p. 225-235.   *In* Food Preservation by Modified Atmospheres, Calderon, M. and Barkai-Golan, R. (eds.). CRC Press, Boca Raton, Florida.
10. Labuza, T.P. and Breene, W.M.   1989.   Applications of "active packaging" for improvement of shelf-life and nutritional quality of fresh and extended shelf-life foods.   J. Food Proc. Preserv. 13: 1-69.
11. Seideman, S.C., Smith, G.C., Carpenter, Z.L., Dutson, T.R. and Dill, C.W.   1979.   Modified gas atmospheres and changes in beef during storage. J. Food Sci. 44: 1036-1040.
12. Smith, J.P., Ooraikul, B. and Koersen, W.J.   1987.   Novel approach to modified atmosphere packaging of bakery products, pp. 332-343.   *In* Cereals in a European Context - First European Conference on Food Science and Technology.   Morton, I.D. (ed.). Ellis Horwood Ltd., England and VCH, Germany.
13. Wolfe, S.K. 1980.   Use of CO- and CO_2-enriched atmospheres for meats, fish, and produce.   Food Technol. 34(3): 55-58, 63.

Chapter 3

TECHNOLOGICAL CONSIDERATIONS IN MODIFIED ATMOSPHERE PACKAGING

B. Ooraikul
Department of Food Science, University of Alberta

There are five key elements that must be considered to achieve the optimum effect of MAP: the initial microbiological state of the food; temperature control; the gas mixture; the barrier film; and the packaging equipment [29]. The effects of microbial load and gas atmospheres on shelf life of foods have been discussed in previous chapters. In this chapter, technological developments concerning packaging materials, packaging equipment and systems, the effect of storage temperature, and the concept of shelf-life dating will be discussed.

3.1 Packaging materials

When food is packaged in an atmosphere other than air to suppress microbiological, chemical and physical changes it is necessary to consider the following:

3.1.1 The nature of the product

Fresh produce normally consists of living tissue which is still respiring. Oxygen may be required for the reactions, which are accompanied by the production of CO_2 and water and, in some cases, volatile compounds, such as alcohols and short-chain fatty acids. Microbial growth may also occur which, depending on the number and type of microorganisms present, may or may not require O_2 but may also produce CO_2, water and some volatile compounds. Eventually, therefore, the headspace atmosphere will become saturated with moisture, while O_2 is depleted and CO_2 accumulates. Some chemical reactions may take place which result in texture and colour changes, e.g. breakdown of starch and chlorophyll in plant tissue, and hydrolysis of proteins and lipids or oxidation/reduction of myoglobin in meat. In processed or semi-processed products, in which the animal or plant tissue is no longer respiring, changes are microbiological and chemical in nature with similar end results, i.e. depletion of O_2 and accumulation of moisture, CO_2 and some volatile compounds. Water activity, pH and storage temperature of the food exert a strong influence on these reactions.

3.1.2 Disposition of metabolic products

Part of the metabolites, i.e. CO_2, moisture vapour and volatile compounds, will be absorbed by the product; some may be lost to the outside atmosphere through the packaging material. The absorbed metabolites will change characteristics of the product such as a_w, pH, colour and texture, which may improve or shorten the storage life of the product. Reduction of pH will make the product less suitable for growth of some microorganisms, while an increase in a_w will stimulate microbial growth and some chemical reactions. Increase in headspace CO_2 and volatiles such as ethanol or organic acids is generally beneficial to storage life of the product. The net result of all these changes depends on the type of food product and its initial conditions, the headspace atmosphere and the stage and rate at which these changes take place. English-style crumpets packaged under CO_2 and N_2 may benefit from a moderate rate of bacterial activity in the first week of storage to prevent mould growth by depleting O_2 and producing some CO_2. However, strong yeast growth in cherry cream cheese cake leads to swelling of the package within 14 d due to excessive production of CO_2.

3.1.3 Permeability of the packaging material

Permeability of the packaging material will determine the atmospheric conditions in the headspace over the period of storage and, ultimately, the shelf life of the product. If an atmosphere high in CO_2 and/or low in O_2 is required, the material should be impermeable to these two gases. Vegetables and fruits, on the other hand, require a certain amount of O_2 in the headspace for maintenance of quality; therefore, packaging material for these products should be quite permeable to O_2 to allow atmospheric O_2 to replenish the gas in the package. Transparency of packaging material to light may also be important for some products. Oxidative changes in lipids and some pigments, for example, are accelerated when exposed to light.

Because of the dynamic state of the product and because there are so many factors that affect changes in the product, it is extremely difficult to maintain the atmosphere surrounding the product at the optimum required to prolong its storage life. CA bulk storage, where atmospheric conditions inside the storage compartment are maintained by mechanical means, is possibly the only commercial method available to keep the atmosphere relatively constant. In MAP, where product is packaged in small individual units, continuous mechanical control of the atmosphere is not possible. Once a gas atmosphere has been applied the level and proportion of headspace gases can only be controlled by judicious selection of packaging material with specified permeability characteristics. There is a wide range of materials available for food packaging, but only a small number are suitable for MAP.

3.1.3.1   Metal containers

The metal container or tinplate can was invented in 1810 in England. It has undergone many improvements in materials and manufacturing technology and it remains one of the most popular and complete food containers. The most important characteristics of the metal can are its strength, its complete impermeability to moisture, gases and light, its relatively low cost and good safety record and its recyclability. However, unless conditions inside the container are essentially static and complete impermeability is desired, metal cans are not suitable as packaging material for MAP. A combination of metal container and flexible film, on the other hand, may be desirable where strength is required to prevent physical damage to the product, while flexibility and a certain degree of permeability is required to accommodate changes inside the container.

3.1.3.2   Glass containers

Like metal cans, the glass container provides strength and a complete barrier against moisture and gas and, if properly dyed, against light. It also lacks flexibility and permeability to accommodate in-package changes. Therefore, the glass container in its traditional form is not considered a suitable packaging material for MAP.

3.1.3.3   Rigid/semirigid plastic and paper containers

The rigid or semirigid containers are made from paper or plastic, or composites of these materials [34]. Paperboard cartons have been used extensively for packaging many dairy products and beverages. Plastic trays, bottles, jars, boxes and tubs have been used for a wide variety of products, e.g. meat, soft drink, confectionery, baked goods, ice cream, butter and margarine. Laminated paper tubes have been used for products such as fabricated potato chips and frozen fruit juices. Paperboard may consist of as many as six layers, including a heat sealing layer, adhesion layers, aluminium foil, paperboard and graphic layers. Plastic containers may be made from a variety of plastics such as polyethylene, polypropylene, polystyrene, polyester, polyvinyl chloride, cellulose acetate and phenol formaldehyde. They may be thermoformed or injection moulded. These rigid and semirigid materials provide protection against physical damage during shipping and marketing of food products, but gas permeability may not have been a prime concern with this type of packaging.

In the context of MAP, if the permeability of these materials can be tailored to meet specific requirements, and sealing can be easily and reliably achieved, they may be highly suitable for this purpose, especially for fragile products. However, high cost and lack of convenience in handling may preclude them from serious consideration. Nevertheless, a combination of low-cost rigid containers with a flexible overwrap that meets MAP specifications may be the best packaging for products that require physical protection as well as attractive display of the products. Packaging aids such as moisture absorbents, O_2 scavengers, CO_2 generators and other atmosphere modifiers may also be unobtrusively included with rigid containers.

3.1.3.4   Flexible films

Plastics account for 17% of the packaging material in the U.S. today and they are expected to grow to 30% by the year 2000 [10]. The rapid rate of growth is due largely to the changing consumer demand where convenience, quality, safety and impact on the environment are prime considerations. The greatest increase will be in the food and beverage industries where it is anticipated that plastic packaging will be used increasingly to replace metal cans and glass containers. However, substantial technical development is needed to overcome problems such as economics, package integrity and recyclability of plastic materials before their full potential can be realized.

Plastics represent a family of chemical compounds of very high molecular weight. The high molecular weight is achieved through polymerization of basic "building blocks" or monomers to form "polymers". Naturally occurring polymers include cellulose, rubber, casein, silk, wood and protein [12]. Most commercial plastics are made from synthetic polymers from either wood pulp or petrochemical products.

No single polymer offers all the properties required by any packaging regime. Some plastics may be elastic like rubber while the others may form tough, clear films. Some may have very high melting points and can be moulded into containers of various shapes, while those with lower melting points may be useful as heat-sealable or thermoformable films. Some may be quite permeable to gases and vapours and others may form good protective films against the migration of these molecules. To provide packaging films with a wide range of physical properties, many of these individual films are combined through processes such as lamination and coextrusion. Orientation or stretching of the films during formation may improve their strength or make them heat shrinkable. Orientation is the mechanical alignment of molecular chains, and may be unidirectional or biaxial, depending on the properties required in the films [12].

A vast number of plastic materials with varying strengths, clarity and permeability are available and a great many more are being developed, many of which are designed for specific purposes. For example, Mylar was developed by DuPont for boil-in-bags, Surlyn for high-speed meat packing, polyethylene terephthalate (PET) to replace glass bottles, and Selar high-barrier resin for packaging that requires films with low gas permeabilities and water vapour transmission rates.

For MAP, flexible films with varying degrees of permeability to gases and water vapour are used almost exclusively. Next to permeability, package integrity is the most important property for MAP films. In MAP systems the package is normally evacuated first, followed by backflushing with the desired gas or gas mixture before heat sealing. The package may be preformed or thermoformed in form-fill-seal packaging equipment. The following are examples of gas atmospheres recommended for MAP of various products.

	Gases, % by Volume		
	O_2	N_2	CO_2
Bakery (crumpets)	--	40	60
Apples	2	1	97
Tomatoes	4	4	92
Pasta	--	--	100
Fresh red meat	70	10	20
Thin sliced meat	--	20	80
Processed luncheon meat	--	--	100
Sausages	40	60	--
Fresh poultry	--	50	50
Fresh fatty fish	--	70	30
Fresh white fish	--	60	40

(Adapted from Rothwell, 1986 [33])

Most processed products rely on high CO_2 concentration and absence of O_2 in headspace atmospheres for shelf-life extension. Therefore, the films used in packaging of these products must have excellent barrier properties against these two gases, and, for high moisture products, against water vapour as well. However, owing to different processing and barrier requirements for each application, no single barrier film will meet all of the requirements and be cost-effective in every case.

There are three types of high-oxygen-barrier polymers available at moderate cost: vinyl alcohol, vinylidene chloride, and acrylonitrile [8]. Copolymerization of films made from these polymers have been developed to combine processability and barrier properties. For example, polyvinylidene chloride (PVDC) polymers are copolymers of vinylidene chloride and vinyl chloride or methyl acrylate. Acrylonitrile is polymerized with styrene or methyl acrylate, while vinyl acetate is polymerized with ethylene and saponified to form ethylenevinyl alcohol (EVOH).

Modified barrier resins based on nylons, polyesters, and EVOH have also been developed to improve their strength, impact and chemical resistance and permeation barrier to flavours, aromas and a variety of gases [8].

For MAP of bakery products such as English-style crumpets, films that have been used quite successfully include polyethylene (PE)/nylon/PE, saran/nylon/PE, and aluminium foil/PE.

Fresh produce, on the other hand, respires or undergoes biological changes, hence anaerobic conditions in the headspace are undesirable. Red meat, especially, needs O_2 to maintain its bright red colour. Films used for MAP of produce, therefore, must be selectively permeable, i.e. they must be good barriers against CO_2 and water vapour, but quite permeable to O_2 to allow permeation from the air into the package. There are many moderate-oxygen-barrier polymers, e.g. nylons, polyesters, and unplasticized polyvinyl chloride which may be used for these purposes [8]. Properly coated or laminated PE, polyester or polypropylene have also been used.

There have been reports of attempts to create the so-called "smart films" by developing new polymers or adding appropriate additives to existing polymers to modify their barrier characteristics. These films could conceivably "sense" changes in the atmosphere inside and outside the package and adjust their permeability accordingly. The films would be well suited for MAP of respiring produce where balance between O_2 and CO_2 in the headspace is crucial to the maintenance of quality and shelf life of the products. Considerable work is being done to increase O_2 permeability of low-barrier films for fresh produce, change ratios of CO_2 to O_2 permeability, improve compatibility and sealability of low-barrier materials, and develop materials with permeability sensitive to the surrounding temperature [21]. However, availability of these films may still be in the distant future.

Some products are fragile or soft and require a package which can withstand physical abuse. Most products require packaging films with good clarity for attractive display. Many products, especially those with high moisture content, may become foggy owing to condensation of moisture on the inside surface of the package, especially when storage temperature changes. Some products require heating before serving, and microwavable or dual-ovenable packages may be desirable in these cases. These are some of the special problems that need to be addressed in selecting the films and designing the packages. Packaging systems such as "Gemella" and "Flavaloc" have been developed for these purposes.

In the Gemella system the package is designed to withstand the stresses of handling, shipping, cold storage, and it is also microwavable. The container is made from a combination of board and plastic film, bonded together by thermoform-lamination [6]. The rigid board, which may be printed on, forms the external protection and promotes visual sales appeal for the product, while the transparent film provides barrier performance, tamper-proofing, and attractive presentation. There are two styles of Gemella trays: the "full tray" where the cartonboard provides support to both walls and base, and the "picture frame" where the board provides strength to the walls and flanges. The system has been used successfully for meat pies, fresh meat and cheese slabs, and shows good potential for bakery products, meat products, produce and fish. The films used include PVC/PE, PS/EVOH/PE and PVC/Saranex for trays, and PET/PE and PET/Saranex for lids. The packaging is done on a fully integrated thermoforming/filling/sealing   line.

The Flavaloc system was developed by Garwood Ltd., Australia, as a "truly controlled atmosphere system" for individual packs [24]. It was developed to address some marketing disadvantages of conventional CAP/MAP. These disadvantages include headspace requirements which make the package bulky and preclude it from vertical display, deterioration of the package appearance during transportation and handling due to product movement inside the package, and high costs of the packaging system. The Flavaloc system is similar to conventional MAP except that an intermediate web is introduced between the thermoformed base and the top lidding web. The intermediate web is sealed onto a flange formed around the side of the base tray and stretched over the contents of the package. This stretched web holds the contents firmly to the thermoformed tray and thus forms a separate sealed compartment in the package. The machine used for this purpose is similar to regular form-fill-seal equipment, but an additional gassing and sealing station and a stretch and flange sealing head are incorporated. The package is evacuated and backflushed with a predetermined gas mixture. Thus both the inner compartment containing the product and the empty compartment between it and the tray and the lidding web contain the gas mixture. The intermediate web is chosen for its permeability characteristics to control the rate of gas transfer from the gas reservoir outside it. In this way the system is said to be a truly controlled atmosphere packaging for retail size packages because the atmosphere surrounding the product will be kept relatively constant during storage.

The stretch web must be highly elastic, with good puncture and tear resistance and excellent heat-sealing properties. Anti-fog additive can be applied to the film for good presentation. Film based on an octene copolymer linear low-density polyethylene (LLDPE) has been tested with as good results as the intermediate web. PVC/EVA multilayer sheet has been used as the base web and a four layer structure of PET/PVDC/LDPE/EVA as the lidding material. Work is underway in Australia to develop techniques to produce PVC/LDPE laminates and coextruded stretch webs at lower cost. It is also envisaged that the Flavaloc system may be extended to include a twin compartment for prepared foods. The two compartments, containing two different portions of a meal such as roast beef and potato salad, will be kept separate until served.

About 75% of American households now own microwave ovens and microwavable foods have become a $3 billion a year industry, indicating that the emphasis in packaging is microwavability [31]. The MAP system definitely complements this trend in providing fresh or near-fresh convenience foods. Together with microwavability, MAP products may be the ultimate in quality and convenience. Many packaging films have been developed to meet these dual purposes. For example, polystyrene/polyphenylene oxide (PS/PPO) blends (NORYL® EFC resins) by GE Plastics increase the heat resistance of polystyrene foam packaging for microwavable applications. The PS/PPO blends produce high microwave heat- and grease-resistant materials suitable for form-fill-seal lines. By varying the amount of PPO resin in the blend, the material can be used to produce either foam or solid containers for any specific product requirements. With the addition of a barrier layer, the containers can be used for MAP applications.

Another development, also by GE Plastics, is an amorphous thermoplastic polyetherimide (ULTEM® 1000 resin) which is used in microwave susceptor films to produce higher heat with reduced metal usage [31]. The films facilitate higher cooking temperatures, crisper foods, and shorter cooking times. These films in combination with appropriate barrier layers would be ideal for MAP of products such as pizza, roast beef or any food that requires browning or crisping.

Solid waste-disposal problems associated with plastic packaging materials have been the subject for intense discussion and research in recent years. Development and application of MAP processes will, no doubt, add to these problems. Increasing environmental awareness among consumers and governments, together with landfill shortages, increasing waste-disposal costs and other related concerns have put pressure on industries to solve or minimize these problems before they reach catastrophic proportions. Several solutions have been put forward, e.g. incineration of solid wastes and recovery of heat energy for useful purposes, development of biodegradable plastics, and recycling of plastic packaging materials. Among these, recycling may be the most viable option as it not only reduces the quantity of solid waste but it also conserves non-renewable resources. Research centres such as the Centre for Plastics Recycling Research at Rutgers University, U.S.A., have been established to develop and demonstrate the technical and economic viability of recycling and reuse of plastic materials [25]. Research projects include recycling of certain types of plastic bottles, utilization of commingled plastics, development of collection, sorting and handling systems, new product applications of database on the plastics recycling industry, and investigation of the recyclability and the recycling of polystyrene. Commingled plastics have already been successfully used in the manufacture of non-food plastic containers, strapping materials, and construction materials to replace concrete, wood, metal, etc. With further technological advances and proper quality assurance systems, recycled plastics may even be usable for food packaging. Indeed, plastics recycling may become an economically viable industry that helps conserve limited resources and protect the environment.

3.2 Packaging technology

3.2.1 Atmosphere modification

The headspace atmosphere surrounding a MAP product is chosen for maximum extension of shelf life while maintaining safety and high product quality. Optimum gas atmospheres for each product must be determined through a series of storage trials. A rough guide of atmospheric requirements for each group of commodities may be established by evaluating general characteristics of the commodity with respect to its chemical, microbiological and physical properties as they affect the shelf life and quality of the food. Based on this guide, the optimum atmospheric requirements for individual products must be determined to maximize the benefits of MAP. For example, fresh produce requires high concentrations of CO_2 with relatively low O_2 concentrations; fresh meat may keep best under high concentrations of both CO_2 and O_2; fresh fish can be stored longer under relatively high concentrations of CO_2, with or without O_2; and bakery products keep best without O_2, but with high concentrations of CO_2. Nitrogen is required as a filler gas in most cases. The desired headspace atmosphere may be incorporated into MAP by two methods, i.e. replacing air with the gas or gas mixture, or generating the atmosphere within the package using suitable atmosphere modifiers, such as O_2 absorbents or CO_2 generators.

3.2.1.1 Gas flushing

In this method, the package containing the product is normally first evacuated then backflushed with the gas or gas mixture. The gas mixture may be premixed according to specifications, or pure gases from individual cylinders may be mixed in desired proportions with a gas blender during the packaging operation. Major gas suppliers will premix gases for customers using a partial pressure technique. The accuracy of the mixture is normally about ±5% of the required concentration, and the cost is high compared with a gas blending method [23].

Gas blenders are capable of giving a mixing accuracy of about ±2% or better [23]. There are several types of gas blenders operating on the principles of differential flow rates or differential pressures of gases from individual gas cylinders. Accuracy, capacity, ease of operation and cost of the blenders vary. Ooraikul [26] used a Smith's Proportional Mixer for MAP of English-style crumpets for a mixture of CO_2 and N_2 (3:2). This mixer comes in three models in which two or three gases may be blended to any proportion by adjusting the setting dial. Safeguard mechanisms are available in some models where the mixer will automatically shut down, or an alarm is sounded, if any of the incoming gases falls below a certain pressure. Using gas blending as part of the packaging system is more cost-effective than using premixed gases.

3.2.1.2   Atmosphere modifier

Suitable atmosphere modifiers may be used to provide the desired headspace atmosphere inside the package. The modifiers commercially available include O_2 absorbers or scavengers, CO_2 absorbers or generators, ethylene absorbers, moisture regulators, and ethanol generators [39]. Most of the technologies used in the manufacture of these modifiers originated in Japan and the materials had been used in that country for many years before their introduction into North America.

One of the O_2 absorbers produced by Mitsubishi Gas Chemical Co., Inc., is "Ageless". Its major ingredient is active iron oxide packaged in a permeable sachet. A suitable amount of Ageless is placed inside the package together with the product before sealing. Active iron oxide will react with the headspace O_2 to reduce it to 0.01% (100 ppm) or lower within a few hours. Part of the reaction is as follows:

$$Fe + 3/4\ O_2 + 3/2\ H_2O \longrightarrow Fe(OH)_3 \longrightarrow Fe_2O_3.3H_2O$$

Ageless eliminates problems associated with aerobic condition in headspace gases. It has been used successfully to inhibit oxidative flavour changes in products such as ground coffee, chocolates, cookies and fried snacks, e.g. potato chips. Thus, the need for antioxidants such as butylated hydroxyanisole (BHA), butylated hydroxytoluene (BHT), and sulphites can be eliminated [2]. It has been used to control insect infestation in cereals and grains. Ageless has been shown to inhibit completely growth of aerobic microorganisms, especially moulds, in products such as breads, rolls, pizza crust, sponge cake and salami [16, 37]. It shows no human toxicity, having an LD_{50} exceeding 10 g/kg, and it has been approved by the Food and Drug Administration (FDA) of the United States for use with foods. A packing machine has also been developed to automatically cut and dispense Ageless sachets into food packages.

Microwavable Ageless, "Ageless FM", has been developed [28]. It can be microwaved without giving an off-odour or scorching. It has been tested with microwavable rice, fried potatoes, hamburgers, chicken nuggets, etc. and found to be suitable for most foods.

Another Japanese product, "Ethicap", is an ethanol generator produced by Freund Industrial Co., Ltd., Tokyo. It contains at least

55% ethyl alcohol microencapsulated in silicon dioxide powder and packaged in a small heat-sealed sachet. It may also contain a trace amount of food-grade flavourings such as vanilla or citrus. When packaged with low- or intermediate-moisture food, Ethicap will slowly release ethanol vapour into the headspace. This has been shown to inhibit growth of certain microorganisms, some fungi, especially moulds. Hardening or staling of some bakery products is also retarded by ethanol vapour. The manufacturer claims that shelf life of products such as bread, rice cake stuffed with bean jam, cupcake, casteilla, boiled squid, chocolate-covered sponge cake, and American cake has been extended to between 1 week and 6 months by Ethicap.

Ethicap is available in various sizes. A suitable size for any product is determined based on the length of shelf life desired, a_w, and weight of the product. The longer the shelf life, the higher the a_w and the larger the product, the greater the amount of Ethicap needed.

A similar type of product, "Fretek", is manufactured under license from Japan by Techno International, Inc., New York [30]. It is a paper pulp wafer impregnated with 95% absolute ethanol in glacial acetic acid, sandwiched between layers of flavour-impregnated film and polyolefin film and packaged in a small sachet. Fretek has been used successfully for cereals, a range of dry products, fresh meats, poultry, seafood and baked goods. The manufacturer suggests that the wafer offers certain cost advantages over other MAP approaches because no specialized packaging equipment and gases are required. The wafer form of product is expected to cost less than the encapsulated powder form.

Natural Pak Purifier is a CO_2 and water vapour absorber manufactured by Natural Pak Systems, Alpine, NJ. It is a packet of calcium hydroxide and calcium chloride which, when packaged with fresh fruit or vegetable in a semi-oxygen-permeable film, e.g. PE, will provide an environment suitable for the produce to create its own preservation atmosphere [1]. This is due to the fact that when the packaged produce respires, it gives off CO_2 and water vapour. $Ca(OH)_2$ and $CaCl_2$ absorb CO_2 and water vapour, maintaining the desired levels of these inside the package. Oxygen level, on the other hand, is controlled by the permeability of the film, the respiration rate of the produce, and the absorption rate of CO_2 and water vapour.

The first commercial application of this process was on fresh tomatoes. In-store tests demonstrated shelf life of produce treated in this way and held at 15-18°C to be equal or better than that of refrigerated produce. Successful laboratory trials have also been done with apples, artichokes, asparagus, avocadoes, beans, corn, cucumbers, mangoes, papayas, peppers, pineapples and plums.

Other atmosphere modifiers such as CO_2 generators and ethylene (C_2H_4) absorbers or generators are also available in similar forms. For example, ferrous carbonate reacts with O_2 in moist air to generate CO_2. Potassium permanganate adsorbed on celite, silica gel or alumina pellets can oxidize C_2H_4 to CO_2 and H_2O. Builder-clay powder with cristobalite as the principal component absorbs ethylene and many other gases. The choice of these products is dependent on the atmospheric requirements for storage of the product or produce. It also depends on economic considerations, availability of equipment, government regulations and consumer acceptance.

3.2.1.3   Headspace gas analysis

Analysis of the headspace atmosphere of MAP products must be done routinely to ensure accuracy of the gas composition. This is especially important when the presence of even traces of gas such as O_2 is undesirable, or when a minimum quantity of O_2 is essential for the safety of the products known for their botulinum hazard. A common gas analyser used for this purpose is based on a single or dual-column gas chromatograph with a thermal conductivity detector [26]. Portable or on-line gas analysers are now available. Some types of gas mixers, such as that manufactured by Thermco Instrument Corp., La Porte, IN, U.S.A., has a built-in gas analyser. Some gas analysers are designed to analyse a single gas, e.g. CO_2 or O_2. One such system is a compact and portable oxygen analyser by Teledyne Analytical Instruments, City of Industry, CA, U.S.A. It has a gun-style sampling assembly which can take a small, fixed volume of gas from either flexible or semi-flexible packages with an appropriate syringe. The sample is analysed with an electrochemical sensor and the result is shown on a meter with an expandable scale capable of measuring 0-25% O_2 with the accuracy of ±2% of full scale. On-line oxygen analysers, such as Toray LF-700D by Modern Controls, Inc., Elk River, MN, U.S.A., are also available for continuous monitoring of headspace O_2, but they are much more expensive than portable units.

3.2.2   Packaging equipment for flexible films

All MAP techniques require replacement of residual atmosphere with a specified gas or gas mixture. This can be done by either vacuum or straight flushing [35]. In the vacuum flushing technique, the packed product is placed in a vacuum chamber in which a vacuum is drawn and replaced by the gas or gas mixture. Equipment used for this type of operation may provide a choice of full vacuum, semi-vacuum, atmospheric pressure or positive pressure. The advantage of this technique is that the air is replaced more completely with the chosen atmosphere, especially for porous products where air cells are finely distributed inside the products and cannot be displaced by simple gas flushing. This is very important when O_2 is undesirable. However, with some packaging machines the package collapses tightly around the product during the evacuation cycle and may damage fragile or soft products.

In the straight flushing technique, air in the package is displaced with the chosen gas or gas mixture. Approximately 2.5 to 3 volumes of replacement gas are needed to remove one volume of residual air. This system may be used for fragile products when package collapsing is damaging, or when some O_2 in the headspace is acceptable. It is normally used in flow wrap or vertical form-fill-seal operations.

Packaging machines used for MAP operations may be classified into two broad categories: chamber, and form-fill-seal systems.

3.2.2.1   Chamber packaging system

The chamber packaging system is used with preformed pouches. The pouch is manually opened, filled with the product and placed in the vacuum chamber. When the chamber lid is closed, the vacuum pump is activated, the chamber is evacuated, then backflushed with the gas and the package is sealed. This type of machine is suitable for small operations. The machines are relatively cheap, and the operation is simple, but slow. The following are examples of chamber packaging equipment currently available on the market.

Kramer & Grebe Vacuum Packaging Machine. This machine comes in several models, e.g. Autovac Quick Duo and Variant vacuum/gas flushing machines. Both have twin chambers, but the latter has impulse sealing bars at the front and rear of each chamber for greater output. The chamber evacuates automatically when the lid is closed. The Autovac model also has a trimmer knife to remove automatically excess film from the package after it is sealed. These machines are claimed to attain a vacuum of 99.6-99.9%.

Supervac Double Chamber Packaging Machines. These machines are similar in operation to the previous ones, but they have a few optional features. They have biactive sealing with upper and lower sealing bars for tougher and more reliable seal, even when there are wrinkles, fat or liquid in the seal. Automatic lid swing is available to reduce labour and increase output. A "ready meals" version is also available for packaging hot liquids and products with sauce or gravy.

Multivac Vacuum/Flushing Packaging Machines. These machines are available in both table and floor models. They are made as either single or double chamber models, with one or two seal bars.

RMF Snorkel-Vac Modified Atmosphere Packaging System. This system operates without a chamber. These machines have three seal bars with corresponding gas flushing snorkels. When packages are attached to the unit and the cycle is activated, the outer bar clamps the packages in place, the snorkels descend and draw the vacuum, the selected amount of gas is injected, the snorkels withdraw, the impulse bars seal the packages and the packages are released once the seal has cooled. There is a detection mechanism for leaking packages which prevents the machine from completing the cycle if the required vacuum is not reached. There is almost no limit to the size of the package that can be processed on this machine. However, because the vacuum is drawn outside a chamber, under atmospheric pressure, delicate products cannot be packaged using this system.

RMF also manufactures the MACH-54 chamber model. It is designed to package dry powders, liquids, and fragile products. It is claimed to be capable of drawing a high vacuum, and to provide low O_2 residue of 0.15%. It uses a programmable controller to allow the user to adjust cycles to fit various packaging applications.

CVP Fresh-Vac System. This system is similar to the RMF Snorkel-Vac system but it only has two snorkels, and a double bar heat sealer. It comes in both table and floor models.

3.2.2.2   Form-fill-seal packaging system

This system is by far the most popular of the two because of its flexibility and speed. It is widely used in the food industry and for packaging of products such as sterilized medical equipment, pharmaceutical products, electronic components, e.g. circuit boards, etc. [35].

The machines can form their own flexible or semi-rigid containers from a base film in the forming station. Heat is applied to soften the film before it is drawn by vacuum into moulds of the required shape and size. The formed containers are then advanced to the filling station where they are loaded with the product either manually or automatically. The filled containers advance to the vacuum/gas chamber where they are covered with lidstock web. At this stage they are evacuated, flushed and heat-sealed before progressing to the cutting section where the sealed packages are separated. Stamping or additional labelling may be done at this point. The system can be fully automated, hence it can be highly productive, cost-effective and labour-efficient.

The following are examples of a range of form-fill-seal machines for vacuum and/or gas-flushed packaging applications that are currently available on the market.

Tiromat from T.W. Kutter Inc., Avon, MA. This machine consists of six integrated sub-assemblies or stations, viz. web feed, heating, forming, loading, sealing, and package separation. All of these sections may be modified at the factory with standard or custom options to develop the desired final configuration. An optional sandwich preheater is available in the heating station when forming thicker rigid materials. Containers are normally formed in the female part of the mould using vacuum to draw the heated film, or a combination of vacuum and compressed air, when forming complex shapes or sizes. Positive male forming, using a combination of compressed air and vacuum, is also available with unusual draw characteristics, sharp contours and very complex shapes. Computer control of packaging variables is possible with Kutter SMART® Control and Process Validation System. Tiromat has been used for vacuum packaging or MAP of meat, dairy, and bakery products.

Ross Thermoform-Fill-Seal Model 550. These machines from Ross Industries, Inc., for pouch or tray packaging, are microprocessor-controlled and feature self diagnostics and interchangeable snap-in modules. They can handle *sous vide* as well as MAP products.

Klöckner-Hooper Packaging Machines. These thermoform-fill-seal machines can be used for products such as sliced luncheon meats, hot dogs, bacon strips, cheese products, cooked microwave-ready sandwiches, etc.

Other form-fill-seal equipment includes CVP System (Fresh-Vac), Multivac, and Coster. The operations of these machines are basically the same; the major differences are in the designs and the control mechanisms. However, the efficiency in evacuation and gas flushing may vary from one machine to the other. Therefore, samples should be taken regularly for analysis and quality control inspection. Quality control should include inspection of seal and package integrity. Weak or incomplete seals, pin holes, and excessively stretched film, especially around the corners of drawn containers, are among the major causes of MAP failure. Composition of headspace gas must be routinely analysed to ensure its accuracy. Headspace volume must also be adequate to provide the quantities of gases necessary for the desired shelf life. Storage and serving instructions must be legible, especially if a specific storage temperature is crucial to product safety. Storage and handling procedures of leftover product, after the package has been opened, should also be specified to prevent consumer abuse.

3.3 Storage temperature

The main objective of MAP is to extend the shelf life of perishable foods while maintaining their fresh or near-fresh quality. This must be accomplished without sacrificing product safety. Therefore, with the possible exception of baked goods and some products with low a_w and/or pH, MAP products must be refrigerated. As a general rule, the effectiveness of MAP decreases as the storage temperature increases because of the decreased solubility of CO_2, which serves as the major biostat in the product. Respiration rate also increases with increasing storage temperature, resulting in a decreased shelf life [15].

Generally, the reaction rate increases two- to three-fold for every 10°C rise in temperature. However, each type of fruit or vegetable product has a lower temperature limit below which chilling injury may occur, which results in faster rates of respiration and senescence. Hence, optimum storage temperatures must be determined for these commodities. It has been found that reduced O_2 and elevated CO_2 in the headspace can overcome the impact of low temperature injury in some commodities [22]. Therefore, a combination of MAP, using suitably permeable films, and proper storage temperature management can be effective in extending shelf life of fresh produce.

Permeability of packaging films is also a function of temperature. In general, permeability increases as temperature increases, and permeability to CO_2 increases faster than that to O_2 [40]. Therefore, a film which is suitable for MAP at one temperature may not be suitable at other temperatures. This reinforces the importance of careful temperature management of MAP products throughout the food chain. The ultimate, of course, is to develop "smart films" that can vary permeability to maintain desirable headspace atmosphere with changes in temperature.

MAP, through the biostatic activity of CO_2, inhibits the growth of a wide range of microorganisms. These include some pathogens such as *Staphylococcus aureus, Salmonella* and Enterobacteriaceae such as *Yersinia enterocolitica* and *Escherichia coli*. However, the effectiveness of CO_2 inhibition is strongly related to storage temperature. For example, *S. aureus* was completely inhibited in an atmosphere of 75% CO_2 and up to 25% O_2 at 12.8°C, while in air numbers increased 2-3.5 logs in 33 d [13]. Growth of *Salmonella* on chicken thighs and minced chicken was inhibited by 80-100% CO_2 at 10°C or lower [3, 11]. Counts of Enterococci on ground beef packaged in 60% CO_2 and stored at 10°C for 10 d were 3 log less than those packaged in air [36].

However, the major health hazard concern associated with MAP products is the possible growth of and toxin production by *Clostridium botulinum*. Of particular concern is the fact that these bacteria, especially the non-proteolytic strains, may grow and produce toxin on the products before detectable spoilage occurs. There have been many studies on the effect of MAs, especially of CO_2 and O_2, and storage temperature on the growth and toxin production of *C. botulinum* [3, 4, 7, 17, 27, 36, 38]. Though the results are far from conclusive, it may be generalized that MAP cannot guarantee the inhibition of *C. botulinum* unless the storage temperature is sufficiently low. Carbon dioxide is not found to be inhibitory to these organisms; in fact it may have a protective or even enhancing effect on them [3, 9, 14, 18], while O_2 is detrimental to growth [3].

Low storage temperature and presence of O_2 in the headspace are very important in preventing botulinal hazard in MAP products. Some non-proteolytic strains of *C. botulinum* grow at temperatures as low as 3.3°C, and atmospheres containing <2% O_2 favour the growth of *C. perfringens* [13]. For products which may be contaminated with these organisms, packaging in an atmosphere containing >2% O_2 and storage at close to 0°C provides adequate safety against their growth and toxin production. Unfortunately, temperature abuse in the food chain, especially at the retail and consumer levels, is not uncommon. Therefore, applicatios of the MAP system to products susceptible to contamination with *C. botulinum* must be done with utmost care.

3.4   Shelf-life dating of MAP food

3.4.1   Shelf-life determination

Shelf-life dating is an essential feature in marketing MAP products. It serves two main purposes: it informs the consumers the length of time they can expect to keep the products under specified storage conditions without significant changes in quality, and it cautions the consumers of possible spoilage and/or health hazard if kept beyond that date. While shelf-life dating is designed to protect both the industry and the consumers, determination of appropriate shelf life for MAP products represents one of the major obstacles and challenges in the application of MAP technology. Shelf life of foods may need to be redefined vis-à-vis MAP, and regulations governing shelf-life dating revised for this particular class of food.

In general shelf life of a food product is determined by the rate of deterioration of quality parameters such as nutrients, colour, flavour, texture, etc., as affected by storage conditions, particularly temperature. When a minimum acceptable quality of a food is defined, its shelf life at various temperatures may be predicted if the deterioration rates at those temperatures are known.

Mathematically, the loss of food quality for most foods can be represented by the following equation [19]:

$$\frac{dA}{d\theta} = kA^n$$

where: A = the quality parameter measured

θ = time

k = a constant which depends on temperature and a_w

n = order of the reaction

$\frac{dA}{d\theta}$ = the rate of change of A with time, θ. A negative sign is used for this term if the change represents a loss of A, and a positive sign if it represents a production of undesirable end product, e.g. microbial growth.

The value of n may be 0 (zero-order reaction) if the rate of change is constant with time, for example enzymatic degradation, nonenzymatic browning and lipid oxidation. It may be 1 (first-order reaction) when the rate of change is exponential, for example microbial growth and nutrient loss. It may, but very rarely, be other values such as 2, as in the degradation of vitamin C in liquid foods [19].

Shelf-life testing at several temperatures should be done to obtain values of the rate constant, k, at various temperature. This may be accomplished on the basis of the Arrhenius relationship between log k and reciprocal of absolute temperature (K). The plot of this relationship will also help establish the Q_{10}, which is the ratio of shelf life at two temperatures 10°C apart. Knowing the Q_{10} values, shelf-life data collected at high temperature can be used to predict shelf life at lower temperatures. This is the principle of Accelerated Shelf-Life Testing of foods where the testing may be conducted at high temperatures in a much shorter period of time to predict shelf life at lower temperatures [20].

These prediction models work relatively well with traditional foods where the state of the materials, the packaging conditions, and the storage environments are relatively stable or predictable. Even the shelf life of fresh produce may be estimated to a certain degree if the storage conditions are known. However, with MAP products the packaging conditions, atmosphere and storage environment become major variables affecting shelf life. To complicate the matter, deterioration of organoleptic and nutritional quality is no longer the only consideration in some MAP products. The potential for development of hazardous microorganisms under MAP environments has become the most important factor that must be taken into account in shelf-life dating of these products.

In shelf-life testing of MAP products there are at least three quality parameters that need to be evaluated at various temperatures over the test period:
1. Microbial safety
2. Microbial spoilage
3. Organoleptic changes

Much has been discussed about microbial safety and spoilage of MAP products [13, 21]. The major concern is the possible growth of anaerobic nonproteolytic toxin producers before signs of spoilage are detected. Meats and seafoods are particularly susceptible to this problem. Appropriate MAP regimes can be designed to minimize microbial hazards in these products. However, even if all the necessary precautions have been taken by the manufacturers, mishandling and thermal abuse through the distribution system, and by the consumers themselves, are possibilities that cannot be ignored. Therefore, in shelf-life testing based on microbial changes in MAP products, the designs of the tests must take these factors into account.

When microbial hazards are minimal, tests for spoilage and organoleptic changes take precedence in shelf-life determination. This may be accomplished by monitoring microbial growth and changes in colour, odour, texture and overall acceptability of the product. Organoleptic changes are often caused by spoilage microorganisms. Therefore, trained sensory panels can be used to determine the termination of shelf life of a product.

Proper labelling will do much to alleviate problems that may be encountered in marketing MAP products. The label should include a "pull" date as well as handling instructions. The pull date may be indicated by words such as "Sell by", "Use by", or "Best before", followed by a date by which the product should be withdrawn from the market or sold at a reduced price. The latter should be allowed only when consumption after that date does not pose a health problem. It has been estimated that suppliers or producers have 20% of the total distribution time, retailers 60%, and consumers 20% [19]. Therefore, the pull date should be set at no more than 80% of the realistic minimum shelf life obtained from a comprehensive shelf-life study.

The label should also have handling instructions such as "Keep Refrigerated", "Store at" or "Heat to ...°C   for ...min before serving". This will help to reduce product mishandling and development of health hazards.

A well-designed quality assurance system, such as the Hazard Analysis Critical Control Points (HACCP) should be an integral part of the overall processing and packaging operation. This system will help identify and monitor critical points and weak links in the operation where special care should be taken to minimize possible failures.

3.4.2 Regulations governing shelf life dating

In the U.S., there is no uniform or universally accepted open dating system for foods [19]. Some states require some form of shelf-life dating on food supplies, and some states may actually enforce these regulations, while others have no such regulations at all. Where shelf-life dating is used it is applied primarily on perishable or semi-perishable foods such as dairy products and meats. Perishable items are those food items with a shelf life of less than three months, and semi-perishable are those with a shelf life of three months to more than two years [19].

In Canada, Food and Drug Regulations require that prepackaged products with a durable life of 90 days or less and packaged at a place other than the retail premises must have on the label the durable-life date and instructions for proper storage if it differs from room temperature. If the product is packaged on the premises, the label must show the packaging date and the durable life of the food. These regulations appear to be uniformly enforced as there have been cases where retailers were prosecuted for having products on the shelves beyond their pull dates.

Notwithstanding the above, the shelf-life dating laws in North America are, at best, inconsistent and confusing. This may stem from the fact that:
1. There is very limited information available on shelf life of various types of food.

2. There are so many factors affecting shelf life that generalization of shelf life of any type of food is almost impossible.

3. Shelf-life testing must be done on individual products, which takes time and costs money.

4. There are conflicting views between industry, consumers and government as to the best way to apply shelf-life dating. Arguments focus on what dates to use: i.e. the "pack" date which indicates the date the food is manufactured, or the "display" date when the food is first put on the retail shelf, or the "pull" date which indicates when the product must be removed from the store shelf, or the "use by" date which indicates the date when the product should be either consumed or discarded [19].

Nevertheless, it is now generally agreed by all parties that some form of open dating should be used, not only to safeguard the consumer but also to standardize practices in the food industry. This need has been made more urgent by the introduction and growth of CAP, MAP, and vacuum-packaged products. When safety becomes a major consideration in marketing the product every effort must be made to determine accurately its safe life, and the information must be conveyed to the consumer in the most effective way. It is the responsibility of industry and government that this is done, because it will take only one serious incident connected with the consumption of these products to destroy or set back the further development of this technology.

For MAP products, the best shelf-life dating method is to use the "pull" date or "use by" date, together with clear storage and handling instructions. These dates must be based on thorough scientific studies of the shelf life of individual products. If regulations have not been promulgated, industry would be well advised to do it voluntarily. Bernard [5] stated that the success of this new generation of foods depends on the ability of industry to mass-market safe foods with a sufficient shelf life, with the knowledge of potential microbiological problems and the means to minimize them. Ronk *et al.* [32] urged both government and industry to do more research into the safety of food processing and packaging techniques, and the regulatory agencies to enhance their ability to monitor new technologies. They pointed out that our past knowledge and experience have shown that we must balance the need to adequately protect the public with the need to encourage innovation.

References

1. Anon. 1985. Controls ripening rate of packaged fruits/vegetables. Food Proc. 46(7): 84.
2. Anon. 1989. 'True CAP' with oxygen absorbers. Food Proc. 50(10):66.
3. Baker, R.C., Qureshi, R.A., and Hotchkiss, J.H. 1986. Effect of an elevated level of carbondioxide-containing atmosphere on the growth of spoilage and pathogenic bacteria at 2°, 7°, and 13°C. Poultry Sci. 65: 729-737.
4. Banwart, G.J. 1989. "Basic Food Microbiology", 2nd ed. AVI, Westport, CT.
5. Bernard, D. 1987. Issues of ultimate regulatory concern on refrigerated salads and precooked foods. CAP '87. Proceedings of the Third International Conference on Controlled/Modified Atmosphere/Vacuum Packaging. Schotland Business Research, Inc., Princeton, NJ, pp. 171-182.
6. Chazal, G. and Piepois, B. 1987. The Gemella Packaging System: Plastic/paperboard modified atmosphere packaging. CAP '87. Proceedings of the Third International Conference on Controlled/Modified Atmosphere/Vacuum Packaging. Schotland Business Research, Inc., Princeton, NJ, pp. 47-54.
7. Coyne, F.P. 1933. The effect of carbon dioxide on bacterial growth. Proc. Roy. Soc. (London) B113: 196-217.
8. Eidman, R.A.L. 1989. Advances in barrier plastics. Food Technol. 43(12): 91-92.
9. Foegeding, P.M. and Busta, F.F. 1983. Effect of carbon dioxide, nitrogen, and hydrogen gases on germination of *Clostridium botulinum* spores. J. Food Prot. 46: 987-989.
10. Fox, R.A. 1989. Plastic packaging - the consumer preference of tomorrow. Food Technol. 43(12): 84-85.
11. Gray, R.J.H., Elliott, P.H., and Tomlins, R.I. 1984. Control of two major pathogens on fresh poultry using a combination potassium sorbate/carbon dioxide packaging treatment. J. Food Sci. 49: 142-145, 179.
12. Griffin, R.C., Sacharow, S. and Brody, A.L. 1985. Principles of Packaging Development. AVI Publ. Co., Inc. Westport, CT.
13. Hintlian, C.B. and Hotchkiss, J.H. 1986. The safety of modified atmosphere packaging: A review. Food Technol. 40(12): 70-76.
14. Holland, D., Baker, A.N., and Wolf, J. 1970. The effect of carbon dioxide on spore germination in some clostridia. J. Appl. Bacteriol. 33: 274-284.
15. Hotchkiss, J.H. 1988. Experimental approaches to determining the safety of food packaged in modified atmospheres. Food Technol. 42(9): 55-64.
16. Idol, R.C. 1987. The American debut of oxygen absorbers from Japan. CAP '87. Proceedings of the Third International Conference on Controlled/Modified Atmosphere/Vacuum Packaging. Schotland Business Research, Inc., Princeton, NJ, pp. 305-310.
17. Kautter, D.A., Lynt, R.K., Lilly, R. Jr. and Solomon, H.M. 1981. Evaluation of the botulism hazard from nitrogen-packed sandwiches. J. Food Prot. 44: 59-61.

18. King, W.L. and Gould, G.W. 1971. Mechanism of stimulation of germination of *Clostridium sporogenes* spores by bicarbonate. *In* "Spore Research 1971," Barker, A.N., Gould, G.W. and Wolf, J. (Ed.), Academic Press, London.

19. Labuza, T.P. 1982. Shelf Life Dating of Foods. Food & Nutrition Press, Inc., Westport, CT.

20. Labuza, T.P. and Schmidl, M.K. 1985. Accelerated shelf-life testing of foods. Food Technol. 39(9): 57-64, 134.

21. Lioutas, T. 1988. Challenges of controlled and modified atmosphere packaging: a food company's perspective. Food Technol. 42(9): 78-86.

22. Lyons, J.M. and Breidenbach, R.W. 1987. Chilling injury. *In* "Postharvest Physiology of Vegetables," Weichman, J. (Ed.), Marcel Dekker, Inc., New York.

23. Mathieu, R.J. 1987. Outline monitoring of controlled atmosphere packaging with reliable time & indicator. CAP '87. Proceedings of the Third International Conference on Controlled/Modified Atmosphere/Vacuum Packaging. Schotland Business Research, Inc., Princeton, NJ, pp. 55-64.

24. McLean, C. and Warne, D. 1987. Flavaloc - the next step in CA packaging. CAP '87. Proceedings of the Third International Conference on Controlled/Modified Atmosphere/Vacuum Packaging. Schotland Business Research, Inc., Princeton, NJ, pp. 75-89.

25. Morrow, D.R. 1989. Recycling of plastic packaging materials. Food Technol. 43(12): 89-90.

26. Ooraikul, B. 1982. Gas packaging for a bakery product. Can. Inst. Food Sci. Technol. J. 15: 313-315.

27. Post, L.S., Lee, D., Solberg, M., Furgang, D., Specchio, J. and Graham, C. 1985. Development of botulinal toxin and sensory deterioration during storage of vacuum- and modified atmosphere-packaged fish fillets. J. Food Sci. 50(4): 990-996.

28. PPS. 1989. Microwavable oxygen scavenger. Packaging Trends Japan. No. 89-2. PPS Inc., Tokyo, Japan.: 6.

29. Rice, J. 1987. Gas-pack technology requires close attention to tricky variables. Food Proc. 48(6): 64-65.

30. Rice, J. 1989. Gas-emitting wafer - a cost-effective MAP approach. Food Proc. 50(10): 42.

31. Rice, J. 1989. Microwave packaging technology advances. Food Proc. 50(10): 40-41.

32. Ronk, R.J., Carson, K.L. and Thompson, P. 1989. Processing, packaging, and regulation of minimally processed fruits and vegetables. Food Technol. 43(2): 136-139.

33. Rothwell, T.T. 1986. An overview of modified/controlled atmosphere markets in England and Western Europe. CAP '86. Proceedings of the Second International Conference and Exhibition on Controlled Atmosphere Packaging. Schotland Business Research, Inc., Princeton, NJ, pp. 75-90.

34. Sacharow, S. and Griffin, R.C. Jr. 1973. Basic Guide to Plastics in Packaging. Cahners' Practical Plastics Series. Cahners Books, Div. of Cahners Publ. Co., Inc., Boston, MS.

35. Sacks, B. and Gore, A. 1988. Foodpak: Techniques and trends. Food Technol. N.Z. 23(1): 19-21.

36. Silliker, J.H. and Wolfe, S.K. 1980. Microbiological safety consid-
    erations in controlled atmosphere storage of meats. Food
    Technol. 34(3): 59-63.
37. Smith, J.P., Ooraikul, B., Koersen, W.J., Jackson, E.D. and
    Lawrence, R.A. 1986. Novel approach to oxygen control in modi-
    fied atmosphere packaging of bakery products. Food Microbiol. 3:
    315-320.
38. Stier, R.F., Bell, L., Ito, K.A., Shafer, B.D., Brown, L.A., Seeger,
    M.L., Allen, B.H., Porcuna, M.N. and Lerke, P.A. 1981. Effect of
    modified atmosphere storage on *C. botulinum* toxigenesis and the
    spoilage microflora of salmon fillets. J. Food Sci. 46: 1639-1642.
39. Wagner, B.F. and Vaylen, N.E. 1990. Getting to know the packag-
    ing activities: A comprehensive view of absorbers, scavengers,
    getters and emitters, and their kin for food preservation. CAP 90.
    Proceedings of the Fifth International Conference on
    Controlled/Modified Atmosphere/Vacuum Packaging. Schotland
    Business Research, Inc., Princeton, NJ, pp. 81-93.
40. Zagory, D. and Kader, A.A. 1988. Modified atmosphere packaging
    of fresh produce. Food Technol. 42(9): 70-77.

Chapter 4

MODIFIED ATMOSPHERE PACKAGING OF BAKERY PRODUCTS

B. Ooraikul
Department of Food Science, University of Alberta

4.1 Introduction

The bakery industry in the U.S. shipped about $22 billion worth of bakery products in 1987, representing approximately 7% of total value of food and kindred products in the same year [36]. In Canada, with population about 10% that of the U.S., the bakery industry shipped $1.33 billion worth of products, representing about 4% of total food industry shipments in the same year [4]. Bread forms about half of the total value of bakery products shipped, while pies, cakes and cookies make up one quarter, plain rolls and buns about 14% and the remaining products are doughnuts, fruit buns and sweet goods, pizza pies and crusts, frozen pre-cooked pies, frozen dough, meat pies, etc. [13]. The U.S. Department of Commerce forecasts aggregate shipments of all baked products to rise 1.5% annually between 1987 and 1992 [36].

The bulk of bakery products are marketed fresh at both wholesale and retail levels, but some, including many of the cooked and uncooked pies, pizzas, waffles, cream or fruit-filled cakes, are marketed frozen. Products are generally frozen to achieve a longer shelf life, especially for distant markets. With the increased use of microwave ovens, even more baked products may appear in the frozen food section of the retail food store.

Most fresh baked products have only a few days' shelf life at ambient temperature. Microbial growth on the products is the major cause of economic loss in the bakery industry. It is estimated that in the U.S. alone losses due to microbial spoilage are 1-3%, or over 90 million kg of product each year [21]. This estimate is based on in-plant and in-store spoilage only. Therefore, if losses at consumer level are also taken into account, total loss could reach a staggering proportion.

Moulds are the most common types of spoilage organisms associated with baked products. However, bacteria and yeasts are also implicated in the spoilage of some products where souring or fermentation occurs. Among the spoilage bacteria *Bacillus mesentericus* and *Serratia marcescens* are of particular interest [24]. The former causes "ropy bread" where the crumb structure becomes soft and sticky, with brown discoloration, due to protein and starch digestion during growth of the bacteria. The latter causes "bleeding bread", characterized by red spots in the bread. These types of spoilage are no longer a major problem, owing to improved plant sanitation and the use of preservatives such as propionates and sodium diacetate.

Bacterial food poisoning outbreaks attributable directly to bakery products are quite rare. They are most often traced to cream-filled cakes, in which *Staphylococcus aureus* is the most common causative agent [35]. Other bacteria that may be responsible for food poisoning in baked goods are *Bacillus cereus* and *Salmonella*. Todd *et al.* [47] reported that outbreaks of foodborne disease involving cream-type pies in Canada accounted for 0.6% of all foodborne incidents between 1973 and 1977, and 0.2% of outbreaks in the U.S. during the same period. Most of these problems are caused by contamination from workers and improperly cleaned equipment, and process failure in the plants producing the responsible ingredients, e.g. cream, milk powder, chocolate, egg, desiccated coconut, etc. Therefore, improved standards of personal and plant hygiene and stricter process control in those plants can contribute significantly to minimizing the problems. With the introduction of synthetic ingredients, e.g. imitation cream, and less-contaminated ingredients e.g. pasteurized frozen egg or dried egg and skim milk powder, the opportunity for microbial contamination should be even less.

Because yeasts generally require high water activity (a_w) for growth, they rarely cause spoilage of confectionery products made with flour. However, osmophilic yeasts, e.g. *Saccharomyces baillii var. osmophilus*, have been implicated in fermentation of products with high sugar content or fillings such as jam, fondant, fruit, marzipan, etc. [35]. During fermentation, the contaminating yeasts produce CO_2 and other metabolites such as alcohols, causing bubbling and cracking of the product. Metabolites also cause a typical yeasty off-odour and flavour which would be rejected by consumers.

Fermentation-free shelf life of some baked products can be increased up to 50% when their a_w and initial yeast load is reduced [35]. Lowering of storage temperature will further improve fermentation-free shelf life. However, in jam- or fruit-filled products, a considerable amount of soluble solids will migrate from jam or fruit to other parts of the product. Thus, shelf life of these products is normally shorter than for the fillings by themselves.

Contamination of products with osmophilic yeasts normally results from unclean utensils or equipment. Therefore, maintaining strict plant sanitation will minimize contamination and initial yeast load. Addition of preservatives such as sorbates, benzoates, and parabens is quite effective in retarding the growth of yeasts [24, 35].

Mould spoilage is most frequently encountered in the bakery industry. In many cases, mould growth determines the shelf life of the products. Apart from osmophilic yeasts, moulds are the most resistant of all microorganisms to low a_w [35]. Some can grow at a_w as low as 0.65. Therefore, they can cause spoilage of a wide range of products.

Water activity and storage temperature have been shown to be the two most important factors governing mould-free shelf life of cakes [35]: the lower they are the longer the shelf life of the products. However, there is a limit to how much a_w can be reduced without affecting the organoleptic properties of the product. Preservatives such as sorbates also inhibit mould growth but there is a limit to the amount of the preservatives that can be added without creating off odours and flavours. More effective methods for control of mould growth are needed. A promising development to achieve this is MAP using a mixture of CO_2 and N_2 in the gas atmosphere.

Early applications of CO_2 atmospheres for the packaging of bakery products to prevent aerobic spoilage, especially mould growth, were reported by Seiler [35] and Bogadke [6]. A 10% increase in CO_2 concentration in the atmosphere inside the package, together with a 10°F (5.5°C) reduction in storage temperature, doubled the mould-free shelf life of bread and cake. Outside of some European countries, this technique has not found much commercial application and as a result it has not been widely studied or fully developed.

In Canada, a university-industry collaborative study on MAP of English-style crumpets was initiated in 1978. By 1982, the MAP technology for crumpets was developed to the point where the company could confidently claim ambient shelf life of the product beyond one month. The MAP crumpets have since been marketed throughout North America. One of the factors that provided impetus to this renewed interest in MAP was the sudden increase in fuel costs in the late 1970s, which made the conventional frozen storage of these products very expensive.

English-style crumpet is a snack-type bakery product made from wheat flour, water, salt, vinegar, yeasts and/or leavening agents, and potassium sorbate as a preservative. The crumpets contain 52% moisture, an a_w of 0.97 and pH about 7. Traditionally, crumpets were consumed within a few days, before mould spoilage occurred, or they were frozen for prolonged storage. In New Zealand, crumpets are considered a winter snack, because there is less difficulty with mould growth during the cool months.

Initial experiments in which crumpets were packaged in 30x30 cm bags of polyethylene-coated nylon film, 90 μ thick, under various gas atmospheres, produced the results shown in Table 4.1 [27]. Potassium sorbate added at 0.07% of the batter did not prevent mould growth when the product was packaged in air. The product developed visible mould growth after 1 week of storage at 25°C. Longer storage resulted in swelling of the package due to CO_2 production from microbial activity. After 2 weeks of storage the CO_2 content in the headspace increased from a trace amount to 38%, by volume, while most of the O_2 was consumed. The product pH also dropped slightly over the same time period, most likely due to the absorption of CO_2 and production of some organic acids.

N_2 by itself appeared to inhibit mould growth, but had no effect on anaerobes which continued to grow and produce CO_2 and other metabolites to swell the package. CO_2 inhibited mould growth but it took a minimum of 41% CO_2 to retard other microbial activity and delay swelling of the package by 2 weeks.

Increasing CO_2 in the headspace to 50% appeared to be adequate for suppression of mould growth and other microbial activities for more than 2 weeks. A pure CO_2 atmosphere, however, resulted in shrinkage of the package, due to absorption of the gas by the product as well as some loss through permeation to the outside atmosphere. Therefore, a small quantity of N_2 in the headspace is necessary to prevent excessive shrinkage of the package as N_2 is not absorbed by the product.

Commercial application of $CO_2:N_2$ (1:1) atmosphere met with partial success. Occasional mould growth and swelling occurred in some packages within 2 weeks of storage at 25°C. Spotty mould growth was observed; this was later found to be due to the fact that with the commercial vacuum/gas-packaging equipment evacuation of air was often incomplete, leaving O_2 residues in the headspace as high as 3%. This level of O_2 was sufficient for growth of mould within 7 d. The short delay in growth could be due to the low O_2 pressure as well as the presence of potassium sorbate. However, the mould colonies remained small because the O_2 in the headspace was

Table 4.1 Changes on crumpets packaged in different gas atmospheres during storage at 25°C.[1]

| Storage time (days) | Gas composition | | | | | | | | Notes |
| | Original condition | | | | After storage | | | | |
	% CO_2[2]	% N_2	% O_2	pH	% CO_2	% N_2	% O_2	pH	
7	Trace	78.0	21.0	7.0	26.0	73.0	1.0	6.8	Mould growth Package started to swell
14	Trace	78.0	21.0	7.0	38.0	61.1	0.9	6.4	Package swollen
7	0.0	100.0	0.0	7.0	68.5	31.5	0.0	6.2	Package swollen No mould growth
7	31.0	67.7	1.3	7.0	66.0	34.0	0.0	-[3]	Package swollen No mould growth
14	41.0	58.0	1.0	7.0	63.0	37.0	0.0	-	Package swollen No mould growth
14	50.0	50.0	0.0	7.0	64.0	36.0	0.0	6.7	No change in product or package
14	100.0	0.0	0.0	7.0	87.0	13.0	0.0	-	Product remained acceptable but package shrunken

[1] Average of six replicates with SD between 0.04-0.2 for pH and 0.2-2.8 for gases
[2] % by volume
[3] No measurement

depleted. Packages with prolific mould growth, on the other hand, were invariably found to be leakers. The leaks were most often found on the heat seal areas, or on the bottom corners of the packages where the film was stretched during formation of the package.

These preliminary studies indicated that prevention of mould growth during the first week of storage is of paramount importance if mould-free shelf life of MAP crumpets is to be achieved. Removal of the traces of O_2, therefore, reduces mould problems. Conceivably, this could be accomplished in three ways: by mechanical or chemical means, providing conditions suitable for other organisms to compete with moulds for O_2, or by adding an inhibitor to prevent mould growth during the first 7 d while allowing other organisms to deplete the O_2.

In packages that showed no visible changes after 2 weeks of storage, CO_2 level in the headspace decreased gradually, in some cases to as low as 35%. The gradual decrease of CO_2 was attributed to the absorption of the gas into the water phase of the product [3, 9, 34] and permeation through the packaging film. The CO_2 permeability of the film at 25°C and 100% R.H. was 155 $cm^3/m^2/24h$. As a safeguard against inadequate CO_2 in the headspace, and to compensate for these losses, the ratio of $CO_2:N_2$ for crumpet packaging was subsequently raised to 3:2.

Swelling of the packages during storage was caused by the production of CO_2 by the anaerobic microorganisms in the product. In addition to CO_2, other metabolites such as lactic and succinic acids, ethanol, propanol and butanol were detected in the product. These metabolites gave the product a distinct fruity odour which was detected when the swollen packages were opened [38]. Microbial studies of the crumpets indicated that the predominant organisms were facultative anaerobes, *Leuconostoc mesenteroides* and *Bacillus licheniformis*, the former being the major producer of CO_2. The data in Figure 4.1 show total anaerobic, lactic acid bacteria (LAB), and mould counts of crumpets packaged in $CO_2:N_2$ (3:2) and stored at 25°C for 21 d [38]. Total anaerobic count increased rapidly, from 10^4 to 10^7 CFU/g, in the first week, then slowly to 10^8 CFU/g during the next 2 weeks. Most of the initial count consisted of *Bacillus* sp. However, after 7 d LAB became the predominant microorganism. Mould counts never exceeded 10^3 CFU/g throughout the 21-d storage period, and visible mould growth was not observed.

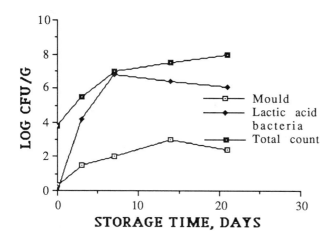

Figure 4.1.   Microbial counts for MAP crumpets stored at 25°C. Total
               and lactic acid bacteria counts were determined under
               anaerobic   conditions,   and   mould   counts   were
               determined under aerobic conditions.

The growth patterns of LAB, the major CO_2 producer, corre-
sponded with the change in package volume. The package began to
swell after 12 d of storage at 30°C due to the increase in CO_2 in the
headspace, following its initial decrease owing to absorption by the
product [39]. The onset of swelling was markedly delayed if the product was stored at 24°C, and no swelling occurred at 22°C or lower. The
effect of lower storage temperatures is two-fold: microbial growth is
retarded while CO_2 absorption is increased. The former was confirmed by the decrease in amounts of metabolites, e.g. lactic acid and
alcohols, produced at lower temperatures, and the latter by the reduction of product pH.

To extend the shelf life of MAP crumpets beyond 2 weeks it
was necessary to conduct detailed studies of various factors influencing shelf life of the product. Using response surface methodology
(RSM) as an experimental design, the effects of a_w, pH, microbial
load, concentrations of potassium sorbate in the product and CO_2 in
the headspace, and storage temperature on shelf life of the product
were investigated [29, 40]. The goal was to identify the most important factors and to optimize them to obtain the desired shelf life.

First- and second-order polynomial equations generated by
the experiments showed that a_w, pH and storage temperature are the
most important factors controlling the growth of LAB, hence CO_2
production, whereas a_w, storage temperature, and headspace CO_2
concentration controlled mould growth. Contour plots generated by
RSM revealed, for example, that CO_2 production could be reduced by
33% if a_w of the product was reduced from 0.990 to 0.976, and pH from
7.7 to 6.8, with the storage temperature of 20°C. Likewise, 25 d mould-

free shelf life could be attained if a_w was reduced to 0.965 and headspace CO_2 increased to 70%, with 20°C storage.

Crumpets formulated with the aid of linear programming to simulate the properties indicated by the RSM attained shelf life very similar to the predicted values [41]. However, organoleptic changes in the product limited the extent to which it could be reformulated. The most successful commercial practice to date is to use the regular product formula, with the addition of 0.34% sorbic acid and 0.43% glucono-δ-lactone (wet batter basis). This reduces a_w to about 0.96 and pH to 6.5, without altering sensory properties of the product. When packaged in $CO_2:N_2$ (3:2), with improved evacuation, packaging film strength and gas barrier characteristics, ambient shelf life of six weeks or more is not uncommon.

4.2 Comparative economy of MAP vs freezing

4.2.1 Process energy requirements

One of the major aims in developing MAP as an alternative to freezing preservation is to reduce energy consumption without adversely affecting the quality of the product during storage and marketing. Making the product shelf-stable would expand its market to include small corner stores which may not have adequate cold storage facilities. Furthermore, elimination of the need for refrigeration would make handling and display of the product easier in the bigger retail outlets. This means that refrigeration capacity would no longer be a limiting factor and stores would be able to carry larger volumes of the product.

Aboagye *et al.* [1] compared frozen and MAP crumpets with respect to energy consumption in production, storage and distribution of the products. It was assumed that the production costs, up to packaging operations, were the same for both methods of preservation. However, packaging requirements for MAP were quite different from frozen crumpets because different types of packaging films are used, with additional cost for gases in the MAP operation. The freezing line requires freezing facilities, frozen storage facilities in trucks and wholesale/retail warehouses, which would not be needed for MAP crumpets. Figure 4.2 shows a general processing and distribution flow chart for both MAP and frozen crumpets. The rectangle shows the boundary of the system over which the energy audit was taken, i.e. from mixing of the batter to the warehousing of the products.

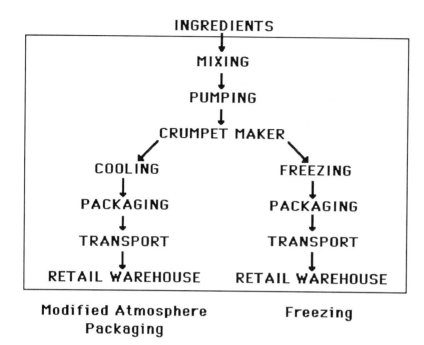

INGREDIENTS

MIXING

PUMPING

CRUMPET MAKER

COOLING	FREEZING
PACKAGING	PACKAGING
TRANSPORT	TRANSPORT
RETAIL WAREHOUSE	RETAIL WAREHOUSE

Modified Atmosphere   Freezing
Packaging

Figure 4.2.   Processing and distribution flowchart for MAP and frozen crumpets.

   In-plant data were collected on the time and length of operation of all equipment performing the different unit operations. The rates of consumption of fuel and power were either measured directly while the lines were operating, or they were estimated from data on equipment specifications and factory records. For calculating energy use, an average power factor of 0.8 and a motor efficiency rate of 80% were assumed [26]. The effects of start-up and shut-down were ignored. Also, energy used for general bakery overheads such as lighting, space heating and provision of sundry facilities was excluded.

   Energy for transportation was estimated from (i) energy required for the vehicle to travel and (ii) energy required to maintain the interior environment of the vehicle at the desired conditions. The average distribution radius of the product was estimated at 1000 km, and transportation was by diesel truck. Travel time by truck was estimated by assuming an average road speed of 100 km/h.

   Cost components considered in this study included: (i) in-plant costs involving only fuel and power, (ii) cost of energy consumed during storage for 30 d, and (iii) cost of packaging materials. Prices for the different commodities are shown in Table 4.2.

Table 4.2.   Data for calculating costs (Based on 1983 costs in Edmonton, AB)

	Energy value/unit	Cost/unit ($)
Electricity	3.6 MJ/kWh	0.0555/kWh
Natural gas	37.3 MJ/1000 L	2.67/GJ
Diesel fuel	38.0 MJ/L	0.30/L
Freezer bag	-	0.077/bag
Gas-barrier bag	-	0.075/bag
Nitrogen gas	-	1.70/kg
Carbon dioxide	-	0.83/kg

Data on energy consumption for the two systems of crumpet production are presented in Table 4.3. The first five unit operations were identical in electrical energy utilization. Electricity for mixing and pumping of batter and power for running associated motors for the griddle, cooling units and packaging machine together totalled 92 kWh for every 1000 kg (120 cases) of crumpets produced.

Table 4.3.   Energy use for the processing of crumpets

Unit operation	Energy Use (per 1000 kg)		
	Electrical (kWh)	Thermal (MJ)	Total Thermal Equivalents (MJ)
Common requirements (a):			
Mixing	28.50	-	102.60
Pumping	18.01	-	64.80
Baking	14.48	1050.00	1102.13
Cooling	1.76	160.00	166.34
Packaging	29.35	-	105.66
Additional MAP requirements (b):			
Transportation (per 1000 km)	-	1019.86	1019.86
MAP total (a+b)	92.10	2229.86	2561.39
Additional frozen storage requirements (c):			
Freezing	3.68	442.12	455.36
Storage (30 d)	7.44	642.82	669.60
Transportation (per 1000 km)	-	1053.36	1053.36
Frozen storage total (a+c)	103.22	3348.30	3719.85

Thermal energy requirements for process heating (baking) and cooling of 1000 kg crumpets were 1050 MJ and 160 MJ, respectively. Thus, fuel for baking was the largest fuel input. Fuel from different sources was converted to thermal equivalents (MJ) in order to provide a common basis for comparison. The results (Table 4.3) indicated that the freezing process requires 455 MJ, an additional 29% of the energy used to produce and cool the crumpets.

No additional direct energy costs were associated with storage of the MAP product; however storage of 1000 kg of frozen crumpets for 30 d used 7.4 kWh of electricity and 643 MJ of refrigeration. In terms of total thermal equivalents, frozen storage added approximately 43% to the consumption of process energy for the crumpets.

The transport distance of 1000 km is the weighted average distance travelled by truck between the factories and the points of retail warehousing. Diesel fuel consumption per 1000 kg crumpets was 1020 MJ for MAP crumpets. Refrigeration for frozen crumpets required an additional 34 MJ. Thus, transportation of frozen crumpets was about 3% more energy-intensive than transportation of MAP crumpets.

The total energy used from electricity, natural gas and diesel fuel was 2561 MJ/1000 kg and 3720 MJ/1000 kg for MAP and frozen crumpets, respectively. Hence, overall, the production, storage and transportation of frozen crumpets was 45% more energy-intensive than of MAP crumpets.

Table 4.4.   Costs of energy for processing and distribution of crumpets

		Cost ($1000/1 mill. kg)	
		MAP	Frozen
Initial processing:			
Mixing		1.58	1.58
Pumping		1.00	1.00
Baking		3.61	3.61
Cooling		2.48	2.48
Freezing		-	7.02
	Subtotal	8.67	15.69
Packaging:			
Bags		214.28	214.28
Gas		10.22	0.00
Energy		1.70	1.70
	Subtotal	226.20	215.98
Freezing storage (30 d)		-	37.66
Transportation (1000 km)		8.05	8.32
	Grand total	242.92	277.65

Table 4.4 shows process energy costs for MAP and frozen crumpets. The corresponding costs associated with the first four unit

operations were the same ($8.67/1000 kg) for both production systems. The freezing operation added approximately 80% to the cost of energy for the production of crumpets up to, but excluding, packaging.

Generally, the cost of packaging depends on the type of container used and the special requirements of the particular product. The data presented in Table 4.4 show packaging to be the most costly operation. The total cost for MAP of $243/1000 kg includes the costs of gas-impermeable bags and of N_2 and CO_2 gases. Most of the packaging cost may be attributed to the bags ($214/1000 kg crumpets). The total cost of N_2 and CO_2 was $10/1000 kg crumpets. The least expensive component was electricity, which only cost $1.70/1000 kg product.

Similarly, the packaging cost for frozen crumpets included freezer bags and electricity. Note that because a form-fill-seal operation was used in the industry for either MAP or frozen storage, the same film was, therefore, used in the cost analyses of both systems. The data in Table 4.4 show the overall cost of processing, transportation and storage of MAP crumpets to be $34.73/1000 kg less than that of frozen crumpets. Breakdown of the costs reveals the actual process energy costs for MAP and freezing lines to be only $18.42/1000 kg and $63.37/1000 kg, respectively. However, because of the overwhelming cost of the packaging compared with process energy cost (over 30 times), the three-fold process energy cost savings of MAP is reduced to an overall cost advantage of approximately 14% over freezing. Thus, for a bakery with an annual production rate of 1,000,000 kg crumpets, a saving of $34,730 can be achieved by using MAP instead of freezing (Table 4.5).

Table 4.5.   Energy consumption and process energy cost calculated for production of 1 million kg of crumpets

System	Energy Consumption (GJ)	Process cost ($)
Freezing	3,724	277,650
MAP	2,561	242,920
Saving	1,162	34,730

It is conceivable that the economic advantage of MAP would be higher if equipment depreciation and maintenance, labour and some intangible costs were included in the audit. For example, the depreciation expense of an IQF (individual quick freezing) unit would add more than $5 to the cost of producing 1000 kg of frozen crumpets. Moreover, the freezer unit would add about $500,000 to the initial capital investment.

Intangible benefits of MAP such as fast turnover due to on-shelf rather than in-freezer marketing should also be considered. These are important benefits which may not be easily measured in terms of production volume or dollar value. Nevertheless, a clear illustration of these benefits is the fact that Forcrest Foods Ltd. of Calgary, Canada increased its production volume four-fold and· expanded its market into the U.S. since its adoption of MAP. Therefore, the overall financial advantage of MAP over freezing system for bakery products may be far greater than the estimated 14%.

4.2.2 Advantages and disadvantages

In comparison with the freezing system, the advantages and disadvantages of MAP may be summarized as follows:

Advantages:

1. MAP provides products with fresh or near-fresh quality, with minimum changes in organoleptic properties, especially texture.
2. MAP products can be marketed on shelves rather than in the frozen product section. Therefore, attractive display is possible and, hence, faster turnover.
3. Fluctuating temperature during storage, shipment and marketing does not normally adversely affect shelf life or quality of MAP product, whereas loss of freezing during any stage in the life of frozen products can mean spoilage or loss of quality.
4. An MAP system requires considerably lower initial investment in processing equipment, storage and shipping facilities, thus lower operating, maintenance and depreciation costs.
5. Production, storage (30 d) and distribution of MAP products is about 45% less energy-intensive. Even when packaging costs are taken into account, an overall cost saving of approximately 14% can be realized with an MAP system, and when the intangible costs and benefits are considered, the cost advantage of MAP may be several-fold greater.

Disadvantages:

1. Ambient shelf life of MAP product is limited. For crumpets it is normally about 6-8 weeks.
2. Integrity of the package is crucial for MAP product.
3. Colour and texture changes through dynamic interchange of components within the product may be a problem for some products.
4. Packaging films with certain transmission characteristics may be costly.
5. There is a potential for the development of hazardous microorganisms, though in the case of bakery products the risks are extremely low. Since the application of MAP with crumpets about 10 years ago, not a single case has been reported concerning health hazard as a result of eating this product.

4.3 Application of MAP to other bakery products

Crumpets represent only one in a vast variety of bakery products. Bakery products have an extremely wide range of physical, chemical, microbial and organoleptic characteristics. The success of MAP with crumpets does not imply equal success with other products, unless they possess similar basic ingredients and product attributes. Studies were conducted to determine the suitability of the MAP system developed for crumpets on other products, to evaluate problems that might be encountered in those products, and possible solutions for the problems.

For these studies, bakery products were grouped into four categories, from those requiring simple ingredients and processes to the more complicated products. Some had only wheat flour as the major ingredient, while some contained eggs or dairy products, e.g. milk, skim milk, butter, cream or cheese. Some products contained several major ingredients, beside wheat flour, e.g. fruit, sugar, dairy products, chocolate and shortening. Some were yeast-leavened, while others used chemical leavening agents. A few products contained preservatives such as sodium sorbate or benzoate. The products chosen for the studies included:

    a. Dough or batter products
        1. Crumpet
        2. Waffle
        3. Cake doughnut
        4. Yeast doughnut
        5. Crusty roll
    b. Cake or pastry
        1. Chocolate Danish
        2. Carrot muffin
    c. Layer cake
        1. Strawberry layer cake
        2. Cherry cream cheese cake
    d. Pie or product with fillings
        1. Butter tart
        2. Apple turnover
        3. Mini blueberry pie
        4. Apple pie, both raw and baked

Products representing each category were selected based on their popularity and availability for the study. Waffles were produced in the laboratory. Fresh cake and yeast doughnuts, crusty roll, carrot muffin, mini blueberry pie, apple turnover and apple pies, which usually have an ambient shelf life of only a few days, were obtained from local bakeries. To determine the applicability of the MAP regime developed for Forcrest Foods crumpets to those produced from a different recipe and with different physical characteristics, crumpets from a bakery in Vancouver were obtained for the study. The rest of the products, which had been marketed across Canada in a frozen form, were obtained fresh from a bakery in Toronto.

The products were repackaged in 30x30 cm bags of 30/60 polyethylene-coated nylon film with total film thickness of 90 μ, using a Multivac Type AG500 vacuum/gas packaging equipment (W.R. Grace & Co. of Canada, Ajax, Ont.). The film had the following gas transmission rates: O_2, 8 $cm^3/m^2/24$ h, 75% RH, 4°C; CO_2, 20 $cm^3/m^2/24$ h, 75% RH, 4°C; water vapour, 4 $g/m^2/24$ h, 90% RH, 38°C. The packages were evacuated and back flushed with a mixture of CO_2 and N_2 (approx. 3:2). The proportion of gases from separate cylinders was regulated by a Smith's Proportional Mixer Model No. 299-006-1 (Tescom, Minneapolis, MA). Products packaged in air were used as controls. Storage trials were done at 25°C.

At regular intervals during storage, samples were analysed for pH, moisture content, a_w, headspace O_2, N_2 and CO_2, microbial metabolites, bacteria, mould and yeast counts, as described by Smith *et al.* [38]. The presence and number of some hygiene- and health-related microorganisms, e.g. *Escherichia coli, Salmonella* spp., *Clostridium perfringens, Staphylococcus aureus* and *Bacillus cereus,* in some products were also monitored, using the media and procedures described by Thatcher and Clark [46], Baird-Parker [2], and the FDA [16].

4.3.1   Dough and batter products

4.3.1.1   Crumpets

The major difference between the Forcrest Foods and Vancouver crumpets was that the former were chemically leavened containing sorbic acid and glucono-δ-lactone as preservatives, while the latter were leavened with both yeast and baking powder and contain sodium propionate as a preservative. The Vancouver crumpets were slightly lower in pH (6.00 vs 6.50), moisture (47 vs 52%), and a_w (0.95 vs 0.97). These resulted in less change in the MAP Vancouver crumpets after 21 d at 25°C, compared with Forcrest Foods crumpets. In fact, headspace CO_2 in the former dropped slightly from 60 to 57%, while in the latter it increased to 65%. This was reflected by the fact that total and *Lactobacillus* counts in Forcrest Foods crumpets were generally higher than in Vancouver crumpets after 21 d of storage. However, the latter contained 400 CFU/g of *S. aureus*, but not the former. Nevertheless, according to the British regulations [35] this level of *S. aureus* was within the maximum of 500 CFU/g allowed for bakery products.

This study confirmed the results of the previous experiments that the MAP system developed was suitable for crumpets.

4.3.1.2   Waffles

Waffles are a simple baked product with wheat flour as the main ingredient. They are normally marketed in frozen form, packaged in a wax-coated paper carton. For this study waffles were produced in the laboratory using the following recipe:

Bleached, enriched flour	4.2 kg
Baking powder	504.0 g
Sugar	336.0 g
Salt	84.0 g
Eggs	2.3 doz
Milk	4.0 L
Salad oil	1.05 L

Fresh waffles had a moisture content of 66%, a_w of 0.94 and pH of 7.2. After 24 d the moisture content of waffles packaged in air dropped to 54%, and those in CO_2/N_2 dropped to 59% at 20°C and 61% at 25°C. The a_w did not change significantly but the pH of the MAP waffles dropped to 6.1 after 24 d at 20°C.

Changes in headspace gases and microbial contents are shown in Figures 4.3 to 4.6. The pattern of change in headspace gases was similar to that of crumpets [39]. Headspace O_2 in air-packaged waffles was depleted after 7 d at 25°C and after 14 d at 20°C, accompanied by an increase of CO_2 to 31% at 20°C and 44% at 25°C at the end of 24 d storage (Figure 4.3). This change was paralleled by a steady growth of microorganisms in the product. Total count increased from $3x10^4$ to $1x10^7$ at 20°C and to $4x10^9$ CFU/g at 25°C after 1 week, while mould increased from 10 to $6x10^4$ and $5x10^3$, respectively over the same period (Figures 4.5 and 4.6). Lactic acid bacteria were <10 CFU/g in the air packaged product during 24 d storage. Total count at 20°C did not change significantly but declined markedly at 25°C. Mould count rose to $8x10^5$ after 2 weeks storage at both temperatures and it remained at that level for the next 10 d. Yeast was not detected on any samples, whether packaged in air or MAP.

In MAP waffles, with a CO_2/N_2 ratio of 2:3, CO_2 increased gradually to 62% after 24 d at 25°C, while at 20°C it was stable at 40% (Figure 4.4). Total count increased at a much slower rate than in air-packaged samples, reaching $8x10^5$ CFU/g at 20°C and $2x10^8$ CFU/g at 25°C after 7 d. Mould counts declined at both temperatures (Figures 4.5 and 4.6). Lactic acid bacteria, on the other hand, increased rapidly from <10 to $9x10^3$ CFU/g at 20°C after 7 d, then gradually to $4x10^4$ CFU/g after 24 d, while at 25°C there was no significant change until 7 d when they started to increase rapidly to $1x10^4$ CFU/g after 2 weeks with a further small increase after 24 d.

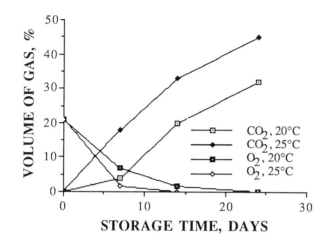

Figure 4.3.   Changes in headspace CO_2 and O_2 of air-packaged
waffles stored at 20 and 25°C.

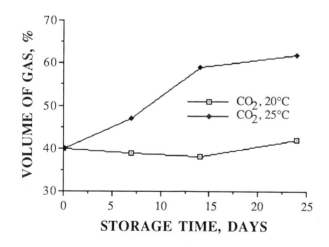

Figure 4.4.   Changes in headspace CO_2 of MAP waffles stored at 20
and 25°C.

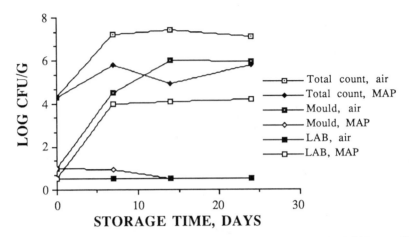

Figure 4.5.   Microbial counts of air-packaged and MAP waffles stored at 20°C.

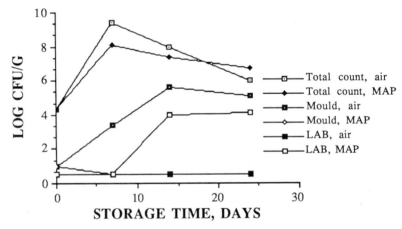

Figure 4.6.   Microbial counts of air-packaged and MAP waffles stored at 25°C.

Less than 10 CFU/g of coliform were detected in fresh waffles, and the count remained unchanged over 24 d storage at both temperatures whether the product was packaged in air or MAP.

Air-packaged waffles developed visible mould growth after a few days at both temperatures. MAP samples remained mould-free throughout the 24 d storage period but developed a slight fruity odour typical of that noted with MAP crumpets when the package had swollen. No noticeable swelling of packages of MAP waffles was observed. Apart from being somewhat crumbly in texture and slightly darker in colour, MAP waffles were organoleptically acceptable after 24 d at either 20 or 25°C. It appeared, therefore, that an MAP technique similar to that applied to crumpets could be used to extend the ambient shelf life of waffles.

4.3.1.3   Cake doughnuts

Cake doughnuts contain wheat flour, soybean flour, skim milk powder, egg yolk powder, shortening, baking powder, baking soda, nutmeg, salt, artificial vanilla and butter flavourings, and sodium propionate as a preservative. When packaged in air and stored at ambient temperature the product developed moulds in about 3-5 d. Tables 4.6 and 4.7 present the results of the chemical, physical and microbial analyses when MAP was applied to this product.

No significant chemical or physical change occurred on MAP doughnuts after 21 d at 25°C, except that sugar granules on the surface were dissolved or absorbed into the doughnut, making it sticky to touch. Undesirable flavour was not detected when the packages were opened at the end of the storage period.

Table 4.6. Chemical and physical properties of MAP cake doughnuts
          stored for 21 d at 25°C

Storage Time Days	pH	a_w	Moisture	Headspace Gas, % (v/v)[1]	
				CO_2	O_2
0	6.61	0.82	17.85	60.8	trace
3	6.56	0.80	16.14	60.5	trace
7	6.60	0.81	17.81	60.9	trace
14	6.52	0.81	15.98	60.7	trace
21	6.54	0.81	16.44	56.4	trace

[1]Average of 10-15 packages

Microbial changes in MAP cake doughnuts during 21 d storage at 25°C were insignificant, which explained the stability of the chemical and physical properties of the product over the same period. The microorganisms present had low initial counts and, in all cases, the counts declined to barely detectable after 21 d storage. Coliforms, clostridia and salmonellae were not present in the fresh product nor were they detected during the storage period (Table 4.7). Again, the study showed that the MAP technique as applied to crumpets could be used to extend the ambient shelf life of cake doughnuts.

Table 4.7. Microbial content of MAP cake doughnuts stored for 21 d at 25°C

Microorganism	Microbial Counts, CFU/g[1]				
	Time of Storage, Days				
	0	3	7	14	21
Total aerobes	Sp[2]	3.3×10^2	4.6×10^2	3.0×10^2	<100
Total anaerobes	Sp	6.0×10^2	2.6×10^2	Sp	<100
Moulds/yeasts	ND[3]	ND	ND	3.9×10^2	<100
Lactic bacteria	ND	690	280	140	ND
Staphylococci	<100	<100	<100	<100	<100
Bacilli	3.8×10^2	3.9×10^2	4.7×10^2	2.9×10^2	<100

[1] Average of two samples, each with duplicate plates
[2] Spreaders, unable to count
[3] Not detected in 10^{-1} dilution

4.3.1.4   Yeast doughnuts

Yeast doughnuts contain cake and bread flours, water, yeasts, granulated sugar, salt, skim milk powder, baking powder, eggs, shortening and butter flavouring. No preservative was used, and when the product was packaged in air its ambient shelf life was only 3 d owing to mould and/or yeast growth.

Chemical and physical properties and microbial content of MAP yeast doughnuts stored at 25°C for 21 d are shown in Tables 4.8 and 4.9.

Table 4.8. Chemical and physical properties of MAP yeast doughnuts stored for 21 d at 25°C

Storage Time Days	pH	a_w	Moisture	Headspace Gas, % (v/v)[1]	
				CO_2	O_2
0	6.48	0.91	27.98	60.1	trace
3	6.45	0.91	27.73	57.9	trace
7	6.45	0.91	27.85	60.7	[2]
14	6.28	0.90	26.56	70.3	-
21	6.06	0.90	27.16	79.8	-

[1] Average of 9-12 packages
[2] Not detectable

There were significant changes in MAP yeast doughnuts after 14 d of storage. The packages started to swell visibly owing to the increase in headspace CO_2 from 60 to 70% (Table 4.8). This was accompanied by a slight drop in pH from 6.48 to 6.28, and the disappearance of traces of headspace O_2. After 21 d of storage headspace CO_2 increased to about 80%. The doughnuts wrinkled slightly during storage, and after 21 d they had developed a slight fruity odour.

The increase in headspace CO_2 was caused by microbial activities, especially those of LAB, bacilli and yeasts, as evidenced by the proliferation of these microorganisms during the 21 d storage (Table 4.9). Obviously, moulds were outgrown by these microorganisms. Yeasts, on the other hand, increased from $2.5x10^2$ to $4.3x10^6$ after 14 d. It was not possible to determine whether the yeasts survived the frying process, or they were the result of recontamination. Nevertheless, they were most likely the major responsible agent for the production of CO_2 and fruity aroma.

Table 4.9. Microbial content of MAP yeast doughnuts stored for 21 d at 25°C

Microorganism	Microbial Counts, CFU/g[1]				
	Time of Storage, Days				
	0	3	7	14	21
Total aerobes	$2.0x10^2$	$1.6x10^4$	$7.3x10^5$	$9.8x10^6$	$1.0x10^7$
Total anaerobes	$1.5x10^2$	$1.5x10^4$	$6.3x10^5$	$6.2x10^6$	$7.3x10^6$
Yeasts	$2.5x10^2$	$6.7x10^3$	$1.1x10^4$	$4.3x10^6$	$2.2x10^6$
Lactic acid bacteria	<100	$9.9x10^3$	$5.6x10^4$	$3.6x10^6$	$1.4x10^5$
Staphylococci	<100	$8.8x10^3$	$3.0x10^5$	$4.6x10^6$	$4.1x10^6$
Bacilli	$1.5x10^2$	$6.6x10^3$	$4.5x10^5$	$4.7x10^6$	$7.4x10^6$

[1]Average of two samples, each with duplicate plates

The proliferation of staphylococci is of concern. The presence of these and other potentially hazardous microorganisms, e.g. *Salmonella* spp., *S. aureus*, *E. coli*, *B. cereus* and *Clostridium* spp. in dough and/or baked goods has been reported by Seiler [35] and ICMSF [23]. However, in yeast doughnuts no coliforms, *Salmonella* or *Clostridium* were detected, and only 1-2% of staphylococci tested was coagulate positive. Nevertheless, yeast doughnuts represent a potentially hazardous product for MAP storage. Proper process and quality control procedures must be observed to prevent contamination of the product with these microorganisms. It appeared, therefore, that yeast doughnut, and/or the MAP technique must be modified before the ambient shelf life of the product could be extended beyond 7 d.

4.3.1.5   Crusty rolls

Crusty rolls contain wheat flour, water, yeast, salt, sugar, skim
milk powder and shortenings. When packaged in air and stored at
ambient temperature the product had a shelf life of 3 d before
moulds started to grow. Chemical and physical properties and
microbial content of MAP crusty rolls stored at 25°C for 30 d are presented in Tables 4.10 and 4.11.

Table 4.10.   Chemical and physical properties of MAP crusty rolls
stored for 30 d at 25°C

| Storage Time | | | | Headspace Gas, % | |
Days	pH	a_w	Moisture, %	CO_2	O_2
0	5.62	0.96	29.28	54.8	trace
3	5.61	-	30.39	51.7	trace
7	5.63	-	31.00	51.5	trace
14	5.60	-	29.01	50.5	trace
21	5.59	-	29.20	46.5	0
30	5.56	0.91	27.58	43.2	0

Table 4.11.   Microbial content of MAP crusty rolls stored for 30 d at
25°C

| | Microbial Counts, CFU/g[1] | | | | | |
| | | | Time of Storage, Days | | | |
Microorganism	0	3	7	14	21	30
Total aerobes	ND[2]	$8.6x10^2$	$1.1x10^4$	$9.6x10^4$	$1.9x10^5$	$6.8x10^5$
Total anaerobes	ND	$4.9x10^2$	$9.5x10^3$	$2.0x10^5$	$1.3x10^5$	$5.5x10^5$
Moulds	ND	ND	ND	$1.4x10^4$	$7.6x10^4$	$4.8x10^4$
Lactic bacteria	ND	$1.5x10^2$	$7.8x10^3$	$1.9x10^5$	$5.6x10^4$	$3.9x10^5$
Staphylococci	ND	$6.3x10^2$	$3.8x10^3$	$4.8x10^4$	$4.8x10^4$	$4.6x10^5$
Bacilli	ND	$1.7x10^2$	$8.5x10^3$	$2.0x10^5$	$1.4x10^5$	$5.0x10^5$

[1]Average of two samples, each with duplicate plates
[2]Not detected in 10^{-1} or 10^{-2} dilution

There was no significant change in pH and moisture content
over a 30 d period, but a_w declined markedly from 0.96 to 0.91. The
important changes were the decrease in headspace CO_2 from about 55
to 43% and disappearance of the trace amount of O_2 after 14 d (Table
4.10).   These latter changes were accompanied by visible mould
growth after 14 d. Total count, LAB, bacilli and staphylococci in-

creased only moderately over a 30 d period, while coliforms, *Clostridium* and *Salmonella* spp. were not detected.

The results indicated that the regular MAP technique could extend the ambient shelf life of crusty rolls only to less than 2 weeks, the major problem being visible mould growth. Because the number of other bacteria was low, moulds could compete effectively for headspace O_2 to initiate growth. Owing to the porous nature of the product, evacuation followed by backflushing with gases in the packaging process might not result in efficient reduction of the headspace O_2. Even with low gas permeability of the packaging film, it was apparently not low enough to prevent the escape of CO_2 and accumulation of headspace O_2 to the level required for the growth of moulds. Unlike crumpets, waffles or doughnuts, in MAP crusty rolls there was no in-package production of CO_2 by microorganisms or, if there was, the rate was too low to even maintain the desirable proportion of headspace CO_2. Therefore, the product and/or MAP technique must be modified to further extend the shelf life of crusty rolls.

4.3.2   Cake or pastry-type products

4.3.2.1   Chocolate Danish

Chocolate Danishes are pastry products with chocolate-based filling, normally marketed in a frozen form and served immediately upon thawing. The ingredients include flour, margarine, icing sugar, water, liquid whole egg, sugar, shortening, invert sugars, glucose, mono- and diglycerides, chocolate, yeast, cocoa, skim milk powder, vegetable oil, whey powder, modified starches, salt, hydrogenated soybean oil, polyoxyethylene stearate, artificial flavour, baking powder, lecithin, baking soda, sodium caseinate, agar and gelatin.

Results of chemical, physical and microbial analyses of air-packaged and MAP chocolate Danish stored for 28 d at 25°C are shown in Tables 4.12 and 4.13. The data in Table 4.12 shows that significant changes in headspace gases of the air-packaged product occurred over the 28-d period at 25°C. Oxygen was almost depleted while CO_2 increased gradually from undetectable to more than 7% at the end of the storage. This was accompanied by a slight decrease in pH from 6.28 to 6.04. MAP product, on the other hand, remained essentially unchanged except for the gradual decline of CO_2 due to permeation through the film and absorption by the product, accompanied by a slight drop in pH.

The chemical and physical changes in the product (Table 4.12) were reflected by the microbial changes shown in Table 4.13. Mould growth was visible on the air-packaged chocolate Danishes around 14 d. Mould and yeast counts increased during 28 d storage from <300 to 7×10^7 and 2.8×10^7 CFU/g, respectively.

Table 4.12. Chemical and physical properties of chocolate Danishes stored for 28 d at 25°C

Packaging/ Storage	Storage Time, Days	pH	a_w	Moisture, %	Headspace Gas, %	
					CO_2	O_2
Air	0	6.28	0.83	22.54	-	20.0
	14	6.17	NT	NT	Tr	15.0
	21	6.01	NT	NT	3.7	12.4
	28	6.04	NT	NT	7.2	7.8
MAP	0	6.28	0.83	22.54	59.0	Tr
	14	6.15	NT	NT	58.8	-
	21	6.08	NT	NT	49.8	-
	28	6.11	0.83	22.03	48.6	-
Frozen	28	6.31	0.84	22.71	NA	NA

NT = Not Tested; Tr = Trace; NA = Not Applicable

Table 4.13. Microbial content of chocolate Danishes stored for 28 d at 25°C

Packaging/ Storage	Microorganism	Microbial Counts, CFU/g			
		Time of Storage, Days			
		0	14	21	28
Air	Total aerobes	<300	4.2×10^4	2.1×10^6	8.7×10^7
	Moulds	<300	3×10^4	2×10^6	7×10^7
	Yeasts	<300	3.7×10^4	3.5×10^6	2.8×10^7
MAP	Total anaerobes	<300	<300	<300	<300
	Lactic acid bacteria	<300	<300	<300	<300
	Moulds and yeasts	<300	<300	<300	<300
Frozen	Total aerobes	<300	NT[2]	NT	<300
	Moulds and yeasts	<300	NT	NT	<300

[1]Average of two samples, each with duplicate plates
[2]Not tested

MAP product, on the other hand, remained virtually unchanged after 28 d at 25°C, similar to the frozen product. Staphylococci, coliforms and clostridia were not detected in chocolate Danishes. Therefore, MAP developed for crumpets can be used to extend the ambient shelf life of chocolate Danishes to at least 28 d.

4.3.2.2   Carrot muffins

Carrot muffins are made from flour, sugar, vegetable oil, shortening, salt, skim milk powder, powdered egg, baking soda, spice, artificial flavour and grated carrots. No preservative is used and their ambient shelf life when packaged in air is about 3 d.

Chemical and physical properties of MAP carrot muffin stored at 25°C for 21 d are shown in Table 4.14. No significant change occurred in pH, a_w or moisture content during storage, except headspace CO_2 which declined from 59 to about 50% while O_2 remained undetectable throughout.

Table 4.14.   Chemical and physical properties of MAP carrot muffins stored for 21 d at 25°C

Storage Time Days	pH	a_w	Moisture	Headspace Gas, % (v/v)[1]	
				CO_2	O_2
0	8.79	0.91	35.14	59.1	ND[2]
3	8.99	0.91	34.60	56.7	ND
7	8.83	0.91	33.84	53.8	ND
14	8.83	0.91	34.98	52.8	ND
21	8.81	0.91	34.45	49.7	ND

[1]Average of 9 - 12 packages
[2]Not detectable

All microbial counts were quite low in MAP carrot muffins and did not increase significantly over the 21-d period (Table 4.15). Neither mould nor yeast growth was detected on the MAP product, while those packaged in air developed luxuriant mould growth after 10 d. Coliforms and clostridia were not detected.

Table 4.15.   Microbial content of MAP carrot muffin stored for 21 d at 25°C

	Microbial Counts, CFU/g[1]				
	Time of Storage, Days				
Microorganism	0	3	7	14	21
Total aerobes	<100	<100	3.7×10^2	4.3×10^5	8.6×10^5
Total anaerobes	ND[2]	<100	2.9×10^2	3.5×10^5	7.9×10^5
Lactic bacteria	ND	ND	ND	2.7×10^3	3.9×10^2
Staphylococci	<100	<100	1.7×10^2	1.3×10^5	6.2×10^4
Bacilli	<100	<100	6.6×10^2	3.0×10^5	7.9×10^5

[1]Average of two samples, each with duplicate plates
[2]No growth detected in the 10^{-1} dilution

After 21 d storage there was no noticeable swelling of the packages, no off-flavour when the packages were opened and the product maintained good texture and colour. The MAP system, therefore, is suitable for the extension of the ambient shelf life of carrot muffins.

4.3.3   Layer   cakes

4.3.3.1   Strawberry   layer   cake

Strawberry layer cake, consisting of three 2-cm layers of cake interspersed with three 0.5-cm layers of fruit-flavoured cream frost-ing, is a more complex product with respect to the ingredients and baking procedure compared with the products previously described. Its ingredients include wheat flour, liquid whole eggs, glucose-fructose, strawberries, margarine, hydrogenated palm kernel oil, modified starch, vegetable oil, modified whey powder, corn syrup solids, baking powder, mono and diglycerides, shortening, salt, arti-ficial flavours and colours, lecithin, polysorbate 60, hydroxypropyl methylcellulose, polyglycerol esters of fatty acids, sodium alginate, sodium benzoate and sorbitan monostearate.

Chemical, physical and microbial changes in strawberry layer cake are presented in Tables 4.16 to 4.19. Fresh product had an overall moisture content of 37.24%, a_w of 0.91 and pH of 6.66 (Table 4.16). However, the three attributes differed considerably between the two layers, i.e. cream frosting and cake, as shown in Table 4.17. Microbial analysis on day 0 showed all counts to be <300 CFU/g, and no coliforms or spore-forming anaerobes were detected (Table 4.18).

Table 4.16.   Chemical and physical properties of strawberry layer cake stored for 21 d at 25°C

Packaging/ Storage	Storage Time, Days	pH	a_w	Moisture, %	Headspace Gas, %	
					CO_2	O_2
Air	0	6.66	0.91	37.24	-	20.8
	7	6.66	NT	NT	Tr	20.9
	14	6.01	NT	NT	30.5	5.0
	21	6.04	NT	NT	49.6	Tr
MAP	0	6.66	0.905	37.24	61.8	Tr
	7	6.71	NT	NT	62.6	Tr
	14	6.52	NT	NT	63.0	Tr
	21	6.10	0.90	35.4	64.3	Tr
Frozen	21	6.63	0.91	36.21	NA	NA

NT = Not Tested; Tr = Trace; NA = Not Applicable

In the air-packaged product moulds and yeasts accounted for practically all the increase in microbial counts during 21 d storage. Mould and yeast counts increased from <300 to 1.9×10^6 and 2.4×10^7, respectively (Table 4.18). Half of the product showed surface mould growth after 7 d and 100% after 14 d, accompanied by package swelling, musty smell and sticky frosting (Table 4.19). Mould growth was prolific after 21 d and the packages swelled to the bursting point. The pH of the product dropped from 6.66 to 6.04 and headspace CO_2 increased from 0 to 50% (Table 4.16).

Table 4.17.   Chemical and physical properties of two layers of strawberry layer cake stored for 21 d at 25°C

Packaging/ Storage	Storage Time, Days	Frosting			Cake		
		pH	a_w	Moisture, %	pH	a_w	Moisture, %
Air	0	6.30	0.91	40.58	7.40	0.89	33.16
	7	6.16	NT	NT	7.00	NT	NT
	14 & 21[1]	NT	NT	NT	NT	NT	NT
MAP	0	6.30	0.91	40.58	7.40	0.89	33.16
	7	6.16	NT	NT	7.00	NT	NT
	14	6.09	NT	NT	6.63	NT	NT
	21	6.04	0.91	36.02	6.62	0.89	32.74
Frozen	0	6.30	0.91	40.58	7.40	0.89	33.16
	21	6.04	0.91	NT	7.39	0.89	NT

[1]Not tested owing to mould growth

Table 4.18.   Microbial content of strawberry layer cake stored for 21 d at 25°C

Packaging/ Storage	Microorganism	Microbial Counts, CFU/g[1]			
			Time of Storage, days		
		0	7	14	21
Air	Total aerobes	<300	$4.3x10^6$	$3.8x10^7$	$3.4x10^7$
	Moulds	<300	$6.0x10^4$	$8.6x10^5$	$1.9x10^6$
	Yeasts	<300	$3.7x10^6$	$2.8x10^7$	$2.4x10^7$
MAP	Total anaerobes	<300	$3.7x10^4$	$2.3x10^6$	$8.1x10^6$
	Lactic acid bacteria	<300	$1.5x10^3$	$7.6x10^5$	$3.8x10^6$
	Yeasts	<300	$3.5x10^4$	$4.7x10^6$	$2.2x10^7$
Frozen	Total aerobes	<300	NT[2]	NT	<300
	Moulds and yeasts	<300	NT	NT	<300

[1]Average of two samples, each with duplicate plates
[2]Not tested

MAP product, on the other hand, showed no mould growth but yeasts increased from <300 to $2.2x10^7$ CFU/g over 21 d (Table 4.18). Lactobacilli also increased from <300 to $3.8x10^6$ CFU/g, while staphylococci and coliforms were <300 CFU/g and anaerobic spore formers were not detected throughout storage. The growth of yeasts and lactobacilli was accompanied by a pH drop from 6.66 to 6.10 and a headspace CO_2 increase from 61.8 to 64.3%. Moisture of the whole cake decreased from 37.24 to 35.40%. When analysed separately, moisture and pH of frosting and cake decreased proportionally (Table 4.17). After 21 d there was a slight swelling of the packages, slight yeasty odour and stickiness of the frosting (Table 4.19). These changes were most likely caused by yeast growth. The frozen samples remained virtually unchanged in all respects after 21 d at -18°C.

Therefore, the MAP system as applied to crumpets can be used to extend ambient shelf life of strawberry layer cake to about 2 weeks. Yeast growth appears to be the major limiting factor.

Table 4.19.   Physical changes in strawberry layer cake stored for 21 d at 25°C

Packaging/ Storage	Days	Package/Product Surface Appearance	Odour	Texture/Colour
Air	0	Normal	Normal	Icing smooth, soft Cake soft, spongy Layers easily separated
	7	Mould on surface of 1/2 of packages, swelling not noticeable	Quite normal	As above
	14	Mould on surface of all packages, considerable swelling	Mouldy musty smell	Icing somewhat sticky Cake texture still spongy
	21	Prolific mould growth on and in product Packages ready to burst	Mouldy musty	Sticky icing texture still okay
MAP	0	Normal	Normal	As Day 0 in Air
	7	No noticeable change	No noticeable change	No noticeable change
	14	No noticeable change	Slight stale "old" odour	No noticeable change
	21	Slight swelling of packages	Slight "yeasty" flavour	Icing slightly sticky
Frozen	21	Normal	Normal	As Day 0

4.3.3.2   Cherry cream cheese cake

Cherry cream cheese cake consists of three distinct layers: a Graham wafer crust which is made of wheat flour and shortening; a cream cheese filling consisting of cheese and whole egg; a cherry topping which is a mixture of fruit, sugar and thickening materials. Each layer poses a very different storage problem, yet they relate to one another as a single product as they are intimately in contact with one another.

The ingredients include: cherries, sugar, cream cheese (milk, water, bacterial culture, salt, carob bean gum), water, baker's cheese, glucose, liquid whole eggs, flour, vegetable oil shortening, Graham flour, skim milk powder, modified starches, hydrogenated soybean oil, molasses, salt, vegetable oil margarine, propylene glycol alginate, xanthan gum, baking soda, ammonium bicarbonate, natural flavour, locust bean gum, cinnamon, ascorbic acid, baking powder, calcium carrageenan and colour.

Chemical, physical and microbial changes in cherry cream cheese cake stored at 25°C for 21 d are presented in Tables 4.20 to 4.23. The cakes as a whole have an average moisture content of 49.9% with a_w of 0.94 and pH 4.51 (Table 4.20). However, these attributes differ considerably among the three layers: cherry topping with 64%, 0.942, 3.82; cream cheese filling with 48%, 0.934, 4.87; and Graham crust with 22.4%, 0.866 and 7.16, respectively (Table 4.21).

Table 4.20.   Chemical and physical properties of cherry cream cheese cake stored for 21 d at 25°C (cross-section of the cake was blended for triplicate analyses)

Packaging/ Storage	Storage Time, Days	pH	a_w	Moisture, %	Headspace Gas, %	
					CO_2	O_2
Air	0	4.51	0.94	49.90	-	20.8
	7	4.43	NT	NT	13.2	16.0
	14	4.39	NT	NT	76.7	2.7
	21	4.33	0.92	53.38	93.7	0
MAP	0	4.51	0.94	49.90	52.0	Tr
	7	4.41	NT	NT	51.8	Tr
	14	4.44	NT	NT	53.5	0
	21	4.37	0.93	49.45	73.7	0
Frozen	21	4.47	0.93	48.11	NA	NA

NT = Not Tested; Tr = Trace; NA = Not Applicable

Day 0 microbial analysis showed all microbial counts to be <300 CFU/g, with no clostridia, coliforms or staphylococci (Table 4.22).

Table 4.21. Chemical and physical properties of three layers of cherry cream cheese cake stored as whole cake for 21 d at 25°C

Packaging /Storage	Storage Day	Topping			Filling			Crust		
		pH	aw	Moisture %	pH	aw	Moisture %	pH	aw	Moisture %
Air	0	3.82	0.94	64.00	4.87	0.93	48.00	7.16	0.87	22.40
	7	4.06	NT[1]	NT	4.60	NT	NT	6.49	NT	NT
	14	4.12	NT	NT	4.47	NT	NT	5.99	NT	NT
	21	4.26	NT	NT	4.45	NT	NT	5.67	NT	NT
MAP	0	3.82	0.94	64.00	4.87	0.93	48.00	7.16	0.87	22.40
	7	4.15	NT	NT	4.57	NT	NT	6.47	NT	NT
	14	4.29	NT	NT	4.41	NT	NT	5.59	NT	NT
	21	4.41	0.93	58.66	4.45	0.93	47.59	5.39	0.91	31.38
Frozen	21	3.91	0.94	57.45	4.90	0.93	46.88	7.19	0.84	20.96

[1]Not tested

Total count in air-packaged cherry cream cheese cake rose rapidly after 7 d from <300 to 1.5×10^6 CFU/g, practically all of which was due to yeasts (Table 4.22). This was accompanied by increase in headspace CO_2 (Table 4.20) and swelling of packages (Table 4.23). The product had an obvious fermented odour and was sticky to touch. Further increase in yeast count and CO_2 production resulted in bursting of many packages after 14 d, with more distinct fruity/fermenting odour, slimy topping, sticky filling and soft crust.

Table 4.22.   Microbial content of cherry cream cheese cake stored for 21 d at 25°C

Packaging/ Storage	Microorganism	Microbial Counts, CFU/g[1]			
			Time of Storage, Days		
		0	7	14	21
Air	Total aerobes	<300	1.5×10^6	1.7×10^7	9.1×10^6
	Yeasts	<300	2.6×10^6	2.4×10^7	1.2×10^7
MAP	Total anaerobes	<300	2.0×10^4	4.1×10^5	1.8×10^6
	Yeasts	<300	2.9×10^4	8.4×10^5	5.8×10^5
Frozen	Total aerobes	NT[2]	NT	NT	<300
	Moulds and yeasts	NT	NT	NT	ND

[1]Average of two samples, each with duplicate plates
[2]Not tested

The rate of microbial growth on MAP product was substantially lower than on the air-packaged product. Total count increased from <300 to 2×10^4 CFU/g after 7 d and 1.8×10^6 CFU/g after 21 d, all of which was due to yeasts. The MAP product remained largely unchanged until 14 d when it started to develop yeasty odour with slimy topping and sticky filling and crust. At 21 d the packages swelled noticeably.

Neither staphylococci nor clostridia were detected on the product and coliforms were present at <300 CFU/g after 21 d at 25°C.

As in the case of strawberry layer cakes, MAP without modification could extend the ambient shelf life of cherry cream cheese cakes to about 2 weeks. Similarly, the major problem was yeast growth which produced CO_2 to swell the packages as well as fermenting odour. Control of yeast growth on this product would be of prime concern in further extension of shelf life.

Table 4.23.   Physical changes in cherry cream cheese cake stored for 21 d at 25°C

Packaging/ Storage	Days	Package/Product Surface Appearance	Odour	Texture/Colour
Air	0	Normal	Normal	Topping: bright red, easily scooped off Filling: fluffy, light, white Crust: crumbly; layers easily separated
	7	Noticeable swelling	Obviously fermented	Sticky to handle Filling: slight yellowing Topping: slight fading
	14	Bursting of 3/8 of packages	Distinct fruity/ fermented	Topping: slimy Crust: moist Filling: sticky, pasty
	21	Remaining packages extremely swollen	As 14 d	Worse than 14 d
MAP	0	Normal	Normal	As Day 0 in air
	7	No noticeable change	Slightly fruity	No obvious change
	14	No noticeable change	Wine-like	Topping: slimy Filling: pastier, yellower than 14 d in air Crust: stickier
	21	Noticeable swelling	Distinctly fruity	More advanced than 14 d
Frozen	21	Normal	Normal	As on Day 0

4.3.4   Pies or products with fillings

4.3.4.1   Butter tarts

Butter tarts consist of: pastry - unbleached pastry flour, water, lard, salt; filling - water, yellow sugar, butter, liquid whole eggs, vegetable oil margarine, raisins, artificial vanilla flavour and vinegar. Fresh product had pH of 5.69, moisture of 19.75% and a_w of 0.784 with total microbial count of <100 CFU/g. These attributes remained essentially unchanged over 21 d storage at 25°C, whether air packaged or MAP. Air-packaged product swelled slightly with slight fruity odour after 21 d. The crust of both MAP and air-packaged butter tarts was crumbly at the end of the storage trial.

This product was the most stable among the bakery products examined, because of its low pH, a_w and moisture. Ooraikul et al. [29] reported pH and a_w to be two of the most important factors controlling the ambient shelf life of crumpets. This fact was well illustrated by the MAP storage of butter tarts. Though it appeared that the product could be stored in air without undue problems, it is prudent to use MAP as a safeguard for lengthy ambient storage.

4.3.4.2   Apple turnovers

Apple turnovers consist of: pastry - water, flour, shortening, salt; filling - water, sugar, dried apple chips, starch, salt, citric acid, spices, dehydrated lemon juice (containing corn syrup solid, lemon juice solid, lemon oil, butylated hydroxyanisole), sodium benzoate and potassium sorbate. The prepared product was dipped in an egg wash (egg water), then granulated sugar before baking.

Chemical, physical and microbial properties of air-packaged and MAP apple turnover stored for 21 d at 25°C are summarized in Tables 4.24 and 4.25. Air-packaged apple turnovers developed visible moulds after 7 d, accompanied by a slight fruity odour and sticky pastry, both of which became more pronounced after 14 d. Headspace O_2 was almost depleted after 21 d, accompanied by the production of CO_2, which reached 19% over the same period. Moisture and a_w remained largely unchanged at about 34% and 0.94, respectively, while pH dropped fractionally from 4.73 to 4.56. This is an expected pattern of spoilage for a product of relatively low pH and moderate a_w, which contains starch, sugar and fruit. The fact that mould growth was delayed for a week could be attributed partially to sodium benzoate and potassium sorbate in the filling.

MAP apple turnovers are much more stable. The only noticeable changes after 21 d at 25°C were a slight fruity odour and the pastry becoming lighter in colour and somewhat sticky to the touch. All microbial counts were never higher than 10^3 CFU/g throughout 21 d of storage, except for yeasts which increased from not detectable in fresh product to 1.5×10^6 CFU/g at the end of the storage test. Though counts on the lactobacilli MRS medium were also high, in parallel with yeast counts, morphological examination of the colonies and cells on the medium identified them as yeasts rather than lactobacilli. Fruity odour and stickiness on the pastry could, therefore, be attributed largely to starch and sugar hydrolysis by yeasts.

None of the health-related microorganisms were of concern as the staphylococcal counts were <100 CFU/g throughout the storage and neither coliforms nor clostridia were detected. Therefore, if yeast growth could be curtailed through modification of either the product or the MAP system, ambient shelf life of apple turnovers could be extended beyond 21 d.

Table 4.24.   Chemical and physical properties of apple turnovers stored for 21 d at 25°C

Packaging/ Storage	Storage Time, Days	pH	a_w	Moisture, %	Headspace Gas, % CO_2	Headspace Gas, % O_2
Air	0	4.73	0.94	34.34	-	20.1
	7	4.63	NT	NT	-	19.1
	14	4.60	NT	NT	2.7	9.4
	21	4.56	0.94	34.63	19.1	1.2
MAP	0	4.73	0.94	35.12	59.6	Tr
	7	4.63	NT	NT	52.1	0
	14	4.60	NT	NT	51.1	0
	21	4.56	0.94	33.58	55.0	0

Table 4.25.   Microbial content of apple turnover stored for 21 d at 25°C

Packaging/ Storage	Microorganism	Microbial Counts, CFU/g[1] — Time of Storage, Days 0	7	14	21
Air	Total aerobes	<100	<100	1.5×10^3	Spreaders
	Moulds	ND	<100	2.4×10^4	5.2×10^5
MAP	Total anaerobes	<100	<100	$<10^3$	Spreaders
	Yeasts	ND	ND	4×10^5	1.5×10^6

[1]Average of two samples, each with duplicate plates

4.3.4.3   Mini blueberry pie

Mini blueberry pie is a product where blueberry filling is contained in a pie shell covered with flaky, firm crust with visible sugar crystals on the surface. The product was made from the following ingredients: crust - flour, water, shortening, salt, sugar; filling - blueberries, water, modified starch, glucose, citric acid, colour, potassium sorbate and sodium benzoate. Chemical, physical and microbial properties of the pie are shown in Tables 4.26 to 4.28.

With low moisture content of 40.2%, a_w of 0.929 and pH of 3.78 mini blueberry pie was considered one of the most stable products investigated. The fact that 17% of the air-packaged pie developed visible moulds after 7 d and 25% after 21 d was attributed to two factors. Firstly, the two parts of the pie, i.e. the crust and the filling, differ quite widely in their chemical and physical properties, the filling having 64.51% moisture, 0.94 a_w and 3.61 pH, while the crust 15.56%, 0.84 and 4.58, respectively (Table 4.27). During storage, the moisture and a_w of the crust increased owing to the equilibration process between the two parts, making the crust more suitable for mould growth. Secondly, mould growth was delayed for 7 d or more because of the presence of potassium sorbate and sodium benzoate.

Table 4.26.   Chemical and physical properties of mini blueberry pie stored for 21 d at 25°C

Packaging/ Storage	Storage Time, Days	pH	a_w	Moisture, %	Headspace Gas, %	
					CO_2	O_2
Air	0	3.78	0.93	40.20	-	20.0
	7	4.02	NT	NT	2.1	16.2
	14	3.97	NT	NT	Tr	15.1
	21	3.95	NT	43.00	3.6	13.3
MAP	0	3.78	0.93	40.20	54.0	Tr
	7	4.08	NT	NT	52.4	Tr
	14	3.91	NT	NT	51.6	Tr
	21	3.91	0.94	43.91	50.7	Tr
Frozen	21	3.94	0.93	42.00	NA	NA

No microbial growth in any category was detected in the MAP mini blueberry pie throughout the storage trial. Therefore, micro-biologically the MAP product was shelf-stable for at least 3 weeks. Physical changes, however, appeared to be the limiting factor in the application of MAP to this type of product. Initially, the crust was flaky and firm with sugar crystals evident on the surface. The odour was typical of freshly baked pies. During storage the crust became progressively softer and stickier. This change was evident by day 7, presumably owing to the migration of moisture from the filling to the crust, and soluble solids such as sugar from the crust to the filling, or simply the sugar being dissolved and absorbed into the crust itself. The odour of the filling after 21 d was essentially un-changed, but that of the crust was slightly stale.

Unless the physical problems of the MAP mini blueberry pie can be corrected the product may have difficulty with consumer acceptance.

Table 4.27.   Chemical and physical properties of the filling and the crust of mini blueberry pie stored as a whole pie for 21 d at 25°C

Packaging/ Storage	Storage Time, Days	Filling			Crust		
		pH	a_w	Moisture, %	pH	a_w	Moisture, %
Air	0	3.62	0.94	64.51	4.58	0.84	15.56
	7	3.93	NT	NT	4.55	NT	NT
	14	3.90	NT	NT	4.36	NT	NT
	21	3.87	NT	NT	4.12	NT	NT
MAP	0	3.62	0.94	64.51	4.58	0.84	15.56
	7	3.96	NT	NT	4.59	NT	NT
	14	3.79	NT	NT	4.19	NT	NT
	21	3.90	0.94	60.67	4.16	0.91	18.06
Frozen	21	3.86	0.94	63.62	4.40	0.84	14.86

Table 4.28.   Microbial content of mini blueberry pie stored for 21 d at 25°C

Packaging/ Storage	Microorganism	Microbial Counts, CFU/g[1]			
			Time of Storage, days		
		0	7	14	21
Air	Total aerobes	ND	<300	3.0×10^5	6.5×10^7
	Moulds and yeasts	ND	<300	4.3×10^5	7.0×10^7
MAP and Frozen:		Not detectable in all categories			

[1]Average of two samples, each with duplicate plates

4.3.4.4   Raw apple pie

Raw apple pie represents the most unstable product with high moisture content and a_w, and low pH. It was included in the study to test the limit of application of the MAP system. The medium size pies, 11 cm diameter, were used in the studies. The ingredients include: crust - flour, water, lard, salt; filling - apple, sugar, modified corn starch, locust bean sugar, citric acid and sodium benzoate. Chemical, physical and microbial changes over 21 d storage at 25°C are summarized in Tables 4.29 to 4.31.

Raw apple pie did not store well, either when packaged in air or MAP. Those packaged in air started swelling after a few days and after 7 d all packages burst, at which point the product had a distinct fruity odour with a distorted appearance and very soft and watery texture. Headspace CO_2 increased from trace to 88%, while O_2 dropped from 22% to 0.

MAP offered a slight improvement in shelf life in that it delayed the swelling of the packages so that they did not burst until after 21 d. The MAP raw apple pie started developing fruity odour after 7 d and became yeasty and putrid after 14 d. Though the product appearance was not as badly distorted as air-packaged products, it became quite watery after 14 d. Moisture content and a_w of the samples were largely unchanged over 21 d, but its pH dropped from 4.26 to 3.64 while headspace CO_2 rose from 60.2 to 82.3%. The product was unsaleable after 7 d.

This study clearly showed that the MAP system would not be suitable for raw bakery products. First of all, the microbial load on raw product would be quite high with wide varieties of microorganisms, depending on the level of contamination of the ingredients and product during processing. Secondly, even if the growth of these microorganisms could be controlled with MAP, the soft texture of the raw product would make it extremely difficult to handle during packaging, storage and shipment. Any microbial growth on the product prior to baking may drastically change the characteristics of the product, as in the case of raw apple pie, to make it unrecognizable or unacceptable after baking.

Table 4.29.   Chemical and physical properties of raw apple pie stored for 21 d at 25°C

Packaging/ Storage	Storage Time, Day	pH	a_w	Moisture, %	Headspace Gas, %	
					CO_2	O_2
Air	0	4.26	0.96	54.75	-	22.0
	7	4.10	NT	NT	88.0	0
	14			All packages burst after 7 d		
MAP	0	4.26	0.96	54.75	60.2	Tr
	7	3.94	NT	NT	73.8	0
	14	3.75	NT	NT	80.2	0
	21	3.64	0.96	54.50	82.3	0

Table 4.30.   Microbial content of raw apple pie stored for 21 d at 25°C

Packaging/ Storage	Microorganism	Microbial Counts, CFU/g[1]			
		Time of Storage, Days			
		0	7	14	21
Air	Total aerobes	5.1×10^3	1.7×10^7	Packages	burst
	Moulds/yeasts	3.6×10^3	8.3×10^6	Packages	burst
MAP	Total anaerobes	3.5×10^3	7.5×10^6	1.3×10^8	2.1×10^8
	Lactic bacteria	3.5×10^3	4.5×10^6	1.1×10^8	5.1×10^7
	Yeasts	3.1×10^3	3.2×10^5	1.1×10^6	3.0×10^6
	Staphylococci	<100	<100	7.0×10^3	<100
	Coliforms	3.1×10^2	ND[2]	ND	ND
	Clostridia	ND	ND	ND	ND

[1]Average of two samples, each with duplicate plates
[2]Not detected in 10^{-1} dilution

Table 4.31.   Physical changes in raw apple pie stored for 21 d at 25°C

Packaging/ Storage	Day	Package/Product Surface Appearance	Odour	Texture
Air	0	Normal	Normal	Normal
	7	Extensive swelling	Distinctly fruity	Appearance distorted; Content oozed liquid
	14	All packages burst		
MAP	0	Normal	Normal	Normal
	7	Swelled slightly	Slight fruity	Content became watery
	14	Extensive swelling	Yeasty & putrid	Content oozed liquid
	21	Extremely swollen	Putrid	Sticky and watery

4.3.4.5   Baked apple pie

Baked apple pies perform quite differently during storage at 25°C from raw pies, especially when the MAP system was applied. Air-packaged pies, with no microorganisms detected on the fresh baked product, developed a soggy surface with visible mould growth and slight fruity odour in 7 d (Table 4.34). The conditions became progressively worse as the storage continued and after 14 d the product was covered with mould with count of $3.1x10^6$ CFU/g, accompanied by a strong mouldy odour. Total count also rose to $4x10^6$ CFU/g over the same period. While its moisture content and a_w were practically unchanged its pH dropped slightly from 4.21 to 4.12 and headspace CO_2 increased from a trace to 13% while O_2 decreased from 22% to 0 after 21 d of storage (Table 4.32).

The MAP baked apple pies, on the other hand, were quite stable throughout storage. No mould growth was detected and moisture and a_w were stable. Headspace CO_2 decreased from 59.8 to 53.9%, presumably through absorption by the product and permeation across the packaging film, and pH dropped fractionally from 4.21 to 4.15 (Table 4.32). A slightly moist appearance and fruity odour developed with some fading of crust colour after 21 d, but the texture remained essentially unchanged (Table 4.34).

Table 4.32.   Chemical and physical properties of baked apple pies stored for 21 d at 25°C

Packaging/ Storage	Storage Time, Days	pH	a_w	Moisture, %	Headspace Gas, %	
					CO_2	O_2
Air	0	4.21	0.95	47.82	-	22.0
	7	4.16	NT	NT	Tr	17.5
	14	4.14	NT	NT	8.0	0
	21	4.12	0.95	47.57	13.0	0
MAP	0	4.21	0.95	47.82	59.8	Tr
	7	4.17	NT	NT	57.4	0
	14	4.14	NT	NT	53.2	0
	21	4.15	0.95	48.55	53.9	0

No growth of any microorganisms was detected in MAP baked apple pies, which had pH 4.12, a_w 0.95 and were packaged in 59.9% CO_2, over 21 d storage (Table 4.33). These results corroborate those on crumpets which showed that pH, a_w and level of headspace CO_2 are the most important factors determining the ambient shelf life of that product [40]. The study on crumpets has shown that CO_2 production by microorganisms in MAP crumpets could be completely inhibited if the product was reformulated to have an a_w of 0.96 - 0.98 and a pH of 5 or less, and it was stored at 20°C.

Table 4.33.   Microbial content of baked apple pies stored for 21 d at 25°C

Packaging/ Storage	Microorganism	Microbial Count, CFU/g[1]			
		0	Time of Storage, Day		
			7	14	21
Air	Total aerobes	ND	Cd[2]	4x10^6	Cd[2]
	Moulds	ND	1x10^6	3.1x10^6	TNTC[3]
MAP	No detectable growth of any microorganism				

[1]Average of two samples, each with duplicate plates
[2]Contaminated
[3]Too numerous to count

Table 4.34.   Physical changes in baked apple pies stored for 21 d at 25°C

Packaging/ Storage	Day	Package/Product Surface Appearance	Odour	Texture
Air	0	Normal	Normal	Normal
	7	Very soggy surface; Visible mould growth	Slightly fruity	Not as intact as MAP
	14	Very soggy; Extensive mould growth	Mouldy	-
	21	Extensive mould growth	Mouldy	-
MAP	0	Normal	Normal	Normal
	7	Slightly moist surface	O.K.	Intact
	14	Slightly moist	Slightly fruity	Intact
	21	Slightly soggy; fading of surface colour	Slightly fruity	Relatively intact

MAP as applied to crumpets is also beneficial for baked apple pies. Physical appearance of the product stored at 25°C for 21 d or longer, i.e. slight soggy surface, may be improved by briefly reheating the product in an oven at about 180°C before serving.

4.3.5 Summary of problems requiring further study

The results of the studies on 14 products showed different responses of each product to the MAP system. Some products, e.g. butter tart, may not require MAP for ambient storage of up to 3 weeks, though the application of MAP would ensure product quality for a longer shelf life. Some products benefit immediately from the application of the MAP, while others show varying degrees of success, and some suffer from different microbial or physical problems. Accordingly, the 14 products may be classified into three categories, based on the types and severity of the problems when the MAP was applied, as shown in Table 4.35.

Group a consists of seven products with extended shelf life for 21 d, viz. butter tarts, cake doughnuts, chocolate Danishes, carrot muffins, mini blueberry pies and waffles. Apart from some minor physical changes during storage the MAP products should be readily accepted by the consumers. However, it appears that, at least in North America, only crumpets have benefitted commercially from the MAP system.

Group b consists of three products, i.e. apple turnover, baked apple pie and crusty roll, which could benefit from the MAP system with, perhaps, slight modification. There was some microbial growth on these products, e.g. yeasts on apple turnover and baked apple pie and moulds on crusty roll, but not to the extent that would cause swelling or major physical changes of the products.

Group c consists of four products, i.e. strawberry layer cake, yeast doughnut, cherry cream cheese cake and raw apple pie, which suffered severe problems under the MAP system. Raw apple pie and probably any raw bakery products are unsuitable for the MAP system. Unacceptable physical and certain microbial and chemical changes during storage would preclude them from benefitting from a further modified system. Other products which had severe problems under the MAP system would require further modification to the system and/or products before their shelf life could be extended with MAP.

The severity of the problems, if any, among the products in the three groups increased with the increase in their a_w. Other factors, i.e. pH or moisture content, did not form a similar discernible pattern. Unfortunately, a_w of a product is a very difficult parameter to manipulate without significantly changing other characteristics of the product. Thus, there were three major problems which required further study so that appropriate measures might be developed to solve them.

Table 4.35.   Classification of the products according to the microbial and physical problems developed when MAP was applied and the product stored for at least 3 weeks at 25°C.

Product	a_w	pH	Moisture, %	Swelling	CO_2 level	Mould	Yeast	Bacteria	Problem*
a. None or only physical problems:									
1. Butter tart	0.78	5.70	19.95	-	-	-	-	-	T-crumbly
2. Cake doughnut	0.82	6.60	17.85	-	-	-	-	-	T-sticky
3. Chocolate Danishes	0.83	6.28	22.54	-	-	-	-	-	None
4. Carrot muffin	0.91	8.70	35.14	-	-	-	-	+	None
5. Mini blueberry pie	0.93	3.78	40.20	-	-	-	-	-	T, C
6. Waffle	0.94	7.20	65.79	-	+	-	-	++	Not much
7. Crumpet	0.97	6.00	47-52.41	-	+	-	-	++	Not much

Table 4.35.   (Continued)

Product	a_w	pH	Moisture, %	Swelling	CO_2 level	Mould	Yeast	Bacteria	Problem*
b. Microbial growth, but no swelling:									
1. Apple turnover	0.94	4.60	35.12	-	-	-	++	-	Y
2. Apple pie baked	0.95	4.21	47.82	-	-	-	+	+	Y,T,C
3. Crusty roll	0.95	5.60	29.28	-	-	++	-	++	M (14 d)
c. Yeast and/or bacterial swelling:									
1. Strawberry layer cake	0.90	6.66	37.24	+	+	-	+++	+	Y, B, S (21 d)
2. Yeast doughnut	0.91	6.40	27.98	+	++	-	+++	+	Y, B, S (14 d)
3. Cherry cream cheese cake	0.94	4.51	49.90	++	++	-	++	-	Y, S (14 d)
4. Apple pie raw (burst 7 d)	0.96	4.26	54.75	+++	+++	-	+++	+	Y, B, S (7 d)

* T = Texture; C = Colour; B = Bacteria; Y = Yeast; M = Mould; S = Swelling (of package)
- = not visible or detectable; + = slight growth/quantity; ++ = medium growth/quantity; +++ = heavy growth/quantity

4.3.5.1   Mould problem

Crusty rolls and similar products such as bread, hot dog buns, etc. are very susceptible to mould growth. This is due largely to the porous texture of the products, which makes it difficult to eliminate all O_2 from the headspace. Even with packaging film of low O_2 permeability, over time it still allows accumulation of the gas to the level sufficient for the growth of moulds, especially when the products do not contain a fungistatic agent. Further studies were, therefore, directed towards solving the headspace O_2 problem, and the application of appropriate fungistats with or without the MAP system to control mould growth.

4.3.5.2   Yeast problem

Products such as strawberry layer cake, yeast doughnut, and cherry cream cheese cake contain live yeasts or fruit or sugar and starch as ingredients and they have a relatively low pH, so they are susceptible to yeast growth, accompanied by swelling of the packages. In some cases, LAB and bacilli also grow, which aggravates the swelling and odour problems. Studies were directed towards developing measures to control yeast growth, either by modifying the MAP system or by using appropriate fungistats with or without the MAP system.

4.3.5.3   Health aspect

The presence and growth of health-related microorganisms under the MA conditions was monitored in most of the products examined. The results indicated in general that microorganisms such as staphylococci, salmonellae, coliforms and clostridia could not compete with other, nonhazardous microorganisms under MA conditions. These potentially hazardous microorganisms were either not detectable or did not grow to a significant level during storage. To be of concern, the initial counts of these microorganisms in the fresh products must be substantial [37], which is unlikely in bakery products. Levels of contamination of the products can be minimized with strict hygiene and a proper quality control program in the bakery. Ooraikul *et al.* [30] showed that a simple portable air sampling device such as the Biotest RCS Air Sampler (Biotest-Serum-Institute GmBH, Frankfurt/Main, West Germany) may be used to monitor air quality in bakeries, which gives an idea of possible level of contamination on the products.

Thus far, there has not been a comprehensive study of health-related microorganisms in MAP bakery products. However, it is expected that as the MAP system becomes more widely adopted by the industry, monitoring of the products for health hazard potential will become part of the quality control functions of the industry as well as responsible Government authorities.

4.4   Studies of mould problems

4.4.1   Types of products susceptible to mould problems

Moulds grow under a wide range of environmental conditions. They grow best at room temperature, but the growth range can be as wide as -5 to 70°C. They can thrive in low moisture foods with an a_w as low as 0.60. They prefer an acidic pH, but they grow between pH 2 and 8. However, they are strict aerobes, therefore they will not grow without some available O_2. Moulds present problems mainly in bakery products, cheese, meat and fruit products in which they produce discoloration, off-flavours, and textural changes, but they do not produce gas [25]. Most bakery products packaged in air develop mould growth after a few days at room temperature.

The MAP system developed for crumpets should solve mould problems because the product is evacuated before flushing with a mixture of CO_2/N_2. However, complete evacuation of air, i.e. total removal of O_2, is not always possible. This may be due to the limitation of the equipment used, or to the highly porous nature of some products e.g. rolls or buns. Moreover, it would be too expensive to use packaging film which is totally impervious to O_2. Hence, there will always be some permeation of gas into the package, albeit at a very low rate. Thus, given adequate time and favourable conditions inside the package, O_2 accumulates to a level sufficient for mould growth.

Two factors are very important in determining a mould-free shelf life of MAP crumpets, i.e. use of fungistats and microbial make up of the product. Crumpets containing 1300 ppm of potassium sorbate or sorbic acid, and a predominant microbial population of facultative anaerobes, *Bacillus licheniformis* and *Leuconostoc mesenteroides* [38] had a mould-free shelf life. The residual O_2 in the headspace of the MAP crumpets could be as low as 0.05% to as high as 1.5%. Permeation of O_2 through the film (40 mL/m^2/24h at 25°C and 100% RH), if allowed to proceed unimpeded, results in visible mould growth on the crumpet surface between 7 and 10 d of storage. However, with added preservative the growth of moulds is inhibited during the first days of storage, allowing enough time for the facultative anaerobes and other aerobes present to use up the residual O_2, or to reduce it to a level such that moulds are unable to grow. The continuing growth of the bacteria, which produce a small quantity of CO_2 [39] to replace the amount lost through film permeability, ensures the continued depletion of O_2 and replenishment of CO_2, thus a "permanent" mould-free shelf life.

Products susceptible to mould problems under the MAP system are those containing no fungistats and/or those having so porous a texture that complete evacuation of O_2 is not possible, especially if they also have low initial bacterial counts. Measures to control mould problems in these products could be (a) use of appropriate preservatives and/or (b) use of appropriate techniques to ensure effective removal of headspace O_2.

4.4.2 Use of preservatives with or without MAP

Three kinds of preservatives were studied [28] with respect to their efficacy in controlling mould growth with or without MAP, viz. Delvocid (GB Fermentation Industries Inc., Charlotte, NC), potassium sorbate and para-hydroxybenzoic acid (PHBA).

Delvocid is an antimycotic agent containing natamycin (pimaricin). It has no colour, odour or taste and is effective against moulds and yeasts but not bacteria [19]. It has been used in naturally ripened cheeses, raw ham and sausages. It is very stable between pH 4 and 7, but it has very low water solubility. Suggested concentrations in foods are 5-20 ppm.

Potassium sorbate is effective against moulds, yeasts and bacteria [15]. It is a white powder with no taste or odour. It is most effective in the pH range of 3.5-6.5 when it is mainly in the dissociated form [5]. Suggested levels of use in foods are 650-1300 ppm.

PHBA is also effective against moulds, yeasts and bacteria. It is a colourless and odourless powder with high water solubility. It is most effective in the pH range of 4-7 with suggested use concentrations in foods of 1000-2000 ppm [5].

A study was conducted using potato dextrose agar (PDA) as the substrate to which varying levels of the preservatives were added and inoculated with *Aspergillus niger*. The plates were packaged in air or in a mixture of CO_2/N_2 (3:2) before incubation at 25°C for 28 d. The results of the study are summarized in Table 4.36.

This study clearly showed the effect of headspace atmosphere, pH and different kinds of preservatives on mould growth. MAP alone extended mould-free shelf life of PDA from a few days to about a week, while pH had a synergetic effect with potassium sorbate, which was more effective at lower pH, especially under MAP. PHBA was not effective against moulds in air but performed better under MAP. The most effective fungistat among the three investigated was Delvocid. At 5 ppm Delvocid delayed mould growth in air for 7 d. The mould-free period was doubled with 20 ppm. A combination of Delvocid and MAP was the best against moulds. There was no mould growth at any pH during the 28-d storage period.

If mould growth is a major problem in a baked product, a combination of 5-20 ppm Delvocid and MAP with CO_2/N_2 (3:2) is one of the most effective measures that can be recommended for use by the baker. Or, if the pH of the product is 5 or lower, a combination of 650-1300 ppm potassium sorbate and MAP may also be used to extend ambient shelf life of the product beyond 28 d.

Table 4.36.   Effect of pH, preservatives and CO_2/N_2 (3:2) atmosphere on mould growth on potato dextrose agar

Preservative	Concentration ppm	Days to Visible Mould Growth[1]					
		In Air			In CO_2/N_2		
		pH			pH		
		5	7	9	5	7	9
Control		1	1	3	7	7	7
Delvocid	5	7	7	10	ng[2]	ng	ng
	20	14	14	-[3]	ng	ng	ng
Potassium sorbate	650	3	2	4	ng	13	14
	1300	7	2	4	ng	16	16
PHBA	1000	2	2	2	5	5	6
	2000	2	2	2	5	5	6

[1] All values are the average of two replicates
[2] No growth after 28 d
[3] Plates contaminated

4.4.3   Removal of headspace oxygen

Experiments with MAP crusty rolls showed that after 16-18 d storage at 25°C all rolls had developed visible mould growth despite being packaged in an "anaerobic environment". Because moulds are strict aerobes, sufficient residual oxygen must be present in the package heasdspace to permit mould growth. This was true with MAP crusty rolls because traces of O_2 were detected in the headspace. The problem is much less serious in products with a less porous texture such as crumpets or fruit pies.

The level of residual O_2 in the gas-packaged product could be due to a number of factors such as (i) oxygen permeability of packaging material, (ii) ability of food to trap air, (iii) leakage of air through poor sealing, and (iv) inadequate evacuation and gas flushing. It has been demonstrated that moulds tolerate and even grow in air with headspace oxygen concentrations as low as 1-2% [20, 45]. Several studies have shown that moulds can grow in low concentrations of O_2 in the presence of elevated levels of CO_2. Tompkins [48] found that the inhibitory effect of CO_2 was independent of the O_2 concentration. He showed that the extent of mould growth reduction in 20, 40 or 60% CO_2 was the same whether the O_2 concentration was 20% or 5%. Dallyn and Everton [10] reported that *Xeromyces bisporus*, a xerophilic mould which causes spoilage of food with low a_w, grew in an atmosphere of 95% CO_2 and 1% O_2. These authors also reported growth of *Aspergillus* sp. in 85% CO_2 and 3% O_2. These results demonstrate that moulds grow at very low concentrations of headspace O_2 even in the presence of high concentrations of CO_2, and that additional control measures are necessary to completely inhibit mould growth in MAP food products. One approach to control of headspace O_2 in MAP is the incorporation of oxygen absorbents into the package to remove completely all traces of O_2 in the package headspace [42]. Products such as Ageless (Mitsubishi Chemical Co., Japan) may be used for this purpose. Ageless contains active iron oxide, which becomes iron hydroxide after absorption of oxygen in the presence of water. It is packed in a small gas-permeable sachet and may be placed in the package with the food. Because this involves a chemical reaction, and not the physical displacement of oxygen as in gas packaging, it completely removes O_2 from the package headspace during the storage as long as package integrity is maintained.

4.4.3.1   Minimum   headspace   oxygen   concentration   for mould growth

Studies were conducted to determine minimum headspace O_2 concentration required for mould growth in a CO_2-enriched atmosphere, using PDA plates inoculated with *Aspergillus niger* and *Penicillium* spp. spores. The plates were packaged with CO_2/N_2 mixture (3:2) in 20x20 cm pouches of polyethylene-coated nylon film. Average permeabilities of the film per $m^2/24$ h were: O_2, 8 cm^3 and CO_2 20 cm^3 at 4°C, 75% RH; N_2, 14 cm^3 at 25°C, 100% RH; water vapour, 4 g at 38°C, 90% RH. The studies revealed that all inoculated PDA plates under MAP, with O_2 content ranging from <0.05% to 10% showed mould growth after 4 to 6 d, while those packaged in air only showed growth after 1 to 1.5 d of storage at 25°C (Table 4.37). Those with initial O_2 between 0 and 0.6% developed growth after 5 to 6 d when headspace O_2 reached 0.4-0.6%, while those with initial O_2 >0.6% showed growth slightly earlier at 4 to 5 d. Obviously, mould growth in the MAP plates was delayed by CO_2, but growth cannot be suppressed as long as O_2 keeps increasing. These results show that the minimum O_2 required for mould growth under the atmosphere of CO_2 and N_2 (3:2) is 0.4%.

4.4.3.2   Use   of   oxygen   absorbent   to   remove   headspace oxygen

Headspace O_2 of air-packaged uninoculated plates remained unchanged at about 21% throughout 35 d of storage, while headspace O_2 under MAP increased at an average rate of 0.06%/d (Figure 4.7). This increase in headspace O_2, which, in the case of inoculated plates, was sufficient to permit mould growth (Table 4.37), may be attributed largely to permeation of O_2 through the film. While an oxygen-impermeable packaging film may be produced the cost of such a film would be prohibitive. A viable alternative is the incorporation of O_2 absorbents such as Ageless into the package. The efficiency of O_2 removal by Ageless is shown in Figure 4.7. When uninoculated plates were packaged in air together with a sachet of

Ageless, headspace O_2 was reduced to <0.05% within 9 h and remained at this level thereafter for at least 30 d.

Table 4.37.   Change in O_2 concentration (%, v/v) in the headspace of air packaged and MAP PDA plates inoculated with mould spores and stored for 7 d at 25°C

Initial Headspace O_2, %	Change in O_2 Conc., %[1] Over Storage Time, days								Days to Visible Growth
	0	1	2	3	4	5	6	7	
0	< 0.05	0.15	0.25	0.35	0.40	0.30	0.10	<0.05	5-6
0.10	0.13	0.20	0.35	0.40	0.50	0.20	<0.05	<0.05	5-6
0.20	0.25	0.35	0.45	0.50	0.55	0.30	<0.05	<0.05	5-6
0.40	0.35	0.45	0.50	0.60	0.50	0.40	<0.05	<0.05	5-6
0.60	0.50	0.60	0.70	0.75	0.60	0.50	0.10	<0.05	5-6
0.80	0.75	0.90	1.00	1.00	0.95	0.45	0.40	<0.05	5
1.00	0.95	1.00	1.10	1.20	1.10	0.50	0.40	0.10	5
2.00	2.10	2.20	2.20	2.10	1.10	0.90	0.10	<0.10	4
10.00	9.60	9.80	9.80	9.50	7.70	1.50	0.50	<0.10	4
20.90 (Air)	20.50	18.50	12.50	1.50	<0.05	<0.05	<0.05	<0.05	1-1.5

[1] Values are an average of 12 replicates with standard deviations of 0.03-0.04

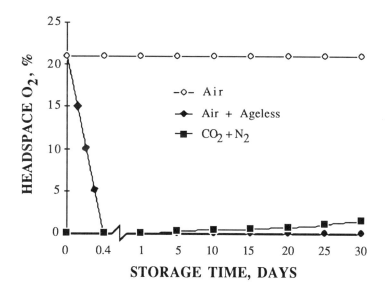

Figure 4.7.   Changes in headspace O_2 of uninoculated PDA plates packaged in air, air + O_2 absorbent 'Ageless', and a mixture of CO_2/N_2 (3:2) over 30-d storage period at 25°C. Each value plotted is an average of 12 replicates with standard deviations ranging between 0.03 and 0.04.

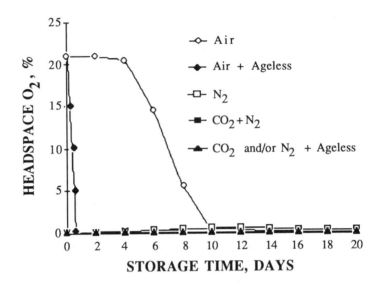

Figure 4.8.   Change in headspace O_2 of crusty rolls packaged in air, N_2, CO_2/N_2 (3:2), air + Ageless, N_2 + Ageless, and CO_2/N_2 (3:2) + Ageless over 20-d storage period at 25°C. Each value plotted is an average of six replicates with standard deviation ranging between 0.03 and 0.14.

4.4.3.3   Application of oxygen absorbent to crusty rolls

When crusty rolls were packaged in the atmospheres described above, similar observations were noted. While mould growth was evident in all air-packaged rolls after 5-6 d, growth was delayed for a further 4-5 d and 11-12 d in rolls packaged in 100% N_2 and CO_2/N_2 (3:2), respectively. Headspace O_2 in rolls packaged in air was depleted to <0.05% after the onset of mould growth (Figure 4.8). While the initial headspace O_2 in crusty rolls packaged in either N_2 alone or CO_2/N_2 was less than 0.05%, the O_2 concentration in both cases steadily increased to approx. 0.6% after 8-9 d. However, mould growth did not appear on the rolls packaged in CO_2/N_2 until after 16-18 d of storage, compared with only 9-11 d for those packaged in 100% N_2. The delayed growth is attributed to the inhibitory effect of CO_2. When Ageless was placed in the packages of MAP rolls, headspace O_2 did not increase beyond 0.05% and the rolls remained mould-free even after 60 d of storage. A similar mould-free shelf life was obtained when Ageless was placed in air- or N_2-packaged crusty rolls.

Thus, an "anaerobic environment" containing elevated levels of CO_2 and/or N_2 is not totally effective in preventing mould growth. Mould growth is only completely prevented when the headspace O_2 can be reduced and maintained at levels less than 0.4%.

While a longer mould-free shelf life is possible by packaging the product in air with Ageless compared with MAP, mould problems occasionally occur in these products, because some products absorb headspace gas, thereby creating a slight vacuum inside the package so that the products become tightly wrapped with the packaging film. This creates localized environments between product surface and the film where the O_2 level may be sufficient to permit mould growth. This indicates the need for a free flow of gas around the product if Ageless is to be effective as an O_2 scavenger. Therefore, a combination of Ageless and MAP may be more effective than either one alone. Nevertheless, this study clearly demonstrated that the use of oxygen absorbents is a viable alternative to gas flushing to extend the mould-free shelf life of bakery products.

4.5 Studies of yeast problems

4.5.1 Types of products susceptible to yeast problems

Yeasts are a major cause of spoilage of products with high sugar content and acidity, such as fruits and their products, because yeasts tolerate low pH and high sugar concentrations. Yeast growth is often accompanied by CO_2, ethanol and acetaldehyde production [32]. In baked goods containing fruit as an ingredient, e.g. strawberry layer cake, cherry cream cheese cake, apple turnover and apple pie, *Saccharomyces cerevisiae* is the prime cause of swelling and fruity odour, especially when the products are packaged in air.

Yeasts are facultative microorganisms, i.e. they can grow with or without O_2, therefore, anaerobic conditions or oxygen scavengers have no effect on their growth. Two methods which could be effective in controlling yeast growth are (a) addition of preservatives, and (b) modification of headspace atmosphere using the vapour of disinfectants such as ethanol.

4.5.2 Use of preservatives with or without MAP

There are three types of preservatives used in Canada to extend shelf life of jams, jellies, fruit spreads and some bakery products, viz. benzoic acid and its Na salt, methyl and propyl *p*-hydroxybenzoic acid (parabens) and their Na salts, and sorbic acid and its Ca, K and Na salts [18].

Benzoic acid and sodium benzoate are often added to fruit products because of their effective inhibition of yeasts and bacteria. Their optimum pH range is 2.5-4.0, they dissolve readily in water and are not deleterious to human health in small doses (<0.5 g/d) [7]. In pie fillings this preservative is used at 0.1% concentration. The main problems associated with the use of benzoates are that yeasts may become resistant to them and they are not as effective against moulds as they are against yeasts [8].

The parabens are also used in fruit products. They are active over a wide range of pH (3-9), have low toxicity, they are active against yeasts, moulds and some bacteria, and they are stable during high-temperature treatment [11]. Methyl and propyl parabens are usually applied in a 3:1 ratio at the levels of 0.03-0.06% in bakery toppings and fillings. The main disadvantage of the parabens is that they are expensive compared with other antimicrobial agents [11]. Ethyl paraben is permitted for direct food use in the U.K., but in the U.S. and Canada it is only permitted for indirect use in adhesives for packaging, transport or holding of food [7].

Sorbates are active in foods with a relatively low pH. They have a high solubility, low acute or chronic toxicity and they are effective inhibitors of yeast growth. However, their solubility decreases in the presence of other solutes, and safety may be of concern because paper, cloth and other absorbent materials may ignite spontaneously when soaked in sorbates [14]. Sorbates are generally added at levels of 0.03-0.30% to pie fillings and toppings [44].

Delvocid is effective over a wide range of pH against all yeasts and moulds. It has been used in cheese, fruit juices and drinks, fresh strawberries and raspberries, and meat products [7, 19]. Yeasts and moulds do not develop resistance to the fungistatic effects of Delvocid even after several years of exposure [12]. This antifungal agent acts by causing the yeast and mould cells to lose cellular components, e.g. nucleic acids, thus preventing reproduction [7]. The minimum inhibitory concentration of Delvocid for yeasts varies between 1.0 and 5.0 μg/mL of media [19].

Preliminary studies showed that even though sorbates are quite effective against moulds, they are not effective against yeasts. Delvocid and ethyl paraben were studied in greater detail to evaluate their effectiveness in controlling yeast growth on PDA and on cherry topping, with and without MAP.

4.5.2.1   Delvocid

PDA with chloramphenicol was inoculated with *S. cerevisiae*, isolated from the cherry topping of a cherry cream cheese cake. An aqueous suspension of Delvocid was either mixed with the sterilized agar before inoculation, or sprayed over the poured, inoculated plates. The concentrations of Delvocid in the medium ranged from 2.5 to 50 ppm. Yeast growth was not observed on any of the plates in which Delvocid was mixed with PDA at the concentrations of 2.5 to 50 ppm, whether air packaged or MAP. However, plates sprayed with Delvocid suspension sustained prolific yeast growth at all concentrations. Therefore, Delvocid mixed directly into the medium was an effective inhibitor of yeast growth at a concentration as low as 2.5 ppm. Spraying did not work because of the extremely low water solubility of Delvocid, the difficulty in controlling the amount and uniformity of the spray, and the additional moisture on the medium surface due to the spray. MAP did not provide further benefit for the control of yeasts with Delvocid.

In cherry topping, Delvocid suspension was mixed directly into the topping at concentrations of 5 to 20 ppm. The results in Table 4.38 indicate that Delvocid did not perform well at 5 or even 10 ppm, especially when packaged in air. With 10 ppm Delvocid, air-packaged topping showed yeast and mould growth after 7 d, while with 20 ppm, the growth was suppressed until day 21. Under the atmosphere of air the main problem is mould growth. Microbial counts on day 28 were substantially lower than those on day 21. This must be due to the depletion of headspace O_2 which decreased from 21% to about 10% on day 14, 7% on day 21 and 3% on day 28.

On the other hand, mould growth did not occur on MAP samples and Delvocid was more effective even at 5 and 10 ppm. However, some replicates at these low concentrations showed some yeast growth. This could be due to the non-uniform mixing of the preservative with the topping. Also, it is likely that higher concentrations of Delvocid were required in cherry topping than in PDA because of the possible lower solubility of the preservative in the substrates containing high soluble solids. Nevertheless, these results indicated that yeast problems in bakery products containing fruit as a major ingredient could be controlled with a combination of MAP and 20 ppm Delvocid.

Table 4.38.   Microbial content of cherry topping with different levels of Delvocid stored for 28 d at 25°C

Packaging/ Delvocid, ppm	Micro-organism	Microbial Counts, CFU/g[1]				
		Time of Storage, Days				
		0	7	14	21	28
Air 0	Moulds & yeasts	ND[2]	4.5×10^6	1.3×10^8	- -	- -
5	Moulds & yeasts	ND	7.9×10^4	9.3×10^6	1.1×10^7	1.2×10^3
10	Moulds & yeasts	ND	1.4×10^5	8.0×10^6	1.9×10^6	<300
20	Moulds & yeasts	ND	ND	ND	1.5×10^5	1.1×10^5
MAP 0	Yeasts	ND	ND	2.0×10^5	7.4×10^6	--
5	Yeasts	ND	ND	ND	ND	1.4×10^7
10	Yeasts	ND	ND	ND	ND	1.6×10^3
20	Yeasts	ND	ND	ND	ND	ND

[1]Average of at least three replicates determined on PDA
[2]Not detectable at 10^{-1} dilution

4.5.2.2   Parabens

Because ethyl paraben is autoclavable it was added directly to the PDA before sterilization to achieve final concentrations of

0.0125-0.2% (w/w). Results showed that ethyl paraben is an effective inhibitor of yeast growth at a concentration of 0.04%, under either air or MAP. At lower concentrations yeast growth was observed with both packaging regimes. At 0.1% and higher the paraben did not dissolve completely in the medium. When added to cherry topping ethyl paraben was effective against both mould and yeast at all concentrations tested (Table 4.39).

Control samples in air showed typical mould and yeast growth during the first 14 d of storage and then declined, owing mainly to the depletion of headspace O_2. Control samples under MAP, on the other hand, showed steady yeast growth over the 28 d storage period. Neither yeast nor mould growth in air was detected when ethyl paraben was added at 0.04% or higher. However, under MAP some yeast growth occurred after 14 d with 0.04% ethyl paraben, but not at higher concentrations. Nevertheless, the yeast counts on days 14, 21 and 28 were all <300 CFU/g and, therefore, would not affect the shelf life of the product. Thus, a minimum concentration of 0.04% ethyl paraben is recommended for the control of yeasts in products containing fruit. The preservative is equally effective against both moulds and yeasts under air or MAP.

Table 4.39.   Microbial content of cherry topping with different levels of ethyl paraben, stored for 28 d at 25°C

Packaging/ % Paraben	Micro-organism	Microbial Count, CFU/g[1]				
				Time of Storage, Days		
		0	7	14	21	28
Air 0	Moulds/yeasts	ND[2]	1.9×10^4	1.0×10^6	7.0×10^5	3.7×10^4
≥0.04	Moulds/yeasts	ND	ND	ND	ND	ND
MAP 0	Yeasts	ND	2.0×10^5	1.0×10^6	3.1×10^4	8.6×10^6
0.04	Yeasts	ND	ND	<300	<300	<300

[1] Average of at least two replicates
[2] Not detected at 10^{-1} dilution

This evidence indicates that ethyl paraben is more effective than Delvocid against moulds or yeasts. This could be due to the very low solubility of Delvocid in water, making it more difficult to incorporate Delvocid than paraben uniformly into the product. Delvocid has a much higher solubility in propylene glycol, glycerol and acetic acid than in water [7]. Dissolving Delvocid in any of these compounds before incorporating it into the product may result in a more uniform distribution and, therefore, greater effectiveness in controlling yeasts and moulds.

4.5.3   Control of yeast growth with ethanol vapour

Ethanol has been widely used as a germicidal agent, especially for equipment and utensils, but little attempt has been made to use it as a food preservative. Only a few studies have evaluated ethanol in modified atmospheres for extension of shelf life of food products. For example, it has been shown that spraying alcohol directly onto the surface of a product prior to packaging and storage can increase the mould-free shelf life of bread [35] and pizza [30].

A novel and innovative way of atmosphere modification with ethanol was recently developed in Japan. A product called Ethicap is produced by encapsulating food-grade ethanol on a fine silicon dioxide powder and packaging it in small, heat-sealed paper/ethyl vinyl acetate (EVA) pouches. The concentration of ethanol on the support material is 55% by weight. When a sachet of Ethicap is packaged with food it absorbs moisture from the food and releases ethanol vapour into the headspace. The size of Ethicap used is dependent on the a_w of the product: the lower the a_w the smaller the amount of Ethicap required, because ethanol is more effective in controlling microbial growth at lower a_w. The Ethicap manufacturer claims that the ethanol atmosphere prevents mould growth and hardening of the product, as well as enriching the flavour, by addition of food-grade flavours to Ethicap.

To evaluate the efficacy of ethanol vapour in shelf-life extension of bakery products, studies were conducted to determine (1) the effect of ethanol concentration in a medium on yeast growth, (2) the effect of a_w on yeast growth and subsequent ethanol and CO_2 production, (3) the effect of a_w on vaporization of ethanol from Ethicap into the package headspace, (4) the relative effect of a_w and ethanol vapour on yeast growth, and (5) the suitability of ethanol vapour for shelf-life extension of bakery products.

4.5.3.1   Effect of ethanol concentration on yeast growth

PDA containing up to 20% (v/v) of ethanol was inoculated with *S. cerevisiae*, packaged in air and stored at 25°C for 21 d. Yeast count, headspace CO_2 and ethanol vapour monitored over the storage period are shown in Table 4.40. It is possible to infer the followimg from these data:

a. Ethanol competes for water in PDA, causing the reduction in the a_w of the medium from 0.99 without alcohol to about 0.92 with 20% ethanol. Because of this it is unclear whether the retardation of yeast growth in the medium is due to the lowering of its a_w or the presence of ethanol, or both.
b. Ethanol added to the medium generates a vapour into the headspace. The amount of vapour generated (initial headspace ethanol) is in direct proportion to the amount of ethanol added to the medium. The final headspace ethanol (i.e. after 21 d) was consistently higher than the initial level, except when 20% ethanol was added, at which concentration the headspace ethanol remained constant. Part of this headspace alcohol is produced by yeasts, as evidenced by the fact that even when no ethanol was added to the medium there was 0.1% (v/v) ethanol in the headspace after 21 d. It is uncertain, however, whether the effectiveness of ethanol against

yeast is due to the vapour in the headspace or the ethanol added to the medium, or both.

c. Ethanol inhibits yeast growth but whether it is ethanol in the headspace or in the medium that performs the function is not clear. Results in Table 4.40 indicated that about 6% (v/v) ethanol in the medium, or 0.43% (v/v) in the headspace, is required to affect growth. To inhibit growth completely, a concentration approaching 20% in the medium or 1.11% in the headspace is needed.

d. Another important yeast metabolite is CO_2. Without any inhibition, yeast can increase headspace CO_2 to about 54% (v/v) after 21 d at 25°C. This renders the product unacceptable, and may cause swelling of the package.

Table 4.40.   Effect of ethanol concentration in PDA on yeast growth during 21 d of storage at 25°C

| Ethanol Conc. in Medium, %(v/v) | A_w | Headspace Ethanol, %(v/v)[1] | | Yeast Growth | | |
		Initial	Final	Days	Extent[2] of Growth	CO_2 Produced[3] (%, v/v)
0	0.99	0.00	0.10	1	+++	54.2
2	0.97	0.08	0.23	1	+++	50.2
4	0.95	0.21	0.35	1	+++	51.0
6	0.95	0.34	0.43	1	++	51.5
8	0.94	0.44	0.56	1	+	49.7
10	0.93	0.51	0.71	2	+	49.9
15	0.92	0.81	0.98	6	+	47.8
20	0.92	1.12	1.11	>21	-	0.0

[1] Average of six replicates with standard deviations of 0.02-0.03
[2] No growth (-); slight growth (+); medium growth (++); heavy growth (+++)
[3] Average of six replicates with standard deviation of 2.45-3.50

4.5.3.2   Effect of a_w on yeast growth and subsequent ethanol and CO_2 production

In order to determine whether it was a_w and/or ethanol that inhibited yeast growth, PDA adjusted to various levels of a_w and inoculated with *S. cerevisiae* was packaged in air and stored at 25°C for 21 d. Glycerol was used to adjust the a_w because it does not have an antimicrobial effect [22].

Results in Table 4.41 show that a_w did not affect yeast growth until it was reduced to 0.90. Even at this level of a_w, yeast still grew to maximum population by 14 d. Not until a_w was reduced to 0.80 was the growth of yeasts completely inhibited. Therefore, the inhibition of yeast growth shown in Table 4.40 is principally due to ethanol.

The amounts of ethanol and CO_2 produced by yeasts parallels the growth of the organisms. The reduction of these metabolites in the headspace as the storage time increased is most likely due to their absorption by the medium.

Table 4.41.   Effect of a_w on yeast growth, ethanol and CO_2 production in PDA plates packaged in air and stored at 25°C

A_w	Days of Storage	Yeast Growth[1]	Headspace Gas	
			Ethanol (%v/v)[2]	CO_2 (%v/v)[3]
0.95	1	++	0.11	15.5
	3	++	0.13	56.3
	7	+++	0.12	52.6
	14	+++	0.12	51.5
	21	+++	0.10	53.0
0.90	1-3	-	-	-
	7	++	0.05	23.7
	14	+++	0.10	32.1
	21	+++	0.05	32.6
0.85	1-7	-	-	-
	14	+	0.01	1.2
	21	++	0.02	8.7
0.80	1-21	-	-	-

[1]No growth (-); slight growth (+); medium growth (++); heavy growth (+++)
[2]Average of six replicates with standard deviations of 0.001-0.01; (-) = not detectable
[3]Average of six replicates with standard deviations of 0.12-1.75; (-) = not detectable

4.5.3.3   Effect of a_w and headspace ethanol on yeast growth

To determine whether ethanol in the substrate or in the headspace is effective against yeasts, PDA plates inoculated with *S. cerevisiae* were packaged with Ethicap of different sizes and stored at 25°C for 21 d. Because ethanol vapour is generated from Ethicap by absorption of moisture, headspace ethanol concentration is influenced by the a_w of the medium. The relationship between the a_w of the substrate and the amount of ethanol in the headspace was also determined so that an appropriate size of Ethicap could be chosen for a product of specified a_w. This was accomplished by packaging uninoculated PDA plates (2 plates/package) with a_w of 0.85, 0.95 and 0.99 (adjusted with glycerol at 41.2, 17.7 and 0%, v/v, respectively) with two sizes of Ethicap, i.e. 1 g (E1) and 4 g (E4), and analysing for headspace ethanol at regular intervals. The results are shown in Figure 4.9.

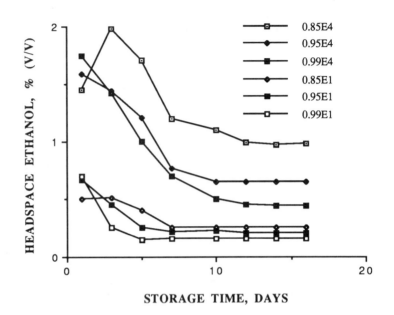

Figure 4.9.   Concentration of headspace ethanol (% v/v) generated
by Ethicap sizes 1 g (E1) and 4 g (E4) over PDA having
a_w of 0.85, 0.95 and 0.99, stored for 16 days at 25°C (Each
value plotted is an average of six replicates with S.D.
ranging between 0.02 and 0.09)

The data in Figure 4.9 clearly demonstrate that the vaporization of ethanol into the headspace and its absorption by the medium
is dependent on the a_w of the system. Headspace ethanol was initially
greater at a_w of 0.99 and 0.95 than at 0.85 (day 1), but it was rapidly
absorbed by water in the PDA, and after 3 d it was higher at lower a_w
values. A greater quantity of Ethicap (E4) generated more ethanol
vapour in the headspace than the smaller quantity (E1). It took about
10-15 d before the exchange between Ethicap, the headspace and the
medium reached an equilibrium. If 0.56% (v/v) headspace ethanol is
taken as a minimum equilibrium concentration that can significantly inhibit yeast growth (Table 4.40), it requires 4 g of Ethicap
(E4) to generate that amount, when packaged with about 50 g product
(2 PDA plates) having a_w of 0.95 and stored at 25°C (Figure 4.9).
However, if complete inhibition is desired the headspace ethanol
must be about 1.1%, and this would require at least 4 g Ethicap for
products with an a_w of 0.92 or lower (Figure 4.9; Table 4.40).

This observation was corroborated by the results in Table 4.42
which show the effect of a_w and headspace ethanol generated by

Ethicap on yeast growth. When two inoculated plates of PDA with a_w 0.95 were packaged with Ethicap size 4 g, yeast growth was slight on day 1 when the initial headspace ethanol was 1.46%. However, as the concentration decreased, owing to absorption by the medium, the yeasts continued to grow and heavy growth was observed on day 3. The growth was completely inhibited for at least 21 d when two PDA plates with a_w 0.90 were packaged with 4 g Ethicap, generating 1.52% initial headspace ethanol which decreased to 0.69% after 21 d. At an a_w of 0.85 or lower only 1 g Ethicap was needed for complete inhibition. CO_2 was detected in the headspace only when yeast growth occurred, and its concentration was proportional to growth. Also, at an a_w of 0.90 and higher headspace ethanol decreased throughout the 21 d storage period owing to absorption by the medium, while at an a_w of 0.85 and lower the concentration remained unchanged after 14 d.

These studies showed that ethanol vapour inhibits yeast growth, and the concentration in the headspace of the package required for inhibition is dependent on the a_w of the system. The inhibitory effect is principally from ethanol vapour rather than absorbed ethanol in the medium, though the latter may affect the growth indirectly through the lowering of the a_w as shown in Table 4.40.

4.5.3.4   Combined effects of Ethicap and MAP or ethyl paraben in controlling yeast growth

The effectiveness of ethanol vapour in controlling yeast growth is strongly dependent on product a_w, and at a_w 0.95 and higher it appears to be very low. Experiments to determine the combined effect of Ethicap and MAP showed that at a_w 0.99 MAP with up to 4 g Ethicap had no effect on yeast growth. Only at a_w 0.95 and lower could MAP with 4 g Ethicap delay yeast growth for about a week.

Use of Ethicap with ethyl paraben did not provide an added benefit. As in the case of cherry topping, 0.04% (w/w) of ethyl paraben was adequate to inhibit completely growth of yeast on agar at a_w 0.99. However, the same concentration of paraben was less effective at a_w 0.95 than at 0.99, perhaps because it was less soluble at low a_w. Ethicap size 1 g with 0.025% paraben was rated slightly more effective at this a_w, because this combination inhibited growth for 21 d. Inhibition was extended to at least 28 d with 4 g Ethicap and 0.04% paraben.

Table 4.42.   Effect of a_w and headspace ethanol generated by Ethicap on yeast growth and CO_2 production in PDA packaged in air and stored for 21 d at 25°C

a_w	Ethicap Size, g	Day Storage	Yeast Growth[1]	Headspace Gas (%v/v) Ethanol[2]	CO_2[3]
0.95	4	1	+	1.46	1.2
		3	+++	1.24	16.3
		7	+++	0.96	43.8
		14	+++	0.60	44.7
		21	+++	0.59	43.2
	6	1	+	1.76	1.2
		3	+++	1.43	21.5
		7	+++	1.28	48.2
		14	+++	0.95	46.7
		21	+++	0.70	45.2
0.90	2	1	-	1.09	-
		3	-	0.85	-
		7	++	0.52	24.8
		14	+++	0.41	38.0
		21	+++	0.39	43.9
	4	1	-	1.52	-
		3	-	1.70	-
		7	-	1.50	-
		14	-	0.87	-
		21	-	0.69	-
0.85	1	1	-	0.56	-
		3	-	0.45	-
		7	-	0.30	-
		14	-	0.20	-
		21	-	0.19	-
	2	1	-	1.15	-
		3	-	0.92	-
		7	-	0.58	-
		14	-	0.46	-
		21	-	0.44	-
0.80	1	1	-	0.61	-
		3	-	0.53	-
		7	-	0.32	-
		14	-	0.21	-
		21	-	0.21	-
	2	1	-	1.10	-
		3	-	0.99	-
		7	-	0.60	-
		14	-	0.44	-
		21	-	0.42	-

[1]No growth (-); slight growth (+); medium growth (++); heavy growth (+++); [2]Average of six replicates with standard deviations of 0.02-0.05; [3]Average of six replicates with standard deviations of 0.12-0.92

4.5.4   Application of Ethicap to bakery products

4.5.4.1   Yeast doughnut

In a previous study it was shown that yeast doughnuts (a_w 0.91) packaged with MAP swelled after 14 d, owing largely to yeast growth and generation of CO_2. Extension of the shelf life beyond 2 weeks requires the control of yeast growth. This could be achieved by the addition of preservatives, e.g. ethyl paraben or Delvocid, to the dough prior to frying. Because paraben can withstand high temperatures, whereas Delvocid is broken down around 100°C, paraben should be the prime candidate for a preservative.

When doughnuts were packaged in air with 1 g Ethicap the growth of yeasts and lactobacilli was significantly suppressed for 28 d, while mould growth was not detected. Only after 2 months' storage was there a marked increase in these microorganisms, possibly because of a loss of headspace ethanol through the packaging film. However, with 2 g Ethicap there was essentially no microbial growth, even after 2 months of storage.

MAP adds little benefit to microbial control when Ethicap is used, though MAP with 1 g Ethicap is practically as effective as 2 g Ethicap in air. Therefore, the choice between these two measures would be based on the economy of operation. Ethicap appeared to be effective against all types of bacteria, moulds and yeasts, while MAP was only effective against moulds and aerobic bacteria. This should be considered when a control measure is chosen for a product.

4.5.4.2   Vanilla layer cake

Previous study with strawberry layer cake indicated that yeast growth was the limiting factor for extending shelf life under MAP. The packages swelled after 21 d owing to CO_2 production by yeasts. Experiments with vanilla layer cake (a_w 0.90), a similar product having vanilla instead of strawberry flavour, showed that the cake packaged in air showed visible mould growth after 7 d. When 120 g of cake was packaged with 3 g Ethicap it did not develop mould growth, but there were signs of swelling after 21 d due to growth of yeast. Packages with 5 g Ethicap remained free of any microbial development throughout the 60 d storage period.

As in the case of yeast doughnuts, MAP did not offer much extra benefit when Ethicap was used, except that with the smaller quantity of Ethicap (3 g) a combination of Ethicap and MAP was more effective than Ethicap alone in controlling microbial growth.

Physical and organoleptic properties of cake packaged with 5 g Ethicap, or together with MAP, remained largely unchanged. In fact, the aroma of the cake so packaged after 60 d was more pleasant than that of the fresh cake, and the taste also remained excellent. This was due to the small quantity of vanilla flavour incorporated into Ethicap.

4.5.4.3   Apple   turnover

Fresh-baked, 100-g apple turnovers (a_w 0.94) were packaged with 4 g Ethicap, with and without MAP. The size of Ethicap used was estimated on the basis of product weight and a_w, using the procedure provided by the manufacturer (Freund Industrial Co., Ltd., Tokyo). The samples were stored at 25°C for 21 d.

The results of the studies confirmed that ethanol vapour can be used effectively to extend the shelf life of bakery products. Yeast counts in the air-packaged and MAP samples increased from not detectable at day 0 to 3.6×10^6 and 4.3×10^5 CFU/g, respectively, after 21 d of storage. There was a concomitant increase in headspace CO_2 of 32% and 19%, respectively, resulting in all packages having a blown appearance. However, for samples packaged with Ethicap, with or without MAP, yeast growth and CO_2 production were inhibited and all packages were normal after 21 d of storage.

Changes in headspace (H) and product (P) ethanol of apple turnovers are shown in Figure 4.10. For samples packaged with Ethicap in air or MAP, headspace ethanol decreased from the initial levels of 1.8 and 1.9% (v/v) to 0.5 and 0.52%, respectively, after 21 d, while headspace ethanol of similarly packaged product without Ethicap increased from 0 to approximately 0.1% over the same storage period. The ethanol content of apple turnovers packaged with Ethicap in air or MAP increased from the initial concentrations of 0.35 and 0.40% (w/w) to about 1.45 and 1.52%, respectively, over 21 d of storage. It was previously shown that the effect of ethanol is principally from the vapour in the headspace rather than from the absorbed ethanol, though the latter may contribute indirectly through its a_w-lowering effect. It would be interesting to determine the extent of the lowering of a_w of the apple turnovers by the absorbed ethanol. Unfortunately, this is not easy to measure owing to the volatile nature of the compound. Smith *et al.* [43] suggested that an estimate of the final a_w of the product can be made using the Ross equation [33] as follows:

$$a_w = a_{wp} \times a_e$$

where a_{wp} = a_w of the product before addition of ethanol, and
          a_e = the a_w-lowering effect of ethanol which can be
calculated from Raoult's Law using the following equation:

$$a_e = n_1/(n_1 + n_2)$$

where n_1 = g water in the product, and n_2 = g ethanol added.

Using these two equations, the final a_w of apple turnovers packaged with Ethicap is estimated to be 0.934-0.935. This is only fractionally lower than the a_w of 0.94 of the fresh product.

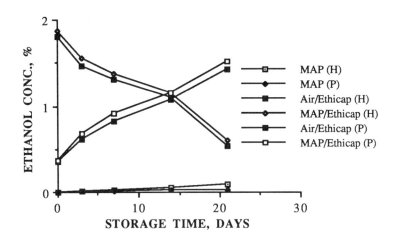

Figure 4.10. Changes in headspace ethanol (H, %v/v) and product ethanol (P, %w/w) of apple turnovers packaged with 4 g Ethicap, with and without MAP, and stored for 21 d at 25°C (Each value plotted is an average of six replicates with S.D. of 0.001-0.03% for H, and 0.001-0.01% for P)

The ethanol content of 1.45-1.52% (w/w) of apple turnovers packaged with Ethicap and stored for 21 d was considerably higher than the 0.05% (w/w) found as a result of yeast growth in the MAP product packaged without Ethicap (Figure 4.10). The absorption of ethanol by the product is a disadvantage of using ethanol vapour for shelf-life extension. Nevertheless, these concentrations of ethanol found in apple turnovers are within the maximum level of 2% by product weight permitted when ethanol is sprayed onto pizza crusts prior to final baking [17]. Preliminary studies showed that the ethanol content of apple turnovers can be reduced to less than 0.1% (w/w) by heating the product at 190°C for 10 min prior to consumption. Therefore, while a longer shelf life may be possible by packaging product with Ethicap, its use as a food preservative may be limited to "brown and serve" type products.

4.5.4.4   Cherry topping

Cherry topping is by far the most susceptible portion of cherry cream cheese cake to growth of yeast, having the highest moisture content (64%), a_w (0.94) and pH (3.82) (Table 4.21). Therefore, effective control of yeast growth in this layer may extend the shelf life of the whole cake.

To evaluate the effectiveness of Ethicap in controlling yeast growth in 40 g samples of cherry topping, samples were packaged with either 2 g or 4 g Ethicap, with or without MAP. The samples were stored at 25°C for 60 d. Samples packaged in air without Ethicap developed heavy yeast growth in 7 d and all packages burst between 7 and 14 d. Under MAP the growth rate was slower, delaying the burst time of the packages to between 21 and 28 d.

Ethicap size 2 g reduced the growth rate slightly in air or MAP, resulting in packages bursting around 14 d in air and after 28 d under MAP. Ethicap size 4 g, on the other hand, was effective against yeast growth under both packaging regimes. In fact, it was somewhat more effective under MAP as yeast growth was reduced from an initial 8.9×10^3 to <300 CFU/g after 7 d. Counts on samples packaged in air were not significantly reduced until after 14 d, though the yeast in the samples was not enough to produce a detectable quantity of CO_2 at any time during the storage.

Physical changes were also observed. After 7 d of storage without Ethicap, either under air or MAP, samples showed signs of fermentation, fading of colour and development of off-odours. In contrast, samples packaged with 4 g Ethicap retained good, bright colouring and pleasant odour over the same storage period. Even after 2 months of storage only slight fading of colour occurred, while the odour remained very pleasant, most likely owing to the small quantity of vanilla in Ethicap.

It was previously shown that preservatives such as ethyl paraben and Delvocid are effective against microbial growth in cherry topping. The choice of control measure for such products would, therefore, be based on factors such as cost, ease of use, Government regulations, and consumer acceptance.

4.5.4.5   Cherry cream cheese cake

Similar studies were done on cherry cream cheese cake by packaging 140 g samples with 6 g Ethicap, with or without MAP. The size of Ethicap, while based roughly on sample weight and overall a_w of 0.94, was limited to 6 g because a smaller size would be inadequate and a bigger size would be impractical owing to cost and visual prominence in relation to the product.

The results of the studies indicate that while Ethicap, especially when used with MAP, suppressed microbial growth, 6 g was clearly inadequate for long-term ambient storage of cherry cream cheese cake. Microorganisms detected were predominantly yeasts. For 140 g of product it would probably require 7 to 8 g Ethicap to effectively inhibit yeast growth. For a product of this nature, i.e. with a sizeable portion of fruit topping, addition of a preservative such as ethyl paraben or Delvocid to the topping, and probably to other portions as well, may be a much more attractive proposition for both the manufacturer and consumers.

4.6   Conclusions   and   recommendations

Studies showed that an MAP system in which $CO_2:N_2$ (3:2) was used to replace air in the headspace can be applied to a number of bakery products. The system uses about 45% less energy than freezing, which has been the mainstay in the industry for long-term storage and marketing of many bakery products. It has also been shown that the overall saving using the MAP system was at least 14%, and could be much more if capital, overhead and maintenance costs of a freezing system, and some intangible benefits of the ambient storage and display of the MAP products are taken into account. The products to which the MAP system could be applied without modification included butter tarts, cake doughnuts, chocolate Danishes, carrot muffins, mini blueberry pies and waffles.

Other products tested with the same MAP system, with the exception of raw apple pies, showed varying degrees of success. Some products developed minor problems, which might be readily solved, or were not serious enough to preclude them from enjoying the benefits of the system. Other products developed more serious problems for which a number of solutions have been developed. These studies indicated that an ambient shelf life of at least 21 d for any baked product can be achieved if an appropriate MAP system and/or product modification is applied.

4.6.1 Minor problems and possible solutions

Minor problems encountered in many products during prolonged storage using MAP were related to texture, colour and taste of the products. Crumpets and waffles, for example, tended to harden after a few weeks of storage, not unlike stale bread or buns. Their colour became slightly pale or greyish. Both of these characteristics recovered almost fully on toasting prior to consumption. The problems were attributed principally to the changes in the starch fractions of the products during storage [49].

The shell of butter tarts, mini blueberry pies and apple pies became crumbly during extended storage. A different flour mixture, type of shortening and surfactants may solve or minimize this problem. Similarly, the sticky problem of cake doughnuts could be solved by using a different form of sugar or sweetening agent and anticaking agent.

4.6.2 Major problems and possible solutions

Major problems encountered with some products related to mould growth, or bacterial and/or yeast growth causing off-odours and swelling of packages.

Mould growth in MAP products is normally associated with those that have low moisture content, low initial microbial counts, and very porous texture. The problem is mainly due to the inability of the packaging system to reduce headspace O_2 to <0.4% and to maintain it at that level throughout the storage. The mould problem could be resolved by

i. using a combination of preservatives, e.g. Delvocid or potassium sorbate, and the MAP system, or paraben without MAP,

ii. packaging the product with an oxygen absorbent, such as Ageless, and an inert gas to ensure headspace O_2 below 0.4%, and

iii. using another form of MA, i.e. ethanol vapour generated by a product such as Ethicap.

Anaerobes such as lactobacilli which cause swelling problems in crumpets or similar products can normally be controlled by MAP. In some other products they are usually outgrown by yeasts, therefore, they are not considered a serious problem.

Yeast growth causes package swelling and off-odours in the product, and it is associated with products having high sugar and starch content, or with added fruit or dairy product such as cream or cheese as a major ingredient. The problem is particularly serious with products having a high a_w and low pH. The control measures developed include

i. the use of preservatives such as Delvocid with MAP, or paraben without MAP, and

ii. the use of Ethicap with or without MAP, depending on the a_w and general characteristics of the product.

4.6.3 Matching products with control measures

The minimum that needs to be done to determine a suitable packaging regime or control measure for any baked product to extend its ambient shelf life to at least 21 d is to study its ingredients, its physical characteristics, moisture content, a_w and pH. With this information the product can be classified to see if it fits any of the categories outlined in Table 4.35. For those having characteristics of group (a), the MAP system may be applied without further modification. In most cases, however, especially those characteristic of group (b) or (c), a series of experiments is recommended before a suitable packaging regime or measure can be chosen. These include:

1. Analysis for moisture content, a_w and pH.

2. Application of the standard MAP (CO_2/N_2; 3:2) to the product and storage at 25°C for at least 3 weeks to observe physical, chemical and microbial changes.

3. If mould growth is observed, an effective method for controlling mould, e.g. the use of a preservative such as Delvocid or sorbate with MAP, or packaging with Ageless and an inert gas, may be chosen.

4. If a swelling problem develops in the packages, microbial analysis is necessary to determine whether the gas produced is due to bacteria or yeasts.

5. If bacterial swelling is indicated modification of the product to slightly lower its a_w and/or pH is necessary before MAP is applied, or the use of paraben or Ethicap may be investigated.

6. If yeast swelling is the problem Delvocid with MAP, or paraben without MAP, or Ethicap with or without MAP may be applied.

References

1. Aboagye, N.Y., Ooraikul, B., Lawrence, R. and Jackson, E.D. 1986. Energy costs in modified atmosphere packaging and freezing processes as applied to a baked product. Proceedings of the Fourth International Congress of Engineering and Food, Edmonton, AB. Food Engineering and Process Applications Vol. 2 Unit Operations. Ed.: Le Maguer, M and Jelen, P. Elsevier Appl. Sci. Publishers, London. pp. 417-425.

2. Baird-Parker, A.C. 1969. The use of Baird-Parker medium for the isolation and enumeration of *S. aureus*. In "Isolation Methods for Microbiologists". Shapton, D.A. and Gould, G.W. (eds.). Academic Press, New York, NY. pp. 1-8.

3. Banks, H., Ranzell, N. and Finne, G. 1980. Shelf-life studies on carbon dioxide packaged finfish from the Gulf of Mexico. J. Food Sci. 45: 156-162.

4. Barratt, R.F. 1988. Economic review and forecast. Food in Canada 48(7): 14-32, 58-62.

5. Barrett, F. 1970. Extending the keeping quality of bakery products. Baker's Digest. 44(4): 48-49, 67.

6. Bogadtke, B. 1979. Use of CO_2 in packaging foods. Ernährungswirtschaft 7/8: 33.

7. Chichester, D.F. and Tanner, F.W. 1975. Antimicrobial food additives. In "Handbook of Food Additives". 2nd Ed. Furia, T.E. (ed.), pp. 115-184. Chemical Rubber Co., Inc. OH.

8. Chipley, J.R. 1983. Sodium benzoate and benzoic acid. In : "Antimicrobials in Foods". Branen, A.L. (ed.), pp. 11-35. Marcel Dekker, New York, N.Y.

9. Clark, D.S. and Takács, J. 1980. Gases as preservatives. *In :* Microbial Ecology of Foods. pp. 170-192. Academic Press, N.Y.

10. Dallyn, H. and Everton, J.R. (1969) The xerophilic mould, *Xeromyces bisporus*, as a spoilage organism. J. Food Technol. 4: 399-403.

11. Davidson, P.M. 1983. Phenolic compounds. In : "Antimicrobials in Foods". Branen, A.L. (ed.), pp. 37-74. Marcel Dekker, New York, N.Y.

12. de Boer, E. and Stolk-Horstuis, M. 1977. Sensitivity to Natamycin (Pimaricin) of fungi isolated in cheese warehouses. J. Food Prot. 40: 533-536.

13. Dickson, R. 1987. Flour and sugar battles keep industry on its toes. Food in Canada 47(2): 18-20.

14. Earle, M.D. and Putt, G. 1984. Sorbates in food - a review. Food Technol. NZ. 19(11): 29-35.

15. Eklund, T. 1983. The antimicrobial action of some food preservatives. J. Appl. Bacteriol. 55: 441-445.

16. FDA. 1978. Bacteriological Analytical Manual. Bureau of Foods, Division of Microbiology. Food and Drug Administration. August, 1978.

17. Federal Register. 1974. 39(185, Sept. 23): 34185.

18. Fondu, M., van Gindertael-Zegers de Beyl, H., Bronkers, G. and Carlton, P. (eds.). 1980. Food Additives Tables. Updated Edition Classes I-IV. Elsevier Scientific Publ. Co. New York, N.Y.

19. GB Fermentation Industries Inc. 1983. Delvocid for the prevention of mould on food products. Product Bull. Del-04/83.09 Am 10.
20. Golding, N.S. 1945. The gas requirements of moulds. IV. A preliminary interpretation of the growth rates of four common mould cultures on the basis of absorbed gases. J. Dairy Sci. 28: 737-750.
21. Hickey, C.S. 1980. Sorbate spray application for protecting yeast-raised bakery products. Bakers Digest 54: 20-22, 36.
22. Hsieh, F.H. 1975. Death kinetics of food pathogens as a function of water activity. Ph.D. thesis, University of Minnesota, MA.
23. ICMSF. 1980. Microbial Ecology of Foods. Vol. II. Food Commodities. pp. 669-730. Silliker, J.H. (ed.). Academic Press. New York, N.Y.
24. King, B.D. 1981. Microbial inhibition in bakery products - a review. Bakers Digest 55: 8-12.
25. Kramer, A. and Twigg, B.A. 1970. Quality Control for the Food Industry. Volume 1 - Fundamentals. 3rd ed. The AVI Publ. Co., Inc. Westport, CT.
26. Mahon, H.P., Kiss, M.G. and Leimer, H.T. 1983. Efficient Energy Management Methods for Improved Commercial and Industrial Productivity. Prentice-Hall, Englewood Cliffs, NJ. pp. 248-269.
27. Ooraikul, B. 1982. Gas packaging for a bakery product. Can. Inst. Food Sci. Technol. J. 15: 313-315.
28. Ooraikul, B. 1988. Modified atmosphere packaging of selected bakery products as an alternative to low temperature preservation. Final Report. Engineering and Statistical Research Institute Report of Contract File #10SC.01916-3-EP16. Agriculture Canada, Ottawa, Ontario. 100 pp.
29. Ooraikul, B., Smith, J.P. and Jackson, E.D. 1983. Evaluation of factors controlling spoilage microbiota from gas packaged crumpets with response surface methodology. Research in Food Science and Nutrition. Vol. 2. Basic Studies in Food Science. McLoughlin, J.V. and McKenna, B.M. (eds.). Proceedings of the Sixth International Congress of Food Science and Technology, Dublin. pp. 92-93.
30. Ooraikul, B., Smith, J.P. and Koersen, W.J. 1987. Air quality in some Alberta bakeries. Can. Inst. Food Sci. Technol. J. 20: 387-389.
31. Plemons, R.F., Staff, C.H. and Cameron, F.R. 1976. Process for retarding mould growth in partially baked pizza crusts and articles produced thereby. U.S. Patent No. 3 979 525.
32. Prescott, S.C. and Dunn, C.G. 1982. Industrial Microbiology. 4th Ed. Reed, G. (ed.). AVI Publ. Co., Inc. Westport, CT.
33. Ross, K.D. 1975. Estimation of water activity in intermediate moisture foods. Food Technol. 29(1): 26-34.
34. Seideman, S.C., Carpenter, Z.L., Smith, G.C., Dill, C.W. and Vanderzant, C. 1979. Physical and sensory characteristics of pork packaged in various gas atmospheres. J. Food Prot. 42: 317-322.
35. Seiler, D.A.L. 1978. The microbiology of cake and its ingredients. Food Trade Rev. 48: 339-344.
36. Semling, Jr., H.V. 1988. 1988 Food Industry Outlook. Food Proc. 49(2): 18-32.
37. Smith, J.P. 1982. A study on the control of microbial spoilage of a gas packaged bakery product introducing the use of response surface methodology. Ph.D. thesis. Department of Food Science, University of Alberta, Edmonton, AB, Canada.

38. Smith, J.P., Jackson, E.D. and Ooraikul, B. 1983a. Microbiological studies on gas packaged crumpets. J. Food Prot. 46: 279-283.
39. Smith, J.P., Jackson, E.D. and Ooraikul, B. 1983b. Storage study of a gas-packaged bakery product. J. Food Sci. 48: 1370-1371, 1375.
40. Smith, J.P., Khanizadeh, S., van de Voort, F.R., Hardin, R., Ooraikul, B. and Jackson, E.D. 1988. Use of response surface methodology in shelf life extension studies of a bakery product. Food Microbiol. 5: 163-176.
41. Smith, J.P., Ooraikul, B. and Jackson, E.D. 1984. Linear programming: a tool in reformulation studies to extend the shelf life of English-style crumpets. Food Technol. in Australia. 36: 454-456, 459.
42. Smith, J.P., Ooraikul, B., Koersen,W.J., Jackson, E.D. and Lawrence, R.A. 1986. Novel approach to oxygen control in modified atmosphere packaging of bakery products. Food Microbiol. 3: 315-320.
43. Smith, J.P., Ooraikul, B., Koersen, W.J., van de Voort, F.R., Jackson, E.D. and Lawrence, R.A. 1987. Shelf life extension of a bakery product using ethanol vapor. Food Microbiol. 4: 329-337.
44. Sofos, J.N. 1983. Sorbates. In : "Antimicrobials in Foods". Branen, A.L. (ed.). pp. 141-175. Marcel Dekker, New York, N.Y.
45. Tabak, H. H. and Cooke, W.B. 1968. The effects of gaseous environments on the growth and metabolism of fungi. Botan. Rev. 34: 126-252.
46. Thatcher, F.S. and Clark, D.S. 1968. Micro-organisms in Foods: Their Significance and Methods of Enumeration. University of Toronto Press, Toronto, Ont.
47. Todd, E.C.D., Jarvis, G.A., Weiss, K.F., Riedel, G.W. and Charbonneau, S. 1983. Microbiological quality of frozen cream-type pies sold in Canada. J. Food Prot. 46: 34-40.
48. Tomkins, R.G. 1932. The inhibition of the growth of meat attacking fungi by carbon dioxide. J. Soc. Chem. Ind. 51P: 261-264.
49. Weatherall, A.E. 1986. Studies on starch in crumpets. Ph.D. thesis. Department of Food Science, University of Alberta, Edmonton, AB, Canada.

Chapter 5

MODIFIED ATMOSPHERE PACKAGING OF MEAT, POULTRY AND THEIR PRODUCTS

M.E. Stiles
Department of Food Science, University of Alberta

5.1 Introduction

Slaughter of animals for food involves the "conversion of muscle to meat." The most significant postmortem change is that the end products of metabolism are no longer removed by the dynamic processes of the body. Nonetheless, meat is a dynamic system in which chemical changes occur, followed by the deteriorative effects of microbial growth. The requirements of humane slaughter should result in adequate levels of glycogen in muscle to allow the post-mortem drop of meat pH to <6.0. "Wholesomeness" of meats ensures that slaughter animals are free of disease so that contaminating microorganisms are **on** (extrinsic) rather than **in** (intrinsic) meat. Tissue of healthy animals is not sterile at the time of slaughter; however, rapid chilling inhibits the growth of intrinsic bacteria, such as *Clostridium perfringens* [49].

Immediately after dressing a meat carcase it is customary to chill it to less than 4°C. Rapid chilling extends the storage life of meats and decreases the possibility of spoilage by bone-taint or bone-sours [83] but it slows the process of ageing. Ageing of meats is

necessary to reverse the toughening effect of *rigor mortis*. This is
of great importance to meat quality because consumers rate meat
tenderness as a very important factor in acceptance of meats [100].
Ultimate tenderness of meats is affected by many factors, from the
age of the animal to the type of meat cut and the method of cooking.
However, adequate ageing allows the optimum tenderness of meat to
be achieved.   Rapid chilling of beef and lamb can result in cold
shortening [112].   This involves   the   more   rapid   shortening
(toughening) of meat fibres at 0°C than at higher temperatures.
Electrical stimulation of carcase meat speeds glycolysis to promote
rigor mortis and avoids cold shortening.   Electrical stimulation is
applied commercially to allow rapid cooling without the effects of
cold shortening.   Other techniques to allow rapid cooling to avoid
holding meats at 15-16°C are available [112].

Retail merchandizing practices for red meats are directed toward maintaining a bright red colour for the maximum possible
time.   This applies especially to beef, but also to pork and lamb.
Redness has little or no meaning for eating quality or freshness of
meat, but it is a major factor in consumer acceptance of meats, especially beef [2, 77].   The degree of redness depends on the state of oxy-
genation or oxidation of the natural pigment of meat, myoglobin, according to the following equation:

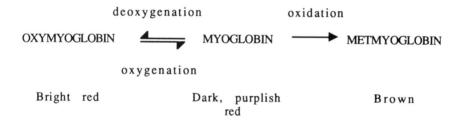

The interconversion of oxymyoglobin and myoglobin depends
on the availability of oxygen.   Ultimately, microbial growth results
in the conversion of myoglobin to metmyoglobin.

Carcase meats have a small surface area to weight ratio.
During refrigerated storage the surface dries and microbial growth
is inhibited.   As the carcase is "broken" into smaller units, the surface area to weight ratio increases.   If meat is ground (comminuted)
then surface contaminants are distributed throughout the meat.
Traditionally, fresh meats were supplied to retailers as sides or
quarters of carcases, and the meat was cut and offered for sale as desired by the customer.   With the advent of self-service in the retail
food trade in the 1950s and the availability of appropriate packaging
materials, retail cuts of meat were wrapped in an oxygen-permeable
film for display in refrigerated cases.   This promotes the retention of
the bright red colour of red meats, while protecting them from ex-

ternal contamination.   However, it does not enhance the case life of the meats.

It was realized that sliced luncheon meats that were packaged anaerobically, by vacuum packaging in gas-impermeable plastic film, had a refrigerated case life of 30 d or more [5, 87]. Because the colour of processed meats is fixed by the reaction of nitrite with myoglobin to form nitrosomyoglobin, the traditional appearance of the sliced meats was not significantly changed under anaerobic conditions.   During the 1960s centralized cutting of beef carcases was developed.   This involves the vacuum packaging of primal (wholesale) cuts in gas-impermeable plastic film as so-called "boxed" or "block-ready" beef.   This eliminates moisture loss, and extends the refrigerated storage life up to 10 to 12 weeks [134], but in the absence of oxygen the cut surfaces of beef are a dark, purplish red colour.   In the retail meat trade, this is not an acceptable colour, but when this meat is cut and exposed to air, it "blooms" to give a bright red colour.

In contrast, vacuum packaging of portion cuts for the hotel and restaurant trade gained rapid acceptance.   Some retail groups are currently selling vacuum-packaged consumer cuts of beef and lamb.   However, this generally requires re-eduction of consumers about the colour of red meats.   The alternative to vacuum packaging of retail cuts of meat to achieve an extended storage life is freezing. Freezing is efficient but expensive.   It has been successfully applied to poultry meats because of the minimal change in visual appearance.   In red meats, especially beef, frozen cuts lack the consumer acceptance of a fresh or chilled meat.   Vacuum packaging is a form of MA in which the on-going respiration of the meat and microbial growth create a carbon dioxide atmosphere which, in turn, markedly affects microbial growth [86].   MA (or vacuum) packaging can provide consumers with a versatile range of fresh (meaning never frozen) meats that have an extended refrigerated storage life.

5.2   Factors influencing storage life of meat

The use of carbon dioxide to inhibit microbial growth was discovered in the 1880s.   It was put into commercial use in the 1930s for the transoceanic shipment of carcase meats from Australia and New Zealand to Britain [39].   However, this CAS technology for meats fell into disuse, and it was only revived in the 1950s when plastics were developed that could serve as a "container" for the gas atmosphere. Impermeable packaging of meats prevents external contamination during transportation and storage, and the elevated CO_2 atmosphere serves as a bacteriostatic agent [25] against a variety of microorganisms, notably the aerobic gram-negative, rod-shaped spoilage bacteria as well as moulds [109, 110].

The extension of storage life of MAP meats and meat products is dramatic.   However, the actual extension of storage life of meat is affected by the initial microbial load, temperature of storage, the gas composition of the MA, and gas permeability of the packaging materials. Unlike controlled atmosphere storage in which the gases are constantly adjusted to desired concentrations, the gas atmosphere of MAP products varies over time as a function of respiration of meat and microorganisms and the gas permeability of the packaging materials. Plastic film with high gas permeability is selected for retail packaging of meats to allow O_2 into the pack which maintains the bright red colour of oxymyoglobin. However, the storage life of aerobically packaged retail cuts of meat with good refrigeration is seldom more than 4 d. Film or foil laminates with low gas permeability are selected for extended storage life of meats. The permeability of the packaging films is expressed by manufacturers in terms of *theoretical* oxygen transmission rates (OTR) in $cm^3/m^2/24$ h at 1 atmosphere pressure and specified (0 or 100%) humidity and temperature (usually 23°C).

5.2.1   Initial microbial load

The importance of maintaining low microbial loads through high standards of hygiene during meat preparation cannot be overemphasized. In hazard analysis of MAP meats, the level of hygiene is a critical control point influencing safety and storage life of the product. High initial microbial contamination of meats negates the benefits of MAP [22]. The effect of initial microbial load on time to spoilage was illustrated by Kraft [91]. Under aerobic conditions, where spoilage is assumed at 10^7 colony forming units (CFU)/cm^2, an initial microbial load of 10^3 as opposed to 10^5 CFU/cm^2 has the potential to double the time to spoilage at a specified storage temperature. Under anaerobic conditions, where spoilage is not directly correlated with microbial count, the time to reach maximum population would be similarly extended but the time to spoilage is more difficult to estimate. Depending on the type of organisms present, off-odours or off-flavours develop 2 to 3 weeks after the maximum population of 10^8 CFU/cm^2 is reached [35]. This depends in part on the composition of the microbial flora and the numbers of microbes initially present that are capable of growth in the MA [106]. The effects of high initial microbial load on storage life of packaged meats have been reported by several researchers [8, 22, 114, 145].

5.2.2   Storage   temperature

Temperature has been cited as the most important factor determining storage life of MAP meats [125]. Numerous studies have demonstrated the effect of storage temperature on storage life. Storage of vacuum-packaged beef cuts at 0°C consistently enhanced storage life compared with 5.5°C [20]. Increasing the storage temperature from 5 to 10°C in an atmosphere of 20% CO_2 in air caused a decrease in storage life from 11 to 6 d [25]. In studies of MAP storage of pork at -1, 4.4 and 10°C in an atmosphere of 40% CO_2 in nitrogen, the storage life in oxygen-impermeable foil laminate was 8, 5 and 2 weeks, respectively [103]. Not only the rate of microbial growth but also the types of microorganisms predominating the meat microflora varied at the different storage temperatures. At -1 and 4.4°C lactic acid bacteria predominated as long as the level of O_2 was minimal; if O_2 increased to above 2 to 3%, *Brochothrix thermosphacta* became an important component of the microflora; and at 10°C Enterobacteriaceae predominated. At storage temperatures of 10°C and above there is an increasing range of microorganisms that can grow on meat, including potentially pathogenic bacteria. *Salmonella* spp. and *Escherichia coli* are pathogens of greatest concern on fresh meats stored at abusive temperatures. Both grow equally well under aerobic and anaerobic conditions [53].

Elevation of storage temperature reduces the colour stability of MAP fresh meat. At temperatures above 3°C myoglobin is more readily oxidized to metmyoglobin [151]. The rate of discoloration of vacuum- and nitrogen-packaged beef increases as storage temperature is increased from -1 to 5°C. The amount of purge or drip loss from MAP fresh meat is also related to storage temperature. At temperatures close to 0°C the water-holding capacity of meat is maintained owing to reduced denaturation of proteins [65]. Storage at 0 or 4°C had little effect on purge volume from vacuum packaged beef [154]. Increasing the storage temperature of nitrogen packaged beef from 0 to 10°C [110] or from 0 to 7°C [137] significantly increased the amount of purge.

5.2.3 Gas atmosphere

The starting point of MAP meats was vacuum packaging. An excellent review of early research on vacuum packaging of beef was published in 1984 [126]. Since Ingram's review on the principles of prepackaging of meat [84], many reports demonstrating the effectiveness of vacuum packaging in prolonging the storage life of fresh meats compared with packaging in air have been published. In vacuum-packaged meat and poultry, levels of CO_2 rapidly rise to 10-20% and reach a maximum of 30% [43]. Vacuum packaging is, therefore, a form of MAP with elevated CO_2. Small residues of air remain in vacuum-packaged meats, but any residual oxygen is readily used by the microorganisms growing on the meat. Vacuum packaging is limited to meat with pH <6.0. Above pH 6.0, for example in DFD (dark, firm, dry) beef or pork, bacteria that produce hydrogen sulphide may grow, resulting in off-odours and formation of sulphmyoglobin, which causes a green discoloration [55, 126, 148].

The gases used for MAP are CO_2, N_2 and O_2, separately or in combination. Carbon monoxide has been considered for application in MAP of red meats because of its potential to overcome discoloration [28, 38, 92, 128, 153]. Unfortunately, the colour retention with CO can mask microbial spoilage [76]. However, some researchers [45, 153] reported that colour deterioration precedes or accompanies spoilage. Carbon monoxide is highly toxic because of its affinity for haemo- and myoglobin. However, the maximum CO concentration of meat would be 8 ppm based on CO saturation of 50% of myoglobin pigment present in meat. If all CO in treated meat was transferred to the blood, which is unlikely, consumption of a normal portion of meat would result in an increase in blood carboxy-haemoglobin of less than 0.1% [153]. The role of CO in extending the storage life of red meat appears to be only the stabilizing of meat colour.

5.2.3.1   Carbon dioxide

This gas is used in the MA storage of meats for its well-documented bacteriostatic effect [11, 25, 26, 27, 57, 80, 97, 147]. Generally, CO_2 is used in combination with N_2 as an inert filler gas and (or) oxygen, although some workers recommend the use of 100% CO_2 [52, 139]. The overall effect of CO_2 is to extend the lag phase and to increase the generation time of sensitive microorganisms [108], as illustrated in Figure 5.1. The efficacy of CO_2 in inhibiting microbial growth depends on (1) the sensitivity of the meat spoilage organisms; (2) the concentration of CO_2; (3) storage temperature; and (4) the growth phase of the microbial population. The inhibitory effect of CO_2 is selective. Moulds and aerobic bacteria such as the pseudomonads are inhibited, but it has only a limited effect on yeasts and anaerobes such as lactic acid bacteria [84, 97], Enterobacteriaceae and B. thermosphacta [57]. Increasing concentrations of CO_2 up to 20 to 30% increases the bacteriostatic effect [26]. Hyperbaric CO_2 pressure decreases the growth rate of lactobacilli growing on pork at 4°C [17]. Decreasing the storage temperature increases the solubility of CO_2 and hence its bacteriostatic effect [152]. Application of CO_2 past the lag phase of growth reduces its bacteriostatic effect [25].

Theories regarding the mechanism of microbial inhibition by CO_2 have been discussed [32]. The displacement of O_2 was first thought to be responsible for the inhibitory effect. However, replacement of air with N_2 does not have the same inhibitory effect as CO_2. When CO_2 is added to the meat atmosphere, it is absorbed by fat and lean tissue, resulting in a decrease in pH due to the formation of carbonic acid [32]. However, CO_2 inhibition of microbial growth occurs in buffered media [89]. The inhibitory effect of CO_2 has not been fully elucidated but it may be related to interference with enzymatic activity in some species [35]; for example, CO_2 may inhibit decarboxylating enzymes such as isocitrate and malate dehydrogenases [10, 80], and it dramatically inhibits the production of extracellular lipase by Pseudomonas fluorescens [119]. Other theories on the mechanism of action of CO_2 include interference in membrane fluidity [40] and toxicity of undissociated carbonic acid.

It is generally agreed that the for MA of fresh meat is 20 to 30%. Most of the studies on the effects of different CO_2 levels on theof MA beef were done in atmospheres containing O_2 concentrations less than those present in air. Systems have been patented which use O_2 levels of 50 to 80% and the balance of the MA filled with CO_2 [146]. These systems prolong the bright red colour of chilled beef for at least one week. In the absence of O_2, beef stored with elevated CO_2 does not discolour. The purple red colour of deoxymyoglobin "blooms" to the cherry red colour of oxymyoglobin when the meat is exposed to air [110]. There is some evidence of a residual bacteriostatic effect when meat stored in CO_2 is exposed to air [41, 128]; however the CO_2 must be present continuously for the on-going inhibition of sensitive microorganisms [26].

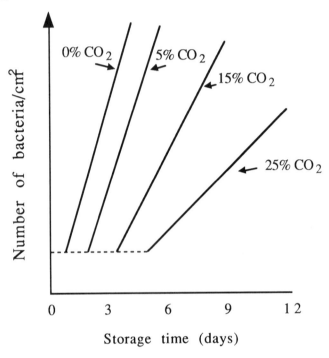

Figure 5.1.   Effect of carbon dioxide on growth curves of bacteria on chicken at 4.4°C. (Adapted from Ogilvy and Ayres [108])

The species-dependent bacteriostatic effect of CO_2 has given cause for concern about product safety. Ground beef samples inoculated with *Salmonella* spp., *Staphylococcus aureus* and Enterobacteriaceae were stored in an atmosphere of 60% CO_2, 15% N_2 and 25% O_2 at 1°C for 4 d. During subsequent temperature abuse at 10 and 20°C, the hazard from *Salmonella* spp. may be reduced, whereas no difference was noted for *S. aureus* or Enterobacteriaceae [98, 99, 135].

5.2.3.2   Nitrogen

This inert gas is commonly used as a filler gas for MA. Originally it was thought that the anaerobic environment created by N_2 would extend the storage life of meat. However, packaging beef in 100% N_2 resulted in greater discoloration of meat compared with CO_2, O_2 or a gas mixture of 70% N_2, 25% CO_2 and 5% O_2 [80]. In contrast, packaging in 100% N_2 caused less discoloration than vacuum packaging for beef [137]. Nitrogen reduces purge as a result of increased water-holding capacity [128, 130, 137]; however it has little or no bacteriostatic effect [126, 128, 129]. Greater survival of *Campylobacter jejuni* was reported on meat packaged in 100% N_2 compared with gas mixture containing CO_2 [141].

5.2.3.3   Oxygen

When O_2 is used in high concentration (greater than 50%) it has two possible functions: (1) to maintain the bright red colour of red meats [12], and (2) to control the growth of some spoilage organisms. A gas atmosphere containing 70 - 80% O_2 and 15% CO_2 extended the odour and colour storage life of MAP beef at 5°C for up to 9 d compared with storage in air [27]. In contrast, other researchers have reported reduced storage life due to metmyoglobin formation and oxidative rancidity [107, 128]. These conflicting results may reflect differences in meat types; for example lamb chops stored in an oxygen-enriched atmosphere developed rancid flavours in 3 weeks [107], whereas lipid oxidation was not a limiting factor in oxygen-enriched storage of pork [6]. Low concentrations of oxygen in MAP (less than 10%), for example 2% O_2, 20% CO_2 and 78% N_2, extended the storage life of beef compared with packaging in air, but vacuum packaging and an MA of 100% CO_2 were more effective [42]. Gas mixtures containing 10% O_2 did not maintain the colour of beef stored at 4°C [12].

5.2.4   Gas permeability of packaging materials

Gas permeability determines the rate of change of the gases in MAP. For retail cuts of meat, where retention of bright red colour is desired, packages with a high oxygen transmission rate are used. However, for cuts of meat, where extended storage life is the primary concern, packages with low gas transmission rates are used [106, 146]. When assessing information on the efficacy of MAP meat, the stage of development of gas-impermeable packages must be taken into account. Foil laminates can exclude virtually all exchange of gases. Plastic films have now been developed that can achieve this. Extension of storage life beyond 6 weeks is generally achieved with storage temperatures of 0 to -2°C, excellent hygiene of carcase meats, 100% CO_2 and gas-impermeable barrier film [51]. Storage life of vacuum-packaged beef, measured by discoloration and off-odours is inversely related to the gas (oxygen) permeability of the wrapping film [106, 128].

5.3 Applicability of MAP to different meat types

Vacuum or MA packaging of meats with elevated CO_2 dramatically affects the storage life of meats. This depends critically on chilled storage at 5°C or below. The applicability of the principles of MAP to all meat types, raw and processed (cooked), must be considered.

5.3.1   Raw chilled meats

Beef was the fresh meat for which vacuum (MA) packaging was developed.   Originally used for the transportation of carcases [39], with the development of plastic films, it is widely used for distribution of wholesale cuts.   Beef cuts represent a reasonably homogeneous surface of muscle tissue for microbial growth to occur.   pH values of beef generally range from 5.5 to 5.8.   High-pH beef (pH >6.6) stored at 4°C in 100% CO_2 develops a microflora dominated by lactic acid bacteria; however, under vacuum or in a gas mixture of 78% N_2, 20% CO_2 and 2% O_2 relatively large numbers of *Brochothrix thermosphacta*, and *Pseudomonas* spp. and *B. thermosphacta* grow with the lactic acid bacteria, respectively [42].   Chilled DFD beef with pH >6.0 stored at 1°C in vacuum packages showed signs of spoilage at 6 weeks whereas normal beef was sound for 9 or more weeks of storage [54].   However, both normal and high-pH beef packaged under CO_2 were not spoiled after 25 weeks of storage.   Under optimum conditions, chilled beef can be stored for more than 24 weeks [52].   Studies on veal showed that MAP in N_2 retards discoloration and exudate loss compared with vacuum packaging, while few differences are detected in bacterial numbers [94].   This is a somewhat surprising result that justifies further study and comparison with CO_2 environments.

Studies on the display life of vacuum packaged beef loin steaks stored for 0, 12 or 24 d and then repackaged as steaks and displayed at 2 or 7°C in either O_2 permeable film for 6 d or in vacuum packages for 30 d [64, 150] revealed the expected differences in meat spoilage. *Pseudomonas* spp. dominated the microflora of the steaks displayed in O_2 permeable film and *Lactobacillus* spp. dominated the microflora of the vacuum-packaged steaks [150].   To maintain product appearance, odour and flavour, steaks must be displayed at or below 2°C [64].   The display life of vacuum-packaged steaks was limited to 10 d by the development of odours and flavours, especially in steaks fabricated from primals which had been stored for 12 or 24 d before repackaging [64].

Development of vacuum and other MA packaging of pork was slower than for beef.   In North America this might be due to the specialized utilization of the hog carcase for curing and processing. Pork loin is the principal primal cut used for retailing.   In other countries where the whole carcase is used for retail this may be different.   Pork that was vacuum packaged and stored at 4°C [41] showed good microbiological and sensory qualities.   Variable results were reported for pork packaged in N_2 [41, 79].   Benefits of N_2 atmospheres for ground beef and veal were not observed for pork [94, 95]; notably there was no difference in meat exudate between samples stored under vacuum or in high nitrogen atmospheres.   Pork cuts stored under vacuum in packaging films with low O_2 transmission rates were grossly spoiled by the growth of *Brochothrix thermosphacta* after 2 weeks' storage at 3°C and 5 weeks' storage at -1.5°C [52].   Extended storage of 18 to 26 weeks was achieved in CO_2 packaged cuts stored at -1.5°C. During prolonged storage under CO_2, the only deteriorative changes occurred with pale, soft, exudative (PSE) pork which showed a distinct loss of colour. This was attributed to the loss of myoglobin with the exudate. Under optimum conditions, chilled pork can be stored for more that 24 weeks [52].

A shorter storage life of 6 to 8 weeks has been reported for vacuum-packaged lamb [85, 107, 132]. Lamb chops stored in air and 20% CO_2 had a storage life 50% greater than in air. Storage in high O_2 and 20% CO_2 gave bright red colour but brown discoloration and rancidity developed in 3 weeks [107]. Much of the surface of lamb is adipose tissue, which has a pH close to neutral and no respiratory activity to reduce the pH. Hence the surface of lamb presents a heterogeneous environment for microbial growth [62]. Storage life of MAP lamb cuts was extended to 12 weeks by packaging in an oxygen-impermeable foil laminate, and holding at 0 to -0.5°C [54]. Comparison of CO_2 and N_2 for head space gas in MAP primal cuts of lamb revealed better storage of retail cuts at 5°C in air for samples stored in CO_2 [63].

The inhibitory effect of CO_2 on slime-forming organisms on chicken was established by Ogilvy and Ayres in 1951 [108]. They reported an increased storage life at 4.4°C with increasing CO_2 concentration, but that the maximum practical level of CO_2 was 25%, because of discoloration of the meat at higher concentrations. However, this may have been a deficiency of the packaging. Subsequently, there have been reports that high-CO_2 atmospheres extend the storage life up to three-fold that of storage in air [9, 78, 120]. In contrast to red meats, poultry does not undergo irreversible discoloration of the meat surface in the presence of oxygen. In a study of broiler carcases packaged under vacuum in film of low O_2 transmission or under CO_2 in gas-impermeable packs [58], storage life was a function of storage temperature, packaging and O_2 availability. Putrid spoilage in gas-impermeable packs after 7 weeks' storage at 3°C or 14 weeks' storage at -1.5°C was attributed to "enterobacteria." In vacuum packages with oxygen transmission rates of 30 to 40 $mL/m^2/24h$, putrid odours were detected after 2 weeks' storage at 3°C and 3 weeks' storage at -1.5°C.

Rabbit meat was also stored under vacuum and MA containing N_2 and CO_2 [44]. While CO_2 atmosphere maintained good microbiological quality of the meat, it caused decreased water-holding capacity which lead to discoloration and toughening of the meat. In contrast, storage under N_2 preserved organoleptic properties but not the microbial quality. Vacuum packaging gave similar results to CO_2 storage. Temperature of storage at 0 or 3°C was cited as a more important factor than gas atmosphere in determining meat quality.

5.3.2   Cooked and cured (processed) meats

The use of anaerobic atmospheres for storage of perishable cooked and cured meats is viewed with concern by Public Health

Officials. The use of nitrite as an additive in cured meats effectively inhibits the outgrowth of *Clostridium botulinum* spores. In recent years there has been an emphasis on reducing use levels of nitrites in processed meats. Nitrite-free processed meats are available in the retail marketplace [149] and increased use of MAP of cooked meats in delicatessens and restaurants creates conditions where the only hurdle to food safety is refrigeration [73]. In studies with sliced roast beef stored in MA with 30, 50 or 70% CO_2 in N_2 at 4.4°C, a predictable lactic acid microflora developed, whereas with cooked hamburgers the development of a microflora could not be predicted in the same way [102]. The safety of nitrite-free processed meats and cooked, uncured meats packaged in anaerobic environments needs extensive study [48].

5.4   Meat microbiology

There have been many reviews detailing the microbiology of meats, including Ingram's review of the microbiological principles of prepackaged meats published in 1962 [84] and more recently chapters by Gill in 1985 [50] and Kraft in 1986 [91]. A combination of intrinsic, processing and extrinsic factors dictate the potential for microbial growth that may result in the hazard of foodborne illness [83]. The intrinsic factors include: the chemical and physical characteristics and nutrient composition of the food; water activity, pH and redox potential; and natural inhibitory substances in foods. The most significant processing parameter is heat treatment. The extrinsic factors include storage temperature, packaging and gas atmosphere, the use of permitted preservatives, as well as cleaning, disinfection and hygiene. These characteristics of a food interact to determine the likelihood that a specific microorganism will survive and (or) grow in the food. The effect of superimposing these limiting factors has been described as the "Hurdle Effect" [96]. This has been succinctly described and illustrated by Scott [124]. She claims that food technologists only have extensive information for predictive hurdle evaluation for *C. botulinum* in cured meats and cheese spreads. The hurdles in the new generation refrigerated MAP foods have not been established to the same extent. For some of these foods refrigeration is the only hurdle assuring product safety. Because adequate refrigeration cannot be guaranteed throughout the food distribution system, this represents a tenuous safety hurdle.

The exterior of animals is heavily soiled so that skinning and opening the body cavity results in surface contamination of the carcase. The extent of contamination depends on the sanitation practiced during slaughter [7]. Despite the wide range and levels of contaminating microorganisms, the microbiology of refrigerated fresh meats is highly predictable [83]. Efficient chilling of carcase meats creates a selective environment which favours the growth of psychrotrophic strains of microorganisms. Psychrotrophs are those bacteria that grow optimally at >20°C, but also grow at refrigeration temperatures below 5°C. Some psychrotrophic bacteria grow at <0°C, as long as there is water available for their growth. Storage of carcase meats under conditions that permit drying, i.e. relative humidity <80%, inhibits bacterial growth but growth of psychrotrophic fungi occurs [83]. However, storage of carcase meats under drying conditions is not favoured because drying results in weight loss and deterioration of sensory quality, which represents a serious economic loss to the meat trade.

The gas atmosphere also creates a selective pressure on the microflora of meats. Aerobic, chilled storage favours the growth of gram-negative, aerobic rod-shaped bacteria including *Pseudomonas*, *Moraxella* and *Acinetobacter* spp. [8]. Many other bacteria are present but *Pseudomonas* spp. predominate and produce off-odours from protein breakdown and amino acid metabolism [21]. In general, off-odours develop at 10^7 CFU of bacteria per square centimeter or gram of meat. The relationship between bacterial numbers and spoilage of meats is illustrated in Figure 5.2. Anaerobic, chilled storage of meat has a fundamental effect on the predominating microflora, more as a function of elevated CO_2 levels than as a result of absence of O_2 [88]. Under anaerobic conditions of chilled storage with elevated levels of CO_2, as occurs with vacuum packaging, growth of the aerobic spoilage microflora is discouraged, while the slower-growing lactic acid bacteria, such as *Lactobacillus*, are encouraged [117, 145]. In the presence of oxygen, growth of *Brochothrix thermosphacta* occurs and may cause spoilage of meats [61]. At poor refrigeration temperatures, for example 10°C, Enterobacteriaceae predominate the microflora of the meats and cause off-odours [43, 103].

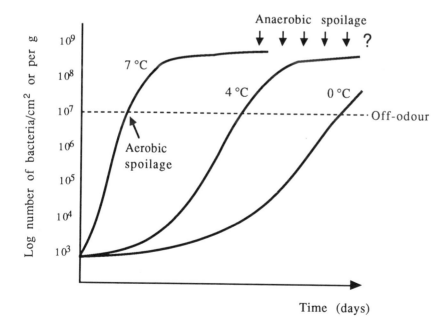

Figure 5.2.   Aerobic and anaerobic spoilage of meat at different storage temperatures.

The change in microflora between aerobically and anaerobically packaged meat is responsible for the extended storage life of vacuum and MA packaged meat with increased levels of CO_2. The meat lactics do not produce off-flavours or off-odours in vacuum-packaged meats stored at 0 to 5°C in the same time frame or at the same microbial load as the aerobic spoilage microflora, and myoglobin is not converted to metmyoglobin [14]. Spoilage by meat lactics is generally secondary to microbial growth, as illustrated in Figure 5.2, unless hydrogen sulphide-producing strains of lactobacilli predominate [133]. Lactobacilli can produce H_2S at low pH, under anaerobic conditions and at low concentrations of carbohydrate [93]. Hydrogen sulphide production by lactobacilli is rare; however, chilled vacuum-packaged meats can develop objectionable sulphide odours with prolonged storage [67, 93].

In a vacuum package the residual oxygen is rapidly consumed by tissue and microbial respiration and CO_2 reaches a high partial pressure [84]. If air is evacuated from the package and replaced with a modified gas atmosphere, extremely low O_2 residues can be achieved. If the gas atmosphere is achieved by gas flushing, the level of residual oxygen depends on the length of time and volume of flushing. Red meats prepared under good standards of hygiene, packaged in saturation levels of 100% CO_2 in a gas-impermeable film using the Captech™ system and stored at -1.5°C have a storage life of up to 6 months [51].

There have been many studies on the lactic acid bacteria (lactics) associated with vacuum-packaged meats. The lactics are a heterogeneous group of microorganisms that produce lactic acid as the end product of carbohydrate metabolism. They are separated into two groups based on the end products of their fermentation. Homofermentative lactics produce almost entirely lactic acid from hexose sugars, but they may produce lactic and acetic acids from pentose sugars [18]. Heterofermentative lactics utilize a different metabolic pathway to break down hexose sugars and yield 50% lactic acid plus CO_2 and a mixture of end products that includes acetic acid, acetaldehyde and ethanol.

The lactic acid bacteria were traditionally subdivided into four genera, *Lactobacillus*, *Leuconostoc*, *Pediococcus* and *Streptococcus*. The genus *Streptococcus* has undergone considerable taxonomic reclassification. The serological group N (lactic) streptococci have been transferred to the genus *Lactococcus*. The lactobacilli are an important group of lactic acid bacteria associated with meats. It is assumed that they can be readily distinguished from other lactics by their rod shape, but some heterofermentative lactobacilli are "coccoid rods" and they are easily confused with leuconostocs. The homofermentative lactobacilli were subdivided by Orla-Jensen into two main groups, the so-called Streptobacteria, which grow at 15°C, and the Thermobacteria, which do not grow at 15°C. The Streptobacteria are important among the lactics on vacuum-packaged sliced cooked meats [104] and on fresh meats [131].

During the 1980s the lactic acid bacteria of vacuum-packaged meats were studied more analytically. Hitchener *et al.* [75] isolated 177 organisms from beef with pH 5.5 to 5.8, stored at 0 to 1°C in packaging with an intermediate oxygen transmission rate. They observed that most of the strains were *Lactobacillus*-type organisms (90%). Based on Orla-Jensen's classification these were largely heterofermentative Betabacteria (65%) or atypical Streptobacteria (20%). The remaining 10% of the strains were identified as leuconostocs. In a study by Shaw and Harding [131], 100 strains were isolated from refrigerated, vacuum-packaged beef, pork, lamb, and bacon. They identified most of the strains as atypical Streptobacteria (88%), which they subdivided into aciduric (57%), which were provisionally identified with *L. sake* or *L. bavaricus,* and nonaciduric (31%), which were not identifiable with any described species. These nonaciduric strains have subsequently been identified as *Carnobacterium* spp. [29]. *Leuconostocs* were also identified, and 6% of the isolates were other strains of lactic acid bacteria. *Leuconostoc* spp. can predominate the microflora of MAP raw meat [121].

In an extensive study on the lactic acid bacteria growing on beef steaks stored at 2°C in high or medium oxygen barrier film [150], the heterofermentative *Lactobacillus cellobiosus* was the most common strain isolated. The homofermentative lactobacilli *L. coryneformis* and *L. curvatus* were also important, accounting for up to 50% of strains isolated at different times of the experiment. *Leuconostoc* spp. did not generally constitute a significant part of the population, except on freshly prepared steaks stored under vacuum.

Schillinger and Lücke [123] studied 229 strains of lactic acid bacteria from fresh and processed meats. They developed their own taxonomic scheme based primarily on easily determined fermentation and physiological characteristics. They identified 13 strains of *Lactobacillus* that they noted as the most common on meats, *L. alimentarius, L. brevis, L. casei* subsp. *casei, L. coryneformis, L. curvatus, L. farciminis, L. halotolerans, L. hilgardii, L. plantarum, L. sake, L. viridescens* and the new species *C. piscicola* (*L. carnis*) and *C. divergens* (*L. divergens*).

Lactic acid bacteria in meats are viewed variously as preservative or spoilage agents. Vacuum- and MA-packaged meats are **fermenting** rather than **spoiling** products. The lactic microflora of MAP meats is not controlled; therefore if a H_2S producer is present in fresh meats or aciduric strains of *Lactobacillus* or *Leuconostoc* strains are present on processed meats with added carbohydrate, then excess acidity (aciduric *Lactobacillus*) or slime (dextran formation from sucrose by *Leuconostoc*) can develop and cause product spoilage. Greening of meats can occur when lactics produce either H_2S or H_2O_2 [127].

Many lactic acid bacteria that predominate the microflora of raw and processed meats are bacteriocin producers [3]. Bacteriocins are proteinaceous antibacterial compounds produced by bacteria, generally with a narrow spectrum of antibacterial activity primarily against closely related bacteria. It is speculated that bacteriocin production may be involved in strain domination in mixed fermentations [90]. Studies have shown that bacteriocin production in many strains is plasmid-mediated [4, 72]. The opportunity exists to develop strains of lactic acid bacteria that dominate the microflora of MAP meats and cause limited or no spoilage defects in the meat for extended periods of refrigerated storage.

Results of studies to extend storage life of meats with lactic acid bacteria have been equivocal. In some cases this may have resulted from use of inappropriate cultures, for example dairy starters [1, 115, 116] or commercial meat starters [113]. Inhibition of psychrotrophic spoilage bacteria by lactic acid bacteria was attributed in part to production of hydrogen peroxide [59]. Growth of *Brochothrix thermosphacta* was inhibited by *Lactobacillus* spp. in vacuum-packaged meat [117]. This was also observed in vacuum-packaged processed meat from which strains of *Lactobacillus brevis* and *Lactobacillus plantarum* were isolated [30]. Inhibition was not due to acidity, lactic acid or hydrogen peroxide [30, 117]. There have been several studies on the effect of lactic acid starter cultures on spoilage and pathogenic bacteria in meats [34, 66, 122, 138]. In cases where dairy starter bacteria were used, large inocula were necessary to achieve the desired effect. Beef steaks inoculated with lactic acid bacteria from vacuum-packaged meats failed to extend storage life at 1 to 3°C [66, 138]. Inoculated steaks had a higher incidence of off-odours, surface discoloration and poor flavour ratings than uninoculated steaks.

In a study of MAP lamb [54], samples were treated with 5% lactic acid or a *Lactobacillus* strain originally isolated from meat. Samples were packaged in plastic film with a moderate oxygen transmission rate and held at -0.5°C for up to 12 weeks. The cuts treated with lactic acid developed an off-odour attributed to growth of "enterobacteria" by 12 weeks. Inoculated cuts developed an unacceptable "dairy" flavour. In contrast, meats treated with lactic acid and stored in foil laminate packages were unspoiled after 12 weeks.

The organism *Brochothrix thermosphacta* (previously known as *Microbacterium thermosphactum*) is an important spoilage organism of refrigerated MAP meats and poultry [82]. It can be classified with the lactic acid bacteria as it is a nonspore-forming, grampositive rod-shaped organism, which produces lactic acid as its major end product of metabolism. It produces acetoin and acetate aerobically [16], and acetoin is oxidized to diacetyl which can cause spoilage defects in meat. Low pH, anaerobic conditions and low temperature should inhibit *B. thermosphacta* [15, 19, 52]. This organism does not compete well in the presence of lactobacilli [30, 117] nor under strict anaerobic conditions [61]. However, *B. thermosphacta* can be a major spoilage organism of MAP meats if the anaerobic environment is not maintained [15, 61, 103] even at a pH as low as 5.4 [19]. Comparison of *B. thermosphacta* and lactobacilli as spoilage agents of vacuum-packaged sliced luncheon meats, showed that *B. thermo-*

sphacta caused rapid spoilage, whereas homofermentative lacto-bacilli caused spoilage much more slowly [36].

Members of the family Enterobacteriaceae may also constitute part of the psychrotrophic microflora of vacuum-packaged meats [67, 68, 69, 105, 144]. Many Enterobacteriaceae, such as *Citrobacter freundii, E. coli, Klebsiella pneumoniae* and *Salmonella* spp., which are commonly found on meats do not grow because the temperature of the chilled storage is below their minimum growth temperatures of 5 to 7°C. In contrast, psychrotrophic *Enterobacter, Hafnia* and *Serratia* spp. and *Y. enterocolitica* can grow at storage temperatures below 5°C [13, 68, 69, 150].

Progressive changes in meat processing and packaging have created concerns of an increased risk of botulism and other foodborne illnesses [149], for example: in the 1940s, vacuum packaging; in the 1950s, perishable canned cured meats; in the 1960s, new process for shelf-stable canned hams; in the 1970s, removing nitrites from cured meats; and in the 1980s, vacuum- and MA- packaging of cooked meats. The anaerobic packaging of meats raises questions about the growth of *Clostridium botulinum* or other microaerophilic or facultative anaerobic pathogens. This applies especially to nonproteolytic strains of *C. botulinum* that grow at temperatures down to 3.3°C [37], and that may not cause spoilage before the meat becomes toxigenic. As well, potentially pathogenic psychrotrophs including *Aeromonas hydrophila, Listeria monocytogenes,* and *Y. enterocolitica* may grow on vacuum- or MA-packaged beef to levels capable of causing foodborne illness [56]. However, specific outbreaks involving vacuum- or MA-packaged refrigerated meats have not been reported.

In vacuum- or MA-packaged meats, the lactic microflora creates a competitive environment in which potential pathogens are not expected to grow. The role of the lactic microflora in relation to growth of selected pathogens needs thorough investigation. *Y. enterocolitica*-like bacteria grow on vacuum-packaged beef and lamb [69]. However, to 1980 no outbreaks of foodborne illness had been traced to *Y. enterocolitica* on vacuum-packaged meats [135]. Enteric bacteria such as *Salmonella* can grow on meat at temperatures >7°C, but growth is reduced in MA with elevated CO_2; *Staphylococcus aureus* do not grow during storage of meat at 10°C for 10 days; it was also suggested that MA storage of fresh meat with elevated CO_2 does not increase the hazard of *C. botulinum* growth [135].

In a review of the microbial safety of MAP fresh meat, Genigeorgis [47] concluded that, despite the predictive possibility of *C. botulinum* growth, with strict temperature control the process is suitable for prolonged fresh storage of meats. However, he questioned the advisability of MA for consumer-size packages. There is surprisingly little information on the growth of *C. botulinum* on anaerobically packaged fresh meats. In pasteurized (cooked) meats there is a distinct difference in predictive microbiology dependent upon the presence or absence of curing salts (especially nitrite). In the case of cured meats such as bologna or ham, extensive studies by Stiles and co-workers [111, 140, 142, 143, 144] revealed that neither *Bacillus cereus* nor *Clostridium perfringens* would grow on these meats even under conditions of severe temperature abuse. *S. aureus* grew when inoculated meats were stored at 21 and 30°C for up to 24 hours. *Salmonella typhimurium* and *E. coli* survived on the meats during extended anaerobic storage. The pathogens were less likely to grow on meats subjected to temperature abuse if they had a developed microflora than if the meats were freshly prepared and had not had time to develop their adventitious microflora. This protective effect was ascribed to the competitive lactic acid microflora that developed on the processed meats during chilled storage [140].

The inhibition of *C. perfringens* can also be extended to other spore-forming bacteria such as *C. botulinum*. Vacuum packaging of luncheon meats inoculated with spores of *C. botulinum* type A and stored at abusive temperatures did not increase the rate of toxin production compared with aerobic storage [23]. However, concerns about nitrosamine formation in cured meats have created a market for nitrite-free products [149]. MAP is also being applied to uncured meats, such as sliced roast beef. In studies of the microbiology of cooked roast beef packaged in MA with 75% CO_2, 15% N_2 and 10% O_2 [73, 74] it was shown that without temperature abuse inoculated pathogens failed to grow [74]. However, with temperature abuse, *S. aureus* did not grow, but *S. typhimurium* grew when the meat was held for 42 days at 4.4°C and then stored at 12.5°C for 7 days [73]. In these studies [73, 74], no reference is made to the growth of a competitive microflora. The fate of *Listeria monocytogenes* inoculated onto processed meats and cooked roast beef was determined during storage in vacuum packages for up to 12 weeks at 4.4°C [60]. *L. monocytogenes* survived but did not grow on summer sausages, grew slightly on roast beef, and grew well on ham, bologna, bratwurst and on sliced chicken and turkey. Growth was best on meats at pH 6 or above, and poor on meats with pH 5 or below.

In a study of MAP sliced roast beef sandwiches [102], a predictable lactic microflora developed. This occurred when packages were prepared with 10% O_2 or without O_2. In the case of commercially prepared, broiled hamburger (minced beef pattie) products, the post-cooking microflora was less predictable. In some samples a lactic microflora developed while in others its development was delayed or it had not developed after 6 weeks of storage. In this case the only "hurdle" to growth of potential pathogens is temperature control. Packages containing 10% O_2 would be the preferred MA, but spoilage of aerobically packaged hamburgers occurred within 2 to 3 weeks owing to oxidative rancidity [102]. The same pathogen challenge studies for cooked meats as those on processed meats [111, 140, 142, 143] have not been done, but the concern for food safety dictates that studies of this nature should be undertaken without delay.

5.5   Packaging   systems   for   meat

The traditional distribution of meats as half or quarter carcases was simple but inconvenient. It resulted in on-going external contamination of the meat and economic loss through evaporation of moisture. Centralized packaging of meats controls both of these problems; however MAP also extends the storage life of meats. The gases of primary importance to extended storage life are CO_2 and O_2. Plastic films are up to four times more permeable to CO_2 than O_2. In gas-permeable (aerobic) films, CO_2 produced by respiration of meat and microorganisms escapes from the package. According to Taylor [146] there are three main methods for vacuum packaging of meat:

(i) The Cryovac method in which meat is vacuum packaged in a heat-shrinkable plastic bag with low oxygen permeability, and sealed with a metal clamp.

(ii) The chamber method in which processed meats are placed in a preformed bag, evacuated and heat sealed. Such packages are suitable for retail use because they allow product labelling on the package. The bags are not heat shrinkable, hence drip can accumulate in the corners of the bags. This can be avoided by using secondary sealing in which an inner layer of surlyn or other heat-sealable film is used so that "free" package space around the meat is eliminated.

(iii) The thermoforming method, which is best adapted to smaller pieces of meat or processed meats. The meat is placed in trays and an upper layer of plastic is heat sealed onto the tray under vacuum. This requires two types of laminated film - a forming film for the base and a nonforming film as a seal.

Vacuum packaging of retail meats is best adapted to processed meats, but application of vacuum- or MA-packaging techniques to fresh meats for retail sale is an obvious attraction. In the current climate of environmentalist concerns about plastics, centralized packaging of retail meats would reduce the amount of plastic material used for marketing fresh meats. For use with red meats, this would require extensive re-education of consumers about meat colour; however, in the case of poultry there should be no restriction to the acceptance of centralized vacuum packaging. Various systems of overwrapping have been developed for red meats, for example the retail cut is packaged in a gas-permeable film and overwrapped in an anaerobic MA containing elevated levels of CO_2. When the overwrap is removed, oxygen crosses the gas-permeable film and oxygenates the myoglobin.

Gas permeability of the packaging film depends on the materials used. Both polyvinylidene chloride (PVDC) and ethylene vinyl alcohol (EVOH) have excellent water vapour and gas barrier characteristics. Meat packaging films are generally multi-layer structures using a variety of polymer resins [46]. For prolonged refrigerated storage of fresh and processed meats a low O_2 transmission rate is required. The O_2 transmission rate is a function of polymer type, thickness, storage temperature, relative humidity, and other factors. Until recently (the end of the 1980s), clear plastics could only achieve O_2 transmission rates of 25-30 mL/m^2/24 h at 23°C and 0% RH. However, lower O_2 transmission rates have been achieved with new clear, composite polymers with O_2 transmission rates <1 mL/m^2/ 24 h.

Forming films require a rigid polymer, such as nylon, which can be softened by heat, formed in a cavity to the desired size and shape, without damaging the sealant layer. Nylon affords a poor moisture barrier, but a relatively good gas barrier. However, the process of forming the film can result in stress and different degrees of thinning that influence the gas and moisture transmission rates of the finished package. Water vapour transmission rates can be reduced with layers of oriented polypropylene and (or) polyethylene (EVA) [46].

Sealing of MA packages is a critical step in forming and maintaining the desired atmosphere in the package. Several sealants can be used. They usually represent the thickest element in the lamination. Surlyn (Ionomer) is widely used, having a broad heat seal range, a lower heat seal temperature and it maintains its strength while hot (hot tack). New developments in sealants are occurring all of the time, so that manufacturers of plastic film have an increasing range of products from which to choose in order to meet customers' needs.

5.6   Research   needs

Vacuum or MA packaging of meats with elevated CO_2 has revolutionized meat marketing, by affording extended refrigerated storage of fresh and processed meats. Many research studies have focused on the effect of the gas atmosphere on extension of storage time. It appears that a minimum of 20% CO_2 is necessary for the inhibitory effect, but some reports indicate that 100% CO_2 is optimum [51]. There may be little or no economic advantage to different amounts of CO_2, hence the optimum concentration of CO_2 should be determined under a wide range of storage conditions.

It is well-known in the trade that the bright red colour of fresh meats is an artifical criterion of meat quality. The obvious solution is to educate consumers about the acceptability of dark red meat. This may require marketing and (or) consumer research for different markets to determine the best education and information strategies. Colour fixing with carbon monoxide to form carboxymyoglobin has been tested [152]; however, this presents problems that could be interpreted as consumer deception, i.e. the colour change to metmyoglobin no longer correlates with bacterial spoilage of fresh meat.

The microbiology of anaerobic MAP meats has been extensively studied, to the point that for raw meats it is known that a predictable lactic acid microflora develops. This is comprised primarily of aciduric and nonaciduric lactobacilli (including carnobacteria) and leuconostocs. Ultimately these organisms, or others such as *Brochothrix thermosphacta*, cause spoilage of the meats. However, the initial fermentation and use of alternative energy sources does not result in immediate spoilage. Some H_2S-producing lactics cause off-odours, which result in rejection of MAP products. More control of the microflora developing on MAP fresh meats is desirable. This requires extensive research on bacterial domination in mixed fermentations. The production of bacteriocins, and the genetic engineering of selected strains of lactic acid bacteria to produce these compounds, could be a route to achieving "controlled" fermentation in a mixed microflora. If bacteriocins do not provide the necessary competitive advantage to the "starter" strains, alternative strategies based on inoculum size and growth rate can be developed.

Bacteriocins, by definition, have a narrow range of antibacterial activity. Some are only active against the same species; others are active against a slightly broader spectrum, for example bacteriocins produced by carnobacteria and leuconostocs [4, 70, 71, 72] have a spectrum of activity that includes *Enterococcus* and *Listeria* spp. Roth and Clark [117] noted that *Brochothrix thermosphacta* is inhibited by lactic acid bacteria. We observed that this could be caused by bacteriocins produced by lactic acid bacteria [Stiles, unpublished data]. One of the main concerns for control of the microbial flora of meats is the inhibition of gram-negative spoilage bacteria. This is not only desirable for control of aerobic spoilage of the meats, but it could also control gram-negative pathogenic bacteria, such as *Salmonella* spp. and enterotoxic or verotoxic *E. coli*. Generally, bacteriocins would not be expected to achieve this but some examples have been cited in the literature [101, 136]

The obvious question for control of the microflora of meats is the use of nisin or nisin-producing *Lactococcus lactis* subsp. *lactis*. *L. lactis* is not adapted to grow on meats, being a mesophilic lactic acid bacterium of dairy origin [81]. Furthermore, nisin is only stable at low pH. At pH 5 and above nisin is inactivated, and it is sensitive to proteases [81]. An alternative to nisin production by *L. lactis* is to transfer the nisin-producing ability into meat lactics. This was reported for leuconostocs, but the genetics of this have not been fully resolved. The post-translational modification of nisin makes it a more complex compound to produce, hence further work will be necessary to utilize nisin-producing strains as a preservation strategy in raw meats. Nisin is used as a food additive in Europe, but in N. America it has only been licensed for use in processed cheese in the United States since 1988. The efficacy of nisin as an antibotulinal agent has been questioned [24]. Alternatives to nisin should be sought among the meat lactics.   This could lead to the discovery of bacteriocins that are better adapted to meats.

Extensive studies were done on processed meats to determine their safety with temperature abuse [111, 140, 142, 143].   With the advent of nitrite-free processed meats and the vacuum or MA packaging of cooked (uncured) meats such as sliced roast beef, there is a need for further trials on the safety of anaerobically packaged, refrigerated meats.   Not only for the possible hazard of *C. botulinum* growth and toxigenesis, but also for the growth of other pathogens during refrigerated storage or with temperature abuse.   During the 1980s a range of pathogenic bacteria capable of growth at refrigerator temperatures was recognized [33], these bacteria should also be tested for their ability to survive and grow on different meat types.

In an age of increasing concern for protection of the environment, the packaging of meats in plastic materials needs critical review and justification.   The advent of plastic films has allowed the dramatic developments in MA storage of meats, with its associated benefits for the meat industry and consumers.   For environmental acceptability, the amount of plastic material used in packages should be minimized, but only to a level that gives the necessary protection against contamination and gas exchange.   The type of plastic material should be selected based on the proposed methods of disposal.   If disposal involves incineration, the chloride-containing polymers should not be used so as to avoid release of hydrochloric acid into the atmosphere.   MAP of meats represents a sufficiently significant advance in meat safety and delay in meat spoilage that it would be unfortunate to lose these advantages because of a misinformed public or lack of environmental awareness by the meat industry. This attitude in itself should speed up the retailing of vacuum- or MA-packaged cuts of red meats.

References

1. Abdel-Bar, N.M. and Harris, N.D.   1984.   Inhibitory effect of *Lactobacillus bulgaricus* on psychrotrophic bacteria in associative cultures in refrigerated foods. J. Food Prot. 47: 61-64.
2. Adams, J.R. and Huffman, D.L.   1972.   Effect of controlled gas atmospheres and temperatures on quality of packaged pork. J. Food Sci. 37: 869-872.
3. Ahn, C. and Stiles, M.E.   1990.   Antibacterial activity of lactic acid bacteria isolated from vacuum-packaged meats.   J. Appl. Bacteriol. 609: 302-310.
4. Ahn, C. and Stiles, M.E.   1990.   Plasmid-associated bacteriocin production by a strain of *Carnobacterium piscicola* from meat. Appl. Environ. Microbiol. 56: 2503-2510.
5. Allen, J.R. and Foster, E.M.   1960.   Spoilage of vacuum-packed sliced processed meats during refrigerated storage.   Food Res. 25: 19-25.
6. Asensio, M.A., Ordonez, J.A. and Sanz, B.   1988.   Effect of carbon dioxide and oxygen enriched atmospheres on the shelf-life of refrigerated pork packed in plastic bags.   J. Food Prot. 51: 356-360.
7. Ayres, J.C.   1955.   Microbiological implications in handling, slaughtering, and dressing meat animals.   Adv. Food Res. 6: 109-161.
8. Ayres, J.C.   1960.   Temperature relationships and some other characteristics of the microbial flora developing on refrigerated beef.   Food Research 25: 1-18.
9. Baker, R.C., Hotchkiss, J.H. and Qureshi, R.A.   1985.   Elevated carbon dioxide atmospheres for packaging poultry.   I.   Effects on ground chicken.   Poult. Sci. 64: 328-332.
10. Bala, K., Marshall, R.T., Stringer, W.C. and Naumann, H.D.   1977. Effect of *Pseudomonas fragi* on the color of beef. J. Food Sci. 42: 1176-1179.
11. Baran, W.L., Kraft, A.A. and Walker, H.W.   1970.   Effects of carbon dioxide and vacuum packaging on color and bacterial counts of meat.   J. Milk Food Technol. 33: 77-82.
12. Bartkowski, L., Dryden, F.D. and Marchello, J.A.   1982.   Quality changes of beef steaks stored in controlled atmospheres containing high or low levels of oxygen.   J. Food Prot. 45: 41-45.
13. Beebe, S.D., Vanderzant, C., Hanna, M.O., Carpenter, Z.L. and Smith, G.C.   1976.   Effect of initial internal temperature and storage temperature on the microbial flora of vacuum packaged beef.   J. Milk Food Technol. 39: 600-605.
14. Benedict, R.C., Strange, E.D., Palumbo, S. and Swift, C.E. 1975.   Use of in-package controlled atmospheres for extending the shelf life of meat products.   J. Agric. Food Chem. 23: 1208-1212.
15. Blickstad, E. and Molin, G.   1983.   Carbon dioxide as a controller of the spoilage flora of pork, with special reference to temperature and sodium chloride.   J. Food Prot. 46: 756-763, 766.
16. Blickstad, E. and Molin, G.   1984.   Growth and end-product formation in fermenter cultures of *Brochothrix thermosphacta* ATCC 11509[T] and two psychrotrophic *Lactobacillus* spp. in different gaseous atmospheres.   J. Appl. Bacteriol. 57: 213-220.

17. Blickstad, E., Enfors, S.-O. and Molin, G. 1981. Effect of hyperbaric carbon dioxide pressure on the microbial flora of pork stored at 4 or 14°C. J. Appl. Bacteriol. 50: 493-504.
18. Bottazzi, V. 1988. An introduction to rod-shaped lactic-acid bacteria. Biochimie 70: 303-315.
19. Campbell, R.J., Egan, A.F., Grau, F.H. and Shay, B.J. 1979. The growth of *Microbacterium thermosphactum* on beef. J. Appl. Bacteriol. 47: 505-509.
20. Carpenter, Z.L., Beebe, S.D., Smith, G.C., Hoke, K.E. and Vanderzant, C. 1976. Quality characteristics of vacuum packaged beef as affected by postmortem chill, storage temperature, and storage interval. J. Milk Food Technol. 39: 592-599.
21. Christopher, F.M., Carpenter, Z.L., Dill, C.W., Smith, G.C. and Vanderzant, C. 1980. Microbiology of beef, pork and lamb stored in vacuum or modified gas atmospheres. J. Food Prot. 43: 259-264.
22. Christopher, F.M., Seideman, S.C., Carpenter, Z.L., Smith, G.C. and Vanderzant, C. 1979. Microbiology of beef packaged in various gas atmospheres. J. Food Prot. 42: 240-244.
23. Christiansen, L.N. and Foster, E.M. 1965. Effect of vacuum packaging on growth of *Clostridium botulinum* and *Staphylococcus aureus* in cured meats. Appl. Microbiol. 13: 1023-1025.
24. Chung, K.-H. 1989. Effects of nisin on growth of bacteria attached to meat. Appl. Environ. Microbiol. 55: 1329-1333.
25. Clark, D.S. and Lentz, C.P. 1969. The effect of carbon dioxide on the growth of slime producing bacteria in fresh beef. Can. Inst. Food Sci. Technol J. 2: 72-75.
26. Clark, D.S. and Lentz, C.P. 1972. Use of carbon dioxide for extending shelf-life of prepackaged beef. Can. Inst. Food Sci. Technol. J. 5: 175-178.
27. Clark, D.S. and Lentz, C.P. 1973. Use of mixtures of carbon dioxide and oxygen for extending shelf-life of prepackaged fresh beef. Can. Inst. Food Sci. Technol. J. 6: 194-196.
28. Clark, D.S. Lentz, C.P., and Roth, L.A. 1976. Use of carbon monoxide for extending shelf-life of prepackaged fresh beef. Can. Inst. Food Sci. and Technol. J. 9: 114-117.
29. Collins, M.D., Farrow, J.A.E., Phillips, B.A., Feruso, S. and Jones, D. 1987. Classification of *Lactobacillus divergens, Lactobacillus piscicola,* and some catalase-negative, asporogenous, rod-shaped bacteria from poultry in a new genus, *Carnobacterium*. Int. J. Syst. Bacteriol. 37: 310-316.
30. Collins-Thompson, D.L. and Lopez, G.R. 1980. Influence of sodium nitrite, temperature, and lactic acid bacteria on the growth of *Brochothrix thermosphacta* under anaerobic conditions. Can. J. Microbiol. 26: 1416-1421.
31. Collins-Thompson, D.L. and Lopez, G.R. 1982. Control of *Brochothrix thermosphacta* by *Lactobacillus* species in vacuum packed bologna. Can. Inst. Food Sci. Technol. J. 15: 307-309.
32. Daniels, J.A., Krishnamurthi, R. and Rizvi, S.S.H. 1985. A review of effects of carbon dioxide on microbial growth and food quality. J. Food Prot. 48: 532-537.
33. Doyle, M.P. 1990. Foodborne Bacterial Pathogens. Marcel Dekker, Inc. New York.

34. Dubois, G., Beaumier, H. and Charbonneau, R. 1979. Inhibition of bacteria isolated from ground meat by Streptococcaceae and Lactobacillaceae. J. Food Sci. 44: 1649-1652.
35. Egan, A.F. and Shay, B.J. 1982. Significance of lactobacilli and film permeability in the spoilage of vacuum-packaged beef. J. Food Sci. 47: 1119-1122, 1126.
36. Egan, A.F., Ford, A.L. and Shay, B.J. 1980. A comparison of *Microbacterium thermosphactum* and lactobacilli as spoilage organisms of vacuum-packaged sliced luncheon meats. J. Food Sci. 45: 1745-1748.
37. Eklund, M.W. 1982. Significance of *Clostridium botulinum* in fishery products preserved short of sterilization. Food Technol. 36(12): 107-112, 115.
38. El-Badawi, A.A., Cain, R.F., Samuels, C.E. and Anglemeier, A.F. 1964. Color and pigment stability of packaged refrigerated beef. Food Technol. 18(5): 159-163.
39. Empey, W.A., Scott, W.J. and Vickery, J.R. 1934. The export of chilled beef - The preparation of the "Idomeneus" shipment at the Brisbane abattoir. Aust. J. Council of Scientific and Industrial Research 7: 73-77.
40. Enfors, S.-O. and Molin, G. 1978. The influence of high concentrations of carbon dioxide on the germination of bacterial spores. J. Appl. Bacteriol. 45: 279-285.
41. Enfors, S.-O., Molin, G. and Ternström, A. 1979. Effect of packaging under carbon dioxide, nitrogen or air on the microbial flora of pork stored at 4°C. J. Appl. Bacteriol. 47: 197-208.
42. Erichsen, I. and Molin, G. 1981. Microbial flora of normal and high pH beef stored at 4°C in different gas environments. J. Food Prot. 44: 866-869.
43. Gardner, G.A., Carson, A.W. and Patton, J. 1967. Bacteriology of prepacked pork with reference to the gas composition within the pack. J. Appl. Bacteriol. 30: 321-332.
44. Gariepy, C., Amiot, J., Simard, R.E., Boudreau, A. and Raymond, D.P. 1986. Effect of vacuum-packaging and storage in nitrogen and carbon dioxide atmospheres on the quality of fresh rabbit. J. Food Qual. 9: 289-309.
45. Gee, D.L. and Brown, W.D. 1978. Extension of shelf-life in refrigerated ground beef stored under an atmosphere containing carbon dioxide and carbon monoxide. J. Agric. Food Chem. 26: 274-276.
46. Gehrke, W.H. 1983. Film properties required for thermo-formed and thermal processed meat packages. Reciprocal Meat Conference Proc. 36: 55-59.
47. Genigeorgis, C.A. 1985. Microbial and safety implications of the use of modified atmospheres to extend the storage life of fresh meat and fish. Int. J. Food Microbiol. 1: 237-251.
48. Genigeorgis, C.A. 1986. Problems associated with perishable processed meats. Food Technol. 40(4): 140-154.
49. Gill, C.O. 1979. A review. Intrinsic bacteria in meat. J. Appl. Bacteriol. 47: 367-378.
50. Gill, C.O. 1985. The control of microbial spoilage in fresh meats. *In* Advances in Meat Research, Vol. 2. Meat and Poultry Microbiology, pp. 49-88. Pearson, A.M. and Dutson, T.R. (eds.). AVI Publishing Co. Inc., Westport, CT.

51. Gill, C.O. 1988. Packaging meat under carbon dioxide: The CAPTECH system. Proceedings of Industry Day, 34th Int. Cong. Meat Sci. Technol., pp. 76-77. Livestock and Meat Authority of Queensland, Brisbane, Australia.

52. Gill, C.O. and Harrison, C.L. 1989. The storage life of chilled pork packaged under carbon dioxide. Meat Sci. 24: 313-324.

53. Gill, C.O. and Newton, K.G. 1980. Growth of bacteria on meat at room temperatures. J. Appl. Bacteriol. 49: 315-323.

54. Gill, C.O. and Penney, N. 1985. Modification of in-pack conditions to extend the storage life of vacuum packaged lamb. Meat Sci. 14: 43-60.

55. Gill, C.O. and Penney, N. 1986. Packaging conditions for extended storage of chilled dark, firm, dry beef. Meat Sci. 18: 41-53.

56. Gill, C.O. and Reichel, M.P. 1989. Growth of the cold tolerant pathogens *Yersinia enterocolitica, Aeromonas hydrophila* and *Listeria monocytogenes* on high-pH beef packaged under vacuum or carbon dioxide. Food Microbiol. 6: 223-230.

57. Gill, C.O. and Tan, K.H. 1980. Effect of carbon dioxide on growth of meat spoilage bacteria. Appl. Environ. Microbiol. 39: 317-319.

58. Gill, C.O., Harrison, J.C.L. and Penney, N. 1990. The storage life of chicken carcasses packaged under carbon dioxide. Int. J. Food Microbiol. 11: 151-157.

59. Gilliland S.E. and Speck, M.L. 1975. Inhibition of psychrotrophic bacteria by lactobacilli and pediococci in nonfermented refrigerated foods. J. Food Sci. 40: 903-905.

60. Glass, K.A. and Doyle, M.P. 1989. Fate of *Listeria monocytogenes* in processed meat products during refrigerated storage. Appl. Environ. Microbiol. 55: 1565-1569.

61. Grau, F.H. 1981. Role of pH, lactate, and anaerobiosis in controlling the growth of some fermentative Gram-negative bacteria on beef. Appl. Environ. Microbiol. 42: 1043-1050.

62. Grau, F.H. 1983. Microbial growth on fat and lean surfaces of vacuum-packaged chilled beef. J. Food Sci. 48: 326-328, 336.

63. Grau, F.H., Eustace, I.J. and Bill, B.A. 1985. Microbial flora of lamb carcasses stored at 0°C in packs flushed with nitrogen or filled with carbon dioxide. J. Food Sci. 50: 482-485, 491.

64. Griffin, D.B., Savell, J.W., Smith, G.C., Vanderzant, C. Terrell, R.N., Lind, K.D. and Galloway, D.E. 1982. Centralized packaging of beef loin steaks with different oxygen-barrier films: Physical and sensory characteristics. J. Food Sci. 47: 1059-1069.

65. Hamm, R. 1975. Water-holding capacity of meat, pp. 321-337. *In* Meat, Cole, D.J.A. and Lawrie, R.A. (eds.). AVI Publishing Co. Inc., Westport, CT.

66. Hanna, M.O., Hall, L.C., Smith, G.C. and Vanderzant, C. 1980. Inoculation of beef steaks with *Lactobacillus* species before vacuum packaging. I. Microbiological considerations. J. Food Prot. 43: 837-841.

67. Hanna, M.O., Savell, J.W., Smith, G.C., Purser, D.E., Gardner, F.A. and Vanderzant, C. 1983. Effect of growth of individual meat bacteria on pH, color and odor of aseptically prepared vacuum-packaged round steaks. J. Food Prot. 46: 216-221, 225.

68. Hanna, M.O., Smith, G.C., Hall, L.C. and Vanderzant, C. 1979. Role of *Hafnia alvei* and *Lactobacillus* species in the spoilage of vacuum-packaged strip loin steaks. J. Food Prot. 42: 569-571.
69. Hanna, M.O., Zink, D.L., Carpenter, Z.L. and Vanderzant, C. 1976. *Yersinia enterocolitica*-like organisms from vacuum-packaged beef and lamb. J. Food Sci. 41: 1254-1256.
70. Harris, L.J., Daeschel, M.A., Stiles, M.E. and Klaenhammer, T.R. 1989. Antimicrobial activity of lactic acid bacteria against *Listeria monocytogenes*. J. Food Prot. 52: 384-387.
71. Harding C.D. and Shaw, B.G. 1990. Antimicrobial activity of *Leuconostoc gelidum* against closely related species and *Listeria monocytogenes*. J. Appl. Bacteriol. 69: 648-654.
72. Hastings, J.W. and Stiles, M.E. 1991. Antibiosis of *Leuconostoc gelidum* isolated from meat. J. Appl. Bacteriol. 70: In press.
73. Hintlian, C.B. and Hotchkiss, J.H. 1987. Comparative growth of spoilage and pathogenic organisms on modified atmosphere-packaged cooked beef. J. Food Prot. 50: 218-223.
74. Hintlian, C.B. and Hotchkiss, J.H. 1987. Microbiological and sensory evaluation of cooked roast beef packaged in a modified atmosphere. J. Food Proc. Preserv. 11: 171-179.
75. Hitchener, B.J., Egan, A.F. and Rogers, P.J. 1982. Characteristics of lactic acid bacteria isolated from vacuum-packaged beef. J. Appl. Bacteriol. 52: 31-37.
76. Holland, G.C. 1980. Modified atmospheres for fresh meat distribution. Proceedings of the Meat Industry Research Conference, pp. 21-39. American Meat Institute, Washington, D.C.
77. Hood, D.E. and Riordan, E.B. 1973. Discolouration in pre-packaged beef: measurement by reflectance spectrophotometry and shopper discrimination. J. Food Technol. 8: 333-343.
78. Hotchkiss, J.H., Baker, R.C. and Qureshi, R.A. 1985. Elevated carbon dioxide atmospheres for packaging poultry. II. Effects of chicken quarters and bulk packages. Poult. Sci. 64: 333-340.
79. Huffman, D.L. 1974. Effect of gas atmospheres on microbial quality of pork. J. Food Sci. 39: 723-725.
80. Huffman, D.L., Davis, K.A., Marple, D.N. and McGuire, J.A. 1975. Effect of gas atmospheres on microbial growth, color and pH of beef. J. Food Sci. 40: 1229-1231.
81. Hurst, A. 1981. Nisin. Adv. Appl. Microbiol. 27: 85-123.
82. Hurst, A. and Collins-Thompson, D.L. 1979. Food as a bacterial habitiat, pp. 79-134. *In* Advances in Microbial Ecology, vol. 3. Alexander, M. (ed.). Plenum Publishing Corp., New York.
83. I.C.M.S.F. 1980. Meats and meat products. *In* Microbial Ecology of Foods, Vol. 2, pp. 333-409. International Commission on Microbiological Specifications for Foods. Academic Press, London.
84. Ingram, M. 1962. Microbiological principles in prepacking meats. J. Appl. Bacteriol. 25: 259-281.
85. Jeremiah, L.E., Smith, G.C. and Carpenter, Z.L. 1972. Vacuum packaging of lamb: Effects of storage, storage time and storage temperature. J. Food Sci. 37: 457-462.
86. Johnson, B.Y. 1974. Chilled vacuum-packed beef. CSIRO Food Res. Quart., 34: 14-20.

87. Kempton, A.G. and Bobier, S.R. 1970. Bacterial growth in re-
     frigerated, vacuum-packed luncheon meats. Can. J. Microbiol.
     16: 287-297.
88. King, A.D. and Nagel, C.W. 1967. Growth inhibition of a
     Pseudomonas by carbon dioxide. J. Food Sci. 32: 575-579.
89. King, A.D. and Nagel, C.W. 1975. Influence of carbon dioxide
     upon the metabolism of *Pseudomonas aeruginosa*. J. Food Sci. 40:
     362-366.
90. Klaenhammer, T.R. 1988. Bacteriocins of lactic acid bacteria.
     Biochimie 70: 337-349.
91. Kraft, A.A. 1986. Meat Microbiology, pp. 239-278. *In* Muscle as
     Food. Bechtel, R.J. (ed.). Academic Press Inc., London.
92. Lanier, T.C., Carpenter, J.A., Toledo, R.T. and Reagan, J.O. 1978.
     Metmyoglobin reduction in beef systems as affected by aerobic,
     anaerobic and carbon monoxide-containing environments. J.
     Food Sci. 43: 1788-1792, 1796.
93. Lee, B.H. and Simard, R.E. 1984. Evaluation of methods for de-
     tecting the production of H_2S, volatile sulfides, and greening by
     lactobacilli. J. Food Sci. 49: 981-983.
94. Lee, B.H., Simard. R.E., Laleye, L.C. and Holley, R.A. 1983.
     Microflora, sensory and exudate changes of vacuum- and nitro-
     gen packed veal chucks under different storage condition. J.
     Food Sci. 45: 1537-1542.
95. Lee, B.H., Simard, R.E., Laleye, L.C. and Holley, R.A. 1985. Effects
     of temperature and storage duration on the microflora, physic-
     ochemical and sensory changes of vacuum- or nitrogen-packed
     pork. Meat Sci. 13: 99-112.
96. Leistner, L. and Rodel, W. 1976. The stability of intermediate
     moisture foods with respect to microorganisms, pp. 120-134. *In*
     Intermediate Moisture Foods. Davies, R., Birch, G.G. and Parker,
     K.J. (ed.). Applied Science Publications, London.
97. Ledward, D.A., Nicol, D.J. and Shaw, M.K. 1971. Microbiological
     and colour changes during ageing of beef. Food Technol. in
     Australia 23: 30-32.
98. Luiten, L.S., Marchello, J.A. and Dryden, F.D. 1982a. Growth of
     Salmonella typhimurium and mesophilic organisms on beef
     steaks as influenced by type of packaging J. Food Prot. 45: 263-
     267.
99. Luiten, L.S., Marchello, J.A. and Dryden, F.D. 1982b. Growth of
     Staphylococcus aureus on beef steaks as influenced by type of
     packaging J. Food Prot. 45: 268-270.
100. McFadyen, S.C., Stiles, M.E., Berg, R.T. and Hawkins, M.H. 1973.
     Factors influencing consumer acceptance of meats. Can. Inst.
     Food Sci. Technol. J. 6: 219-225.
101. McGroarty, J.A. and Reid, G. 1988. Detection of a lactobacillus
     substance that inhibits *Escherichia coli*. Can. J. Microbiol. 34:
     374-378.
102. McMullen, L. and Stiles, M.E. 1989. Storage life of selected meat
     sandwiches at 4°C in modified gas atmospheres J. Food Prot. 52:
     792-798.
103. McMullen, L.M. and Stiles, M.E. 1991. Changes in microbial pa-
     rameters and gas composition during modified atmosphere stor-
     age of fresh pork loin chops. J. Food Prot. In press.

104. Mol, J.H.H., Hietbrink, J.E.A., Mollen, H.W.M. and van Tinteren, J. 1971. Observations on the microflora of vacuum packed sliced cooked meat products. J. Appl. Bacteriol. 34: 377-397.
105. Newton, K.G. and Gill, C.O. 1978. The development of the anaerobic spoilage flora of meat stored at chill temperatures. J. Appl. Bacteriol. 44: 91-95.
106. Newton, K.G. and Rigg, W.J. 1979. The effect of film permeability on the storage life and microbiology of vacuum-packed meat. J. Appl. Bacteriol. 47: 433-441.
107. Newton, K.G., Harrison, J.C.L. and Smith, K.M. 1977. The effect of storage in various gaseous atmospheres on the microflora of lamb chops held at -1°C. J. Appl. Bacteriol. 43: 53-59.
108. Ogilvy, W.S. and Ayres, J.C. 1951. Post-mortem changes in stored meats. II. The effect of atmospheres containing carbon dioxide in prolonging the storage life of cut-up chicken. Food Technol. 5(3): 97-102.
109. O'Keeffe, M. and Hood, D.E. 1980-81. Anoxic storage of fresh beef. 1. Nitrogen and carbon dioxide storage atmospheres. Meat Sci. 5: 27-39.
110. O'Keeffe, M. and Hood, D.E. 1980-81. Anoxic storage of fresh beef. 2. Colour stability and weight loss. Meat Sci. 5: 267-281.
111. Paradis, D.C. and Stiles, M.E. 1978. Food poisoning potential of pathogens inoculated onto bologna in sandwiches. J. Food Prot. 41: 953-956.
112. Pearson, A.M. 1987. Muscle function and postmortem changes, pp. 155-191. In The Science of Meat and Meat Products, 3rd ed. Price, J.F. and Schweigert, B.S. (ed.) Food and Nutrition Press, Inc., Westport, CT.
113. Raccach, M., Baker, R.C., Regenstein, J.M. and Mulnix, E.J. 1979. Potential application of microbial antagonism to extended storage stability of a flesh type food. J. Food Sci. 44: 43-46.
114. Reagan, J.O., Smith, G.C. and Carpenter, Z.L. 1973. Use of ultraviolet light for extending the retail caselife of beef. J. Food Sci. 38: 929-931.
115. Reddy, S.G., Hendrickson, R.L. and Olson, H.C. 1970. The influence of lactic cultures on ground beef quality. J. Food Sci. 35: 787-791.
116. Reddy, S.G., Chen, M.L. and Patel, P.J. 1975. Influence of lactic cultures on the biochemical, bacterial and organoleptic changes in beef. J. Food Sci. 40: 314-318.
117. Roth, L.A. and Clark, D.S. 1972. Studies on the bacterial flora of vacuum-packaged fresh beef. Can. J. Microbiol. 18: 1761-1766.
118. Roth, L.A. and Clark, D.S. 1975. Effect of lactobacilli and carbon dioxide on the growth of Microbacterium thermosphactum on fresh beef. Can. J. Microbiol. 21: 629-632.
119. Rowe, M.T. 1988. Effect of carbon dioxide on growth and extracellular enzyme production by Pseudomonas fluorescens B52. Int. J. Food Microbiol 6: 51-56.
120. Sander, E.H. and Soo, H.-M. 1978. Increasing shelf life by carbon dioxide treatment and low temperature storage of bulk pack fresh chickens packaged in nylon/surlyn film. J. Food Sci. 43: 1519-1523, 1527.
121. Savell, J.W., Hanna, M.O., Vanderzant, C. and Smith, G.C. 1981. An incident of predominance of Leuconostoc sp. in vacuum-

packaged beef strip loins - sensory and microbial profile of steaks stored in O_2 - CO_2 - N_2 atmospheres. J. Food Prot. 44: 742-745.

122. Schillinger, U. and Lücke, F.-K.   1986.   Milchsaurebakterien-Flora auf vakuumverpacktem Fleisch und ihr Einfluss auf die Haltbarkeit.   Fleischwirtschaft 66: 1515-1520.

123. Schillinger, U. and Lücke, F.-K.   1987.   Identification of lacto-bacilli from meat and meat products. Food Microbiol. 4: 199-208.

124. Scott, V.N.   1989.   Interaction of factors to control microbial spoilage of refrigerated foods.   J. Food Prot. 52: 431-435.

125. Sebranek, J.G.   1985.   Stabilizing of properties of meat products with packaging systems.   Meat Industry Research Conf. pp. 150-167.

126. Seideman, S.C. and Durland, P.R.   1984.   Vacuum packaging of fresh beef: A review. J. Food Qual. 6: 26-47.

127. Seideman, S.C. and Durland, P.R.   1984.   The utilization of modi-fied gas atmosphere packaging for fresh meat: A review.   J. Food Qual. 6: 239-252.

128. Seideman, S.C., Smith, G.C., Carpenter, Z.L., Dutson, T.R. and Dill, C.W.   1979.   Modified gas atmospheres and changes in beef dur-ing storage. J. Food Sci. 44: 1036-1040.

129. Seideman, S.C., Carpenter, Z.L., Smith, G.C., Dill, C.W. and Vanderzant, C.   1979.   Physical and sensory characteristics of beef packaged in modified gas atmospheres.   J. Food Prot. 42: 233-239.

130. Seideman, S.C., Carpenter, Z.L., Smith, G.C. and Hoke, K.E. 1979. Effect of degree of vacuum and length of storage on the physi-cal characteristics of vacuum packaged beef wholesale cuts.   J. Food Sci. 41: 732-737.

131. Shaw, B.G. and Harding, C.D. 1984. A numerical taxonomic study of lactic acid bacteria from vacuum-packed beef, pork, lamb and bacon. J. Appl. Bacteriol 56: 25-40.

132. Shaw, B.G., Harding, C.D and Taylor, A.A. 1980. The microbiol-ogy and storage stability of vacuum packed lamb.   J. Food Technol. 15: 397-405.

133. Shay, B.J. and Egan, A.F.   1981.   Hydrogen sulphide production and spoilage of vacuum-packaged beef by a *Lactobacillus*, pp. 241-251.   *In* Psychrotrophic Microoganisms in Spoilage and Pathogenicity,   Roberts, T.A., Hobbs, G., Christian, J.H.B. and Skovgaard, N. (ed.).   Academic Press, London.

134. Shay, B.J. and Egan, A.F.   1987.   The packaging of chilled red meats.   Food Technol. in Australia   39: 283-285.

135. Silliker, J.H. and Wolfe, S.K.   1980.   Microbiological safety con-siderations in controlled-atmosphere storage of meats.   Food Technol. 34(3): 59-63.

136. Silva, M., Jacobs, N.V., Deneke, C. and Gorbach, S.L.   1987. Antimicrobial substance from a human *Lactobacillus*   strain. Antimicrob. Agents Chemother. 31: 1231-1233.

137. Simard, R.E., Lee, B.H., Laleye, C.L. and Holley, R. 1985. Effects of temperature and storage time on the microflora, sensory and exudate changes of vacuum- or nitrogen-packed beef. Can. Inst. Food Sci. Technol. J. 18: 126-132.

138. Smith, G.C., Hall, L.C. and Vanderzant, C. 1980. Inoculation of beef steaks with *Lactobacillus* species before vacuum packaging. II. Effect on meat quality characteristics. J. Food Prot. 43: 842-849.

139. Spahl, A., Reineccius, G. and Tatini, S. 1981. Storage life of pork chops in CO_2-containing atmospheres. J. Food Prot. 44: 670-673.

140. Steele, J.E. and Stiles, M.E. 1981. Food poisoning potential of artificially contaminated vacuum packaged sliced ham in sandwiches. J. Food Prot. 44: 430-434.

141. Stern, N.J., Greenberg, M.D. and Kinsman, D.M. 1986. Survival of *Campylobacter jejuni* in selected gaseous environments. J. Food Sci. 51: 652-654.

142. Stiles, M.E. and Ng, L.-K. 1979. Fate of pathogens inoculated onto vacuum-packaged sliced hams to simulate contamination during packaging. J. Food Prot. 42: 464-469.

143. Stiles, M.E. and Ng, L.-K. 1979. Fate of enteropathogens inoculated onto chopped ham. J. Food Prot. 42: 624-630.

144. Stiles, M.E. and Ng, L.-K. 1981. Enterobacteriaceae associated with meats and meat handling. Appl. Environ. Microbiol. 41: 867-872.

145. Sutherland, J.P., Patterson, J.T. and Murray, J.G. 1975. Changes in the microbiology of vacuum-packaged beef. J. Appl. Bacteriol. 39: 227-237.

146. Taylor, A.A. 1985. Packaging fresh meat. *In* Developments in Meat Science, 3rd ed. Lawrie, R. (ed). Elsevier Applied Science Publishers, London.

147. Taylor, A.A. and MacDougall, D.B. 1973. Fresh beef packed in mixtures of oxygen and carbon dioxide. J. Food Technol. 8: 453-461.

148. Taylor, A.A. and Shaw, B.G. 1977. The effect of meat pH and package permeability on putrefaction and greening in vacuum packed beef. J. Food Technol. 12: 515-521.

149. Tompkin, R.B. 1986. Microbiological safety of processed meat: New products and processes - new problems and solutions. Food Technol. 40(4): 172-176.

150. Vanderzant, C., Hanna, M.O., Ehlers, J.G., Savell, J.W., Smith, G.C., Griffin, D.B., Terrell, R.N., Lind, K.D. and Galloway, D.E. 1982. Centralized packaging of beef loin steaks with different oxygen-barrier films: Microbiological characteristics. J. Food Sci. 47: 1070-1079.

151. Walters, C.L. 1975. Meat colour: the importance of haem chemistry. *In* Meat, pp. 385-401. Cole, D.J.A. and Lawrie, R.A. (eds). AVI Publishing Co. Inc., Westport, CT.

152. Wolfe, S.K. 1980. Use of CO- and CO_2-enriched atmospheres for meats, fish, and produce. Food Technol. 34(3): 55-58, 63.

153. Wolfe, S.K,. Brown, W.D. and Silliker, J.H. 1976. Transfresh shipping of meat. Proceedings of the Meat Industry Research Conference, pp. 137-148. American Meat Institute, Washington, D.C.

154. Zarate, J.R. and Zaritzky, N.E. 1985. Production of weep in packaged refrigerated beef. J. Food Sci. 50: 155-159, 191.

Chapter 6

MODIFIED ATMOSPHERE PACKAGING OF FISH AND FISH PRODUCTS

B.J. Skura
Department of Food Science, University of British Columbia

6.1 Problems

Fish products (fresh and saltwater fish, molluscs and crustaceans) used for food are highly perishable. Spoilage of freshly harvested fish products is usually microbial in nature. Traditionally, ice and refrigerated sea water or brine have been used to delay the onset of microbial spoilage of fish products. This chapter will review published research on the use of modified/controlled atmospheres for extending the storage life of fish products held at low temperatures. The concerns related to the potential growth and toxin production by psychrotrophic *Clostridium botulinum* types B and E in MAP fish products will also be reviewed.

6.2 Past and present developments

To illustrate potential benefits of MAP of fish products, literature reports dealing with wild as well as cultured fish, molluscs and crustaceans will be reviewed. Literature, up to 1981, on the benefits of MAP storage of fishery products was described in reviews by Parkin and Brown [59] and Wilhelm [89]. More recent reviews have been written by Genigeorgis [31], Smith [74], and Skura [73].

Strategies used in Australia in the development of MAP systems for storage-life extension of fish products were recently described by Statham and Bremner [79] and prior to that by Statham [77].

6.2.1 Trout

The storage life of cultured trout (*Salmo gairdneri*) was doubled by storage in an 80% CO_2, 20% N_2 atmosphere [4]. Trout stored in air were spoiled within 12 d while trout stored in MAP at 1.7°C were still of very good quality after 14 d and of fair quality after 20 d. The raw trout were considered marginally acceptable after 25 d storage in MAP, at 1.7°C, but after cooking the trout were rated as good quality by a sensory panel. Treatment of the trout with 2.3% potassium sorbate prior to MAP did not extend the storage life. The post-MA storage life at 1.7°C was about one week.

In another study [16], farmed rainbow trout packaged in vacuum packages or under CO_2 showed much less lipid oxidation than trout stored in air at 1 to 2°C. MAP with elevated CO_2, followed by vacuum packaging and air storage, was most inhibitory to lipid oxidation but also caused the greatest loss of carotenoid content. Cooked flavour of vacuum-packaged trout was better than that of trout packed under CO_2. Vacuum packaging of farmed trout produced in the United Kingdom gave a longer storage life than MAP in 60% CO_2, 40% N_2 at 5°C [14].

Farmed rainbow trout (*Salmo gairdneri*) had a storage life of 2 weeks at 0°C when vacuum packed in 0.075 mm thick polyethylene bags [44]. Treatment of the fish with 1 and 2 kilogray (kGy) of electrons (10 MeV) extended the storage life to 3 and 5 weeks, respectively. Storage life of trout stored at 5°C was 1, 3 and 4 weeks for fish treated with 0, 1 and 2 kGy of 10 MeV electrons, respectively. Clearly, irradiation of the trout extended the storage life of the vacuum-packed trout at 0 and 5°C.

Rainbow trout (head on) vacuum packed in polyethylene film were successfully stored in CO_2 at 1.8 bar for 8 d compared with 7 d in air (1 bar) at 1°C [61]. Longer-term storage in CO_2 caused fading of skin colour and eye discoloration. Storage of the vacuum packed trout, at -12°C, in CO_2 at 1.8 bar did not improve storage life over that attainable by storage of the vacuum-packed trout held in air (1 bar) at -12°C.

Sea trout (*Cynoscion regalis*) packed under CO_2 and distributed through normal channels to the supermarket level showed slower growth of psychrotrophic bacteria when 327 cm^3 CO_2/450 g fillet was used compared with 164 cm^3 CO_2/450 g fillets [36]. No information was reported on the amount of exudate formed at the two levels of CO_2 used.

6.2.2 Salmon

Farmed Atlantic salmon (*Salmo salar*) had a longer storage life at 5°C when packed in a 60% CO_2, 40% N_2 atmosphere than when vacuum packaged [14]. Whole, eviscerated Coho salmon (*Oncorhynchus kisutch*) was stored for 3 weeks in a 90% CO_2, 10% air atmosphere at 0°C [5].

Sockeye salmon (*Oncorhynchus nerka*) fillets had a longer storage life at -1°C than at 1°C when packaged in a CO_2 atmosphere [65]. Pretreating the fillets with an acidic solution (1% citric acid, 1% ascorbic acid, 0.5% calcium chloride) improved fillet colour but did not improve or detract from the sensory properties [65]. Treatment of sockeye salmon with 1% (w/v) potassium sorbate and an antioxidant dip (0.2% sodium erythorbate, 0.2% citric acid and 0.5% sodium chloride) prior to packaging in an MA of 60% CO_2, 35% N_2, 5% O_2 lead to slight improvement in quality after 18 d at 1°C compared with salmon stored in the MA without the potssium sorbate and antioxidant pretreatments [23]. Addition of 1% CO to the gas mixture caused a slight decrease in quality scores of antioxidant- and potassium sorbate-treated sockeye salmon after 18 d storage at 1°C.

Sockeye salmon was stored at -8°C under an MA of CO_2 for over 9 months [65] at which time a consumer panel rated the salmon of good quality. Fat-rich fish normally have a storage life of only 0.8 months at -6.7°C [64].

6.2.3 Cod

Cod fillets were stored in bulk at 0°C in a 25% CO_2, 75% N_2 atmosphere. The MAP cod had a storage life of at least 8 extra days compared with cod stored in air at 0°C [85]. An MA of 60% CO_2, 40% air was somewhat better than a CA of the same composition for storage of cod in 4.5-kg quantities at 1°C [90]. Wang and Ogrydziak [87] showed that CO_2 exerted a residual preservative effect on cod fillets after removal from MA storage.

Cod fillets were stored under vacuum and under 40% CO_2, 30% N_2, 30% O_2 at 0, 5 and 10°C [85]. The fish had an acceptable storage life of 14, 6 and 3 d in MAP and 10, 4 and 2 d under vacuum packaging at 0, 5 and 10°C, respectively. In all cases psychrotrophic populations were over 10^7 CFU/g at the time when cooked flavour of the fillets reached 5.5 on a 10-point rating scale. Cooked flavour deteriorated at the rate of 0.25, 0.60 and 1.00 units/day in MAP and at 0.3, 0.7 and 1.2 units/day under vacuum packaging at 0, 5 and 10°C, respectively.

Cod fillets packed under CO_2 and stored at 4°C had a storage life of 43 d while at 8°C storage life was 23 d [63]. Temperature abuse markedly decreased the storage life of the CO_2-packed fillets at 4°C. This study and the work of Gibson [33] and Ogrydziak and Brown [58] clearly demonstrated the need for strict temperature control during MAP of fish fillets.

Cod fillets dipped in carbonic acid (pH 4.6) at 2°C for 5 to 10 min before packaging in polystyrene trays were overwrapped with permeable or barrier films and vacuum packed prior to storage at 2°C [18]. Growth of psychrotrophic bacteria was most effectively controlled by barrier packaging with an MA containing 25% CO_2. The study showed that dipping of cod fillets in carbonic acid prior to packaging was equal to but not more effective than packaging cod fillets directly in an MA containing 98% CO_2.

Cod fillets were packed under various gas mixtures ranging from air to 100% CO_2 in glass storage vessels with tightly sealed lids [81]. The time taken for the microbial population on the fillets to reach 10^6 CFU/g increased with increasing CO_2 concentration. The ratio of *Lactobacillus* spp. to total microflora increased with increasing CO_2 concentration while the ratio of *Shewanella putrefaciens* (formerly *Alteromonas putrefaciens*, MacDonnell and Colwell, [50]) decreased with increasing CO_2 concentration. An atmosphere of 50% CO_2, 50% O_2 was recommended by Stenström [81] for extended storage life of cod fillets for the following reasons: microbial growth exhibited a 14-d lag phase at 2°C; development of a dominant *Lactobacillus* microflora was favoured; visual appearance of the cod fillets was preserved. Packages of cod with an initial atmosphere of 50% CO_2 50% O_2 contained 43% O_2 after 34 d storage at 2°C while packages with an initial air atmosphere contained 0.8% O_2 and 18% CO_2 after 19 d at 2°C [81].

A hypobaric storage system was evaluated for storage of cod fillets in 4-kg plastic trays [8]. They found that the hypobaric system did not significantly extend the storage life of the cod fillets. In contrast, Haard *et al.* [37] stated that hypobaric storage at 0.02 atmosphere pressure extended the storage life of cod over that attained by storage of cod at one atmosphere pressure in nitrogen. In the study by Haard *et al.* [37], the fish were not stored in bulk containers as they were in the study reported by Bligh *et al.* [8].

Cod fillets were vacuum packed or packed in a 60% CO_2, 40% air atmosphere in barrier bags followed by treatment with 1 kGy gamma radiation and subsequent storage on ice [47]. The CO_2-packaged, irradiated cod fillets retained quality attributes longer than the vacuum-packed product, which retained their quality longer than air-packed fillets.

6.2.4 Snapper

The storage life of snapper (*Chrysophrys auratus*) was doubled by either vacuum packaging or elevated CO_2 atmosphere storage at 3°C compared with storage in air [70]. The authors stated that although MA extended the storage life of the snapper, commercial use of the technique was dubious because of the substantial variability in the quality of the trawler-caught snapper. The study showed that MAP cannot improve the quality of poor-quality fish.

This is in agreement with the results reported by Gray *et al.* [36] and Statham and Bremner [79]. Aquaculture has a substantial advantage over natural fishing in that the quality of the fish and handling procedures prior to processing can be more carefully controlled and it should be possible to deliver a superior quality fish to the processor. In that case, it is conceivable that farmed fish should perform better than wild-caught fish of the same species under MAP, all other factors being equal. This has to be thoroughly evaluated by suitably designed MAP storage trials of cultured and wild fish.

Snapper fillets showed the best storage life when packed in a CO_2 atmosphere and stored at -1°C [71]. This was better than storage in 60% N_2, 40% CO_2 or vacuum packaging. Storage life of the fillets in CO_2 at -1°C was reported to be 2.25 times that in CO_2 at 3°C.

6.2.5   Other   whitefish

A study comparing several potassium sorbate (0, 2.5, 5.0%) treatments and barrier bag permeabilities (low, intermediate, high) with CO_2 storage at 3°C, of Great Lakes whitefish (*Coregonus clupeaformis*) revealed that fillets dipped in 5% potassium sorbate had the potential for longest shelf life [72]. Potassium sorbate treatment gave no advantage if a storage life of 5 d or less was required. The storage life end point of fillets which were not treated with potassium sorbate was between 10 and 14 d, while fillets treated with 2.5 and 5% potassium sorbate had a storage life in excess of 15 d at 3°C.

Perch (*Meronia   americanus*), croaker (*Micropogon   undulatis*) and bluefish (*Pomatomus   salatrix*) were stored in CO_2 MAP at 1.1 and 10°C [36]. Psychrotrophic bacteria grew at a slower rate in CO_2-packed fish than on the air stored fish on ice. The rate of growth and length of lag phase observed for the fish packed in CO_2 varied with the fish species.

Fillets of morwong (*Nemadactylus   maropterus*), a whitefish, had the longest storage life (13 d) when dipped in 10% polyphosphate and 1.2% potassium sorbate solutions followed by storage under 100% CO_2 in barrier bags at 4°C [80]. Individual dips followed by vacuum packaging or sorbate and polyphosphate dips combined with vacuum packaging were not as effective as the combined dips with the CO_2 MAP.

Fillets of trevalla (*Hyperoglyphe   porosa*) packed in 100% CO_2 had a shelf life of 8 to 16 d longer at 4°C than trevalla fillets stored under aerobic conditions [78]. A CO_2 to fish ratio (v/v) of 4:1 was used. Psychrotrophic bacteria grew rapidly in the air-stored trevalla, reaching a population of 10^9 CFU/cm^2 within 8 d while the psychrotrophic population was less than 10^7 CFU/cm^2 at 8 d in the CO_2-packed fillets. Lactic acid bacteria formed the dominant microflora in the CO_2-packed trevalla.

Prior to vacuum packaging and storage at 4°C, sand flathead (*Platycephalus bassensis*) fillets were subjected to a variety of dipping treatments: untreated; citrate buffers (pH 4.8 and 5.4); glucose (50 g/L); and pH-5.4 citrate buffer containing glucose [52]. Dipping of fillets in the pH-4.8 citrate buffer lowered the pH of the fillet surface and inhibited growth of *S. putrefaciens* and the development of sulphide-like odours. The acid treatment, however, caused bleaching of the fillets and development of a milky exudate. Glucose treatment of the fillets did not lead to storage-life extension. This may have been due to the low initial population of lactic acid bacteria on the fillets at the time of packaging. None of the treatments lead to an extension of the storage life of the sand flathead fillets and storage life of the vacuum-packed fillets was not longer than that of the air-stored control. *S. putrefaciens* formed the major portion of the microflora of untreated vacuum-packed sand flathead fillets, while Enterobacteriaceae dominated in treated fillets.

Greenland halibut (*Reinhardtius hippoglossoides*) fillets were stored at 0°C in atmospheres containing the following $N_2:CO_2$ ratios: 100:0, 75:25, 50:50, 25:75, and 0:100 [7]. The best gas mixture was 25% CO_2 - 25% N_2 because texture and pH were least affected and exudate formation was minimal. The greatest exudate formation occurred with fillets stored in 100% CO_2. Samples were stored only six days.

Mitsuda *et al.* [54] showed that dipping fillets of farmed hamachi (*Seriola aurevittata*) in 5% sodium chloride for one minute prior to packaging under CO_2 and storage at 3°C maintained colour and texture of the tissue during a 7-d storage period. Carbon dioxide was found to be better than N_2 for maintenance of flesh texture. Sodium chloride was shown to be better than potassium chloride as the dipping solution.

A CA system was used by Oberlander *et al.* [57] in a study of storage characteristics of swordfish (*Xiphias gladius*) steaks under various CO_2-enriched atmospheres. Lowest psychrotrophic populations were observed on swordfish steaks stored in 100% CO_2 and 70% CO_2, 30% N_2. Psychrotrophic populations were higher on steaks stored in 70% CO_2, 30% O_2 than on steaks stored in 70% CO_2, 30% N_2. All of the CO_2-enriched atmospheres retarded growth of microorganisms for 22 d at 2°C when compared with storage in air. Total volatile nitrogen increased most rapidly in air storage and at the slowest rate in steaks stored in 70% CO_2 - 30% O_2.

Comercially cultured channel catfish (*Ictalurus punctatus*) packed in polyethylene bags with atmospheres of 100% air, 100% CO_2 or 80% CO_2 and 20% air were treated with 0, 0.5 or 1.0 kGy of gamma radiation [66]. The packaged samples were stored at 0 to 2°C for up to 30 d. The lowest psychrotrophic plate counts were obtained with the 100% CO_2 atmosphere although the difference in counts on fish among the gas atmospheres was not significant. The authors reported that their study did not reveal large differences in microbial counts compared with other studies which showed significantly lower microbiological populations in fish stored under elevated CO_2 atmospheres [86]. This was probably due to the use of polyethylene by Przybylski *et al.* [66] which is quite permeable to CO_2 and O_2. There was no significant difference in microbial populations, regardless of packaging atmosphere, when catfish were irradiated with 0.5 or 1.0 kGy of gamma radiation. Increased radiation doses did, however, decrease the microbial population in fish sampled after 20 d storage. The atmosphere in the packages did not have any significant effect on the development of thiobarbituric acid reactive substances (TBARS) in the catfish at each of the three radiation doses. Increasing radiation dose did increase the TBARS value of the packaged fish. MAP had minimal benefit on shelf life of channel catfish packaged in polyethylene film when compared with the effects of gamma radiation.

Rockfish (*Sebastes* spp.) fillets were stored in an MA of 80% CO_2, 20% O_2 for 20 d at 4°C. *In vitro* protein digestibility of the air-stored fillets decreased while that of MAP fillets remained the same as that of the fish at zero time [56]. The computed protein efficiency ratio of the fillets stored in the MA for 14 d was similar to that of fresh fillets.

White hake (*Urophycis tenuis*) fillets treated with 0.5% erythorbic acid and vacuum packaged (in low-barrier film) deteriorated in quality at -7°C more rapidly than fillets stored in air without the erythorbic acid treatment [67]. Erythorbic acid hastened the rate of deterioration of white hake and the effect was magnified by vacuum packaging.

6.2.6 Herring and sardines

Herring fillets stored in CO_2 had a storage life 3.5 times that in air at 2°C, while N_2 extended storage life to 1.5 times that possible in air. *Lactobacillus* species dominated the microflora in CO_2 after 28 d storage, while a mixed flora was found in herring stored in N_2 [55].

Storage life of herring fillets packed in 40% CO_2, 30% N_2, 30% O_2 at 0°C was 8 d, while that for fillets packed in 60% CO_2, 40% N_2 was 3 d. Vacuum-packed herring had a storage life of 13 d at 0°C [13]. At 5°C storage life of the herring fillets was 3 d in the vacuum packages and 2 d in MAP [13]. In contrast to the study reported by Molin *et al.* [55], *Lactobacillus* spp. were not isolated by Cann *et al.* [13].

Sardines (*Sardinops melanostictus*) were stored at 5°C in barrier bags which contained the following atmospheres after sealing: air; 20% N_2 80% CO_2; 80% N_2 20% CO_2 [27]. The atmospheres in the flushed bags changed very little while the atmosphere in the bags containing air (20.5% O_2, 0% CO_2) at the time of sealing contained 13% O_2 and 5.6% CO_2 by day 21. The air-packaged sardines developed a modified atmosphere during storage. Growth of aerobic and anaerobic bacteria was slowest in sardines packed under 80% CO_2 and 20% N_2. Initially microflora were predominantly Vibrionaceae,

Moraxella, and *Acinetobacter* [28].   After 10 d storage in air
Moraxella, *Acinetobacter*, *Lactobacillus*, and *Streptococcus* formed
the majority of the microflora on the sardines, while microflora on
sardines stored under 20% CO_2, 80% N_2 was predominantly
Lactobacillus and *Streptococcus* sp. Microflora on sardines stored for
10 d under 80% CO_2, 20% N_2 was predominantly unidentifiable cocci
as well as *Lactobacillus* and *Streptococcus* sp. [28]. The rate of increase of K-value of the sardines was not affected by the in-package
atmosphere [27].   Increases in trimethylamine (TMA) concentrations
and thiobarbituric acid (TBA) values were slowest in sardines packed
under 80% CO_2, 20% N_2.

6.2.7 Mackerel

Hot-smoked mackerel had a storage life of 8 d when packed in
40% CO_2, 30% N_2, 30% O_2 and 13 d when vacuum packed and stored at
0°C [13]. In a second study by Cann *et al.* [13] hot-smoked mackerel
packed in 60% CO_2, 40% N_2 had a storage life of 16 d compared with 17
d for the vacuum-packed fish stored at 0°C. Clearly O_2 in the gas
mixture contributed to oxidative rancidity of the mackerel. Cold-
smoked jack mackerel (*Trachurus declivir*) was vacuum packed and
stored at 0°C [26]. Vacuum packaging permitted a storage life in excess of 30 d. Fletcher *et al.* [26] hypothesized, based on the microbial
growth rate, that the mackerel would probably spoil during the second month of storage.

Santos and Regenstein [67] showed that treatment of mackerel
(*Scomber scombrus*) fillets with a 0.5% solution of erythorbic acid
prior to vacuum packaging, in a low-barrier film, improved storage
life at -7°C.   Vacuum packaging with the antioxidant treatment
offered better protection for the mackerel than did glazing of the
antioxidant-treated fillets.

6.2.8 Scallops

MAP extended the storage life of scallops (*Pecten alba*) packed
in barrier bags which were evacuated and backflushed with CO_2
prior to storage at 4°C [10]. MAP scallops had a storage life of 22 d
compared with 10 d for those stored in air at 4°C. Bremner and
Statham [9] evaluated the effect of addition of *Lactobacillus plantarum* to scallops (*Pecten alba*) prior to vacuum packaging and storage at 4°C. The lactobacilli did not suppress growth of spoilage bacteria with the result that storage life was not increased.

Shucked scallops (*Pecten maximus*) harvested in Scotland had
a storage life of 7.3 d at 0°C and 4 d at 5°C when packaged in MA of
40% CO_2, 30% N_2, 30% O_2 [15].

6.2.9 Prawns

MAP of cultured fresh water prawns (*Macrobrachium
rosenbergii*) in a CO_2 atmosphere extended the storage life by 2 to 3 d
at 4°C [62].

Spotted shrimp (*Pandalus platyceros*) stored under a CA of 100% CO_2 were acceptable after 14 d at 0°C while shrimp stored in air had become unacceptable [51]. Microbial counts and drip loss were lower in the head-off shrimp than for the head-on shrimp stored in CO_2. Storage in a CO_2 atmosphere was also reported to be effective in controlling black spot formation, possibly through inhibition of phenol oxidase activity in the shrimp flesh [51]. A CA of 100% CO_2 was more effective than an atmosphere containing 50% CO_2 in delaying the initiation of microbial growth on the shrimp, although the 100% CO_2 atmosphere produced a drip loss of up to 13 to 14% after 12 d storage at 0°C [46]. In a recent study with pink prawns (*Pandalus platyceros*) harvested off the coast of British Columbia a longer storage life was attained when CO_2 was used for the MA compared with N_2 [19]. Drip loss was higher but colour retention was better in the CO_2-packed prawns. Sulphide producing bacteria were the primary spoilage causing agent in the N_2-packed prawns [19]. Lannelongue *et al.* [45] showed that microbial growth on brown shrimp (*Penaeus aztecus*) harvested from the Gulf of Mexico was slower as the CO_2 concentration of the MA increased.

Cooked freshwater crayfish (*Pacifastacus leniusculus*) packed under 80% CO_2, 20% air for 21 d at 4°C was similar in sensory properties to freshly cooked crayfish, while aerobically stored crayfish was fishy in odour and flavour after 14 d at 4°C [86]. Ammonia and trimethylamine levels in the aerobic controls increased but remained relatively static in the MAP crayfish.

Cooked, peeled freshwater crayfish (*Procambaris clarkii*) tail meat was stored on ice in air, 100% CO_2 or 80% CO_2 and 20% air for 21 d [32]. The 80% CO_2, 20% air atmosphere led to the lowest microbial counts (aerobic psychrotrophs and anaerobes) at the end of the storage period. Ammonia and trimethylamine concentrations increased most rapidly in meat in air and slowest in 100% CO_2-packed tail meat. Although the 100% CO_2-packed tail meat was of better overall quality, sensory panelists detected y due to carbonic acid and lactic acid production.

6.2.10 Crab

Cooked whole Dungeness crab was stored up to 25 d in 80% CO_2, 20% air at 1.7°C [60]. Total aerobic psychrotrophs on the MA-stored crabs remained below 10^4 cfu/g of tissue, while aerobic control samples held at 1.7°C had psychrotrophic counts exceeding 10^6 cfu/g after 14 d of storage. As in other studies involving CO_2 enriched atmospheres, the pH of the crab meat under MA decreased during storage while that of the aerobic control increased. The highest amount of exudate was formed in the MAP crab meat, probably due to the effect of pH on the water-holding capacity of the muscle proteins.

Cooked crab claws (*Cancer pagurus*) packed in an atmosphere of 40% CO_2, 30% N_2, 30% O_2 had a storage life of 10 d at 0°C and 6 d at

5°C [15]. The air-stored claws had a storage life of 5 d at 0°C and 3 d at 5°C. The MA had only a minor inhibitory effect on the growth of *S. putrefaciens* and *Brochothrix thermosphacta* compared with the air-stored controls [15].

6.3 Synopsis of benefits of MAP

In an interesting series of experiments, Fletcher and Statham [24] found that the end of storage life was reached at a similar time at 4°C for sterile and naturally spoiling yellow-eye mullet (*Aldrichetta forsteri*). They concluded that for this species, at least, inhibition of spoilage microflora by MAP would not lead to substantial increases in storage life owing to the autolytic degradation that occurred in the mullet flesh, and that MAP would not be effective for fish which are rapid producers of hypoxanthine.

From the examples cited in the text it is evident that MAP can extend the storage life of a variety of fish products. However, MAP is not equally effective for extending the storage life of all fish products. Comparison of data from different studies is often difficult because of differences in the starting quality of the harvested fish, the type of microflora associated with the fish, the gas mixtures and packaging materials used, the storage temperatures employed and the types of analyses (chemical, microbiological and sensory) performed.

6.4 Risks associated with MAP of fish products

MAP of fish products may pose a potential danger [20, 31]. The major concern in relation to the safety of MAP fish products is the potential for growth and toxin production by *Clostridium botulinum* type E. *C. botulinum* produces potent neurotoxins that cause the deadly disease botulism. *C. botulinum* is a strict anaerobe, thus the anaerobic conditions often encountered in MAP low acid foods (pH >4.6) can be favourable to growth and toxin production by the bacteria if the suitable storage temperatures prevail. Three types of *C. botulinum*, types A, B and E, have been most commonly associated with human cases of botulism, with type E being most commonly associated with outbreaks related to consumption of seafood [11, 38, 39]. Proteolytic strains of types A and B cannot grow at refrigeration temperatures and therefore do not pose a hazard if the susceptible foods are stored at or below 4°C. Type E and nonproteolytic type B *C. botulinum* are psychrotrophic and can grow and produce toxin at temperatures as low as 3.3°C [41]. *C. botulinum* type E is found in marine and freshwater environments, thus fish products have the potential of being contaminated.

The frequency of contamination of fish products with *C. botulinum* type E was reviewed by Hobbs [41] and Hauschild [38]. A recent study of the incidence of *C. botulinum* in seafoods purchased in retail stores in California showed that, on average, 43% of the seafood samples were contaminated with nonproteolytic, psychrotrophic strains [3]. Although the incidence of *C. botulinum* type

E in various fish products appears to be low and of a sporadic nature, depending on the geographical region, for assurance of food safety it must be assumed that all fish products are contaminated with *C. botulinum* and they must be handled accordingly [38, 75]. Eyles and Warth [22] concluded that the risk of botulism from consumption of fish prepared from vacuum-packaged raw fish produced in Australia is extremely small if reasonable precautions are taken. It appears, however, that the incidence of *C. botulinum* in waters around Australia and New Zealand is low [21, 34] compared with other areas in the Northern Hemisphere [17, 38, 42, 43].

The following events must occur if foodborne botulism is to occur [20]: (a) the food must be contaminated with *C. botulinum*; (b) processing of the food must allow survival of *C. botulinum* or post-processing contamination must occur; (c) the food must be stored under conditions which allow *C. botulinum* to grow and produce toxin; (d) a sufficient amount of the toxin-containing food must be consumed without cooking or after cooking that is insufficient to inactivate the toxin. *C. botulinum* type E and nonproteolytic type B usually do not affect the colour, odour or flavour of food.

Cod, flounder and whiting were inoculated with 50 *C. botulinum* type E spores per gram and stored at 26, 12, 8 and 4°C aerobically, in vacuum packages, or in MAs of N_2 or CO_2 [63]. At 12 and 8°C, MAP and vacuum-packaged flounder spoiled before toxin was detected. Vacuum-packaged or MAP cod, on the other hand, became toxic before it was spoiled during storage at 12, 8 and 4°C. Storage of cod in a CO_2 MA at 4°C with intermittent abuse at 26°C, prior to continuous storage at 4°C, decreased the time required for the fish to become toxic, which was well in advance of the time at which the cod spoiled. Similar results were obtained for whiting stored in vacuum packages and under MAP. These results showed clearly that MAP of cod and whiting extended the storage life, but even with storage at 4°C the fish became toxic before there was evidence of spoilage. Temperature-abused, MAP cod and whiting represented an increased hazard. That study also showed that there are differences in the ability of *C. botulinum* type E to grow and produce toxin on various types of fish. Gola *et al.* [35] showed that cod fillets inoculated with *C. botulinum* spores and stored in vacuum packages at 4°C did not contain toxin after 42 d. Toxin was detected in fillets stored at 8 and 10°C.

Lindroth and Genigeorgis [49] concluded that at 4°C, naturally occurring levels of *C. botulinum* in red snapper (*Sebastes paucipinis*) would not produce toxin prior to 21 d storage under MAP. At higher temperatures, such as 8°C, one spore could germinate and grow with subsequent toxin production prior to the occurrence of spoilage.

Studies of salmon (*Oncorhynchus tshawytscha*) fillets by Garcia *et al.* [30] and flesh homogenates [29] inoculated with spores of *C. botulinum* types B, E and F showed that homogenates became toxic after 60 d at 4°C when vacuum packed but not when packed under 100% CO_2 or 70% CO_2 and 30% air. The fillets did not become toxic at 4°C but at 8°C they became toxic within 6 d in vacuum packages and within 9 d under 100% CO_2, while odour scores indicated that the salmon was of acceptable quality. Although the salmon spoiled before detection of toxin at 4°C, the fact that the salmon became toxic before it was spoiled when stored at 8°C suggests that somewhere between 4 and 8°C the salmon could become toxic before becoming spoiled. These studies showed that C. *botulinum* produce toxin more readily in the vacuum-packaged salmon than in the salmon packaged under 100% CO_2 or 70% CO_2 [30].

Garcia *et al.* [29, 30] derived best-fit equations relating the length of the lag phase and probability of toxin production to storage temperature, length of storage, the MA and C. *botulinum* spore inoculum level. Those equations enabled calculation of the probability that one C. *botulinum* spore would germinate and lead to a population of actively metabolizing cells that would produce detectable levels of toxin. Baker and Genigeorgis [2] developed a mathematical model for prediction of lag time prior to C. *botulinum* toxigenesis. Their model revealed that about 75% of experimental variation was explained by temperature while the size of the C. *botulinum* spore inoculum explained 7.5% of the experimental variation. Collectively, factors such as MA composition, fish species and C. *botulinum* spore type explained only 2.3% of the experimental variation. Their data showed that temperature control of MAP products was critical and that increased C. *botulinum* contamination shortened the length of the lag time and increased the risk of toxin formation by C. *botulinum* in MAP fish products. Their predictive model also produced conservative approximations of the lag time for toxigenesis by C. *botulinum* when applied to qualitative data for a variety of fish, inoculated with C. *botulinum* and packed under a variety of MAs and stored under a variety of temperatures. The work reported by Garcia *et al.* [29, 30] and Baker and Genigeorgis [2] provides a basis upon which statistically based predictive studies can be conducted to fully assess the hazard posed by psychrotrophic C. *botulinum* in MAP fish under a variety of conditions that could potentially be encountered in distribution, retail and food service sectors as well as during handling by the consumer.

Trout fillets inoculated with 10^5 C. *botulinum* type E Beluga spores/g of fish were vacuum packed in 0.75-mm polyethylene film and subjected to irradiation with electrons (10 MeV) at doses of 0, 1 and 2 kGy [44]. No toxin formation was observed at 0°C and the fish became toxic long after spoilage occurred at 5°C. At 10°C irradiated (1 and 2 kGy) fillets became toxic before spoilage was observed, while unirradiated fillets spoiled before they became toxic. Irradiation, in this case, increased the hazard potential of the vacuum-packaged fillets stored at 10°C.

An earlier study by Stier *et al.* [82] with King salmon (*Oncorhynchus tshawytscha*) inoculated with spores or vegetative cells of C. *botulinum* types B and E, stored in an MA of 60% CO_2, 25% O_2 and 15% N_2 did not show any toxin development during storage at

4.4°C. The salmon spoiled within 12 d at 4.4°C and toxin was not detected after 57 d. The salmon was of poorer microbiological quality (aerobic plate count at the time of packaging was 10^6 CFU/g) compared with salmon used in a later study where the aerobic plate count of the salmon was about $10^{2.5}$ CFU/g salmon fillet [29].

Eklund [20] reported that salmon inoculated with 100 *C. botulinum* type E spores/100 g and stored in 60% or 90% CO_2 at 10°C became toxic within 10 d but were not spoiled. There was no toxin detected after 7 d at 10°C. The 90% CO_2 atmosphere was more inhibitory to *C. botulinum* than the 60% CO_2 atmosphere. Toxin was detected earlier at both MAs as the level of the spore inoculum increased. In contrast, Garcia and Genigeorgis [29] showed that an atmosphere of 70% CO_2, 6% O_2, 24% N_2 was more inhibitory to toxin production than an atmosphere of 100% CO_2 at 4, 8 and 12°C. Cann *et al.* [14] showed that salmon and trout inoculated with 100 *C. botulinum* type E/g, stored at 10°C in 60% CO_2, 40% N_2 or in vacuum packages, spoiled before they became toxic. This is in contrast to the data of Garcia *et al.* [30] and Eklund [20].

C. botulinum did not form toxin in vacuum-packaged scallops held at 4 and 10°C [25]. Only *C. botulinum* type A formed toxin in scallops held at 27°C but the scallops were spoiled before toxin was detected.

Lilly and Kauter [48] subjected 1074 samples of commercial vacuum-packaged freshwater and saltwater fish, produced in the United States from 27 types of fish, to mild temperature abuse at 12°C for 12 d. None of the samples were positive for *C. botulinum* toxin. The authors concluded that either the fish in the packages sampled did not contain *C. botulinum* spores or the spores were unable to grow out and produce toxin within the 12 d of temperature abuse.

The role of oxygen as a factor in the risk of *C. botulinum* toxin formation in MAP fish has not been adequately investigated. Schvester [68, 69] stated that proprietary gas mixtures for MAP of seafoods have been developed which contain elevated levels of O_2 to ensure levels of O_2 in MAP fish, after temperature abuse, that would be inhibitory to growth and toxin production by *C. botulinum*. He presented no data on the inhibition of *C. botulinum* by the gas mixtures except to show their effects on the total aerobe and anaerobe populations in the MAP seafood exposed to temperature abuse. Stenström [81] showed that cod packaged in 50% CO_2 and 50% O_2 contained elevated levels of O_2 after 34 d at 2°C, whereas cod packaged in air contained only 0.8% O_2 after 19 d storage. However, the effect of abusive temperatures on O_2 concentrations in the packaging system reported by Stenström [81] has not been determined. Garcia *et al.* [30] showed that salmon fillets packaged in 70% CO_2 and 30% air (3% O_2 in the mixture) contained undetectable levels of O_2 after 21 d at 8°C. Toxin was detectable in the samples of fillets inoculated with *C. botulinum* after storage for 9 d at 8°C when some O_2 was still presumably present in the packages. The time to toxin detection in cod and whiting packaged in barrier bags with elevated CO_2 atmospheres containing 2% and 4% O_2 was shorter than with elevated CO_2 atmospheres that did not contain O_2 [63]. It has been shown that even when foods are exposed to air *C. botulinum* can still grow within parts of the food where a sufficiently low oxidation-reduction potential exists [76]. Hauschild [38] stated that the assumption that growth of *C. botulinum* in food is dependent on oxygen tension is incorrect. It is clear that much more definitive work is needed to ascertain whether elevated levels of O_2 in gas mixtures for MAP seafoods actually provides protection from the risk of toxin development by *C. botulinum* during conditions of mild temperature abuse that would be encountered in retailing and foodservice distribution of these products. This conclusion was also noted by Hintlian and Hotchkiss [40] in a recent review article about the safety of MAP foods.

Eklund [20] and Lilly and Kauter [48] stated that MAP fish products would not pose a potential botulinal hazard if refrigeration below 3.3°C could be assured at all times between packaging and ultimate removal from the package for preparation for consumption by the consumer. However, it has been clearly established that inadequate refrigeration or cooling is the most frequent factor contributing to outbreaks of foodborne disease [1, 12]. Other studies have cited temperature abuse of food products during distribution and sale and also by the consumer [20, 21]. Because MAP fish products do not contain any added antibotulinal agents, the only hurdle that can be controlled to prevent growth of and toxin production by *C. botulinum* type E or nonproteolytic type B is temperature. Because the minimum growth temperature of *C. botulinum* type E is 3.3°C, normal refrigeration temperatures of 4 to 7°C will not inhibit the growth and toxin production of this organism in MAP fish products. Post *et al.* [63] stated that vacuum and MA packaging alone are not capable of providing the safety required for extended storage of fish fillets in the absence of a fail-safe mechanism whereby storage temperature could be maintained at 0°C at all times. Outbreaks of botulism have proven to be very costly in terms of economic losses as well as human suffering [84].

Guidelines for marketing MAP fish at the retail level in the United Kingdom were recently described by Mills [53]. The guidelines stress the critical constraints of time and temperature necessary for maintaining seafood quality and safety. Consumers were to be advised to protect MAP fish products from temperature abuse on the way home from market and to consume the products as soon as possible. Smith [74] concluded that MAP of fish should be conducted only when storage temperature of 2°C or below can be guaranteed. Whether such a marketing system can be developed in other regions such as North America, remains to be seen. A survey by Wyatt [91] showed a shocking lack of awareness of proper and safe food handling practices by a majority of food retail store managers surveyed. Similar problems exist within foodservice institutions, because the foodservice sector is responsible for the greatest proportion of recorded foodborne disease outbreaks in Canada [1], The Netherlands

[6] and the United States [12]. Training programs, levels of expertise and awareness of food safety among retail and foodservice managers would have to be increased.

Histamine production by bacteria in products such as tuna is another area of concern with regard to the safety of fish products. Tuna, spiked with *Klebsiella oxytoca* T2, *Morganella morganii* or *Hafnia alvei* T8 followed by vacuum packaging, contained more histamine than spiked tuna that was not vacuum packaged [87]. Low storage temperature was the most important factor in controlling histamine production in the spiked tuna samples.

It was recently shown that treatment of fish with nisin just prior to packaging in a CO_2 MA delayed the onset of toxin production by *C. botulinum*. Nisin-treated fillets spoiled at the same rate at 26°C and 10°C as untreated fillets. Nisin had no inhibitory effect on spoilage microflora of cod, herring and smoked mackerel fillets packed under the CO_2 MA, even though *C. botulinum* toxigenesis was delayed [83].

Another alternative that should be explored is the development of MAP regimes which ensure that spoilage microorganisms will outgrow pathogenic microorganisms, such as *C. botulinum* type E, if the product undergoes temperature abuse [40].

6.5 Future developments

MAP of fish products has been shown to have potential benefits of extending storage life. In the light of the potential dangers posed by psychrotrophic *C. botulinum*, packaging, distribution and sale of MAP fish products can only be accomplished safely with strict temperature control from the time of packaging to the time at which the consumer prepares the fish for consumption. Clearly the benefits for MAP of fish products would have to far outweigh the risks. The risks can be minimized by developing fail-safe mechanisms should temperature abuse occur. These areas should be the focus of intensive research so that the benefits of extended storage life of fish products afforded by MAP can be used to advantage without the spectre of risks associated with potential growth and toxin production by psychrotrophic strains of *C. botulinum*.

References

1. Anon. 1986. Foodborne and waterborne disease in Canada, Annual Summaries. 1980, 1981, 1982. Health Protection Branch, Health and Welfare Canada, Ottawa, ON.
2. Baker, D. A. and Genigeorgis, C. 1990. Predicting the safe storage of fresh fish under modified atmospheres with respect to *Clostridium botulinum* toxigenesis by modeling length of the lag phase of growth. J. Food Prot. 53: 131-140, 153.
3. Baker, D. A., Genigeorgis, C. and Garcia, G. 1990. Prevalence of *Clostridium botulinum* in seafood and significance of multiple incubation temperatures for determination of its presence and type in fresh retail fish. J. Food Prot. 53: 668-673.
4. Barnett, H.J., Conrad, J. W. and Nelson, R. W. 1987. Use of laminated high and low density polyethylene flexible packaging to store trout (*Salmo gairdneri*) in a modified atmosphere. J. Food Prot. 50: 645-651.
5. Barnett, H. J., Stone, F. E., Roberts, G. C., Hunter, P. J., Nelson, R. W. and Kwok, J. 1982. A study in the use of a high concentration of CO_2 in a modified atmosphere to preserve fresh salmon. Mar. Fish. Rev. 44(3): 7-11.
6. Beckers, H. J. 1988. Incidence of foodborne diseases in the Netherlands: Annual summary 1982 and an overview from 1979 to 1982. J. Food Prot. 51: 327-334.
7. Belleau, L. and Simard, R. E. 1987. Effets d'atmospheres de dioxyde de carbone et d'azote sur des fillets de poisson. Sci. Aliments. 7: 433-436.
8. Bligh, E. G., Woyewoda, A. D., Shaw, S. J. and Hotton, C. H. 1984. Hypobaric storage of Atlantic Cod fillets (*Gadus morhua*). Can. Inst. Food Sci. Technol. J. 17: 266-270.
9. Bremner, H. A. and Statham, J. A. 1983. Spoilage of vacuum-packed chill-stored scallops with added lactobacilli. Food Technol. Aust. 35: 284-287.
10. Bremner, H. A. and Statham, J. A.. 1987. Packaging in CO_2 extends shelf-life of scallops. Food Technol. Aust. 39: 177-179.
11. Bryan, F. L. 1987. Seafood-transmitted infections and intoxications in recent years, p 319-337. *In* Seafood Quality Determination, Kramer, D. and Liston, J. (eds). Elsevier Science Pub., Amsterdam.
12. Bryan, F. L. 1988. Risks of practices, procedures and processes that lead to outbreaks of foodborne diseases. J. Food Prot. 51: 663-673.
13. Cann, D. C., Smith, G. L. and Houston, N. G. 1983. Further studies on marine fish stored under modified atmosphere packaging. Torry Research Station, Aberdeen, U.K.
14. Cann, D. C., Houston, N. C., Taylor, L. Y., Smith, G. L., Thomson, A. B. and Craig, A. 1984. Studies of salmonids packed and stored under a modified atmosphere. Torry Research Station, Aberdeen, U.K.
15. Cann, D. C., Houston, N. C., Taylor, L. Y., Stroud, G., Early, J. C. and Smith, G. L. 1985. Studies of shellfish packed and stored under a modified atmosphere. Torry Research Station, Aberdeen, U.K.

16. Chen, H. C., Meyers, S. P., Hardy, R. W. and Biede, S. L. 1984. Color stability of astaxanthin pigmented rainbow trout under various packaging conditions. J. Food Sci. 49: 1337-1340.

17. Craig, J. M., Hayes, S. and Pilcher, K. S. 1968. Incidence of *Clostridium botulinum* type E in salmon and other marine fish in the Pacific Northwest. Appl. Microbiol. 16: 553-557.

18. Daniels, J. A., Krishnamurthi, R. and Rizvi, S. S. H. 1986. Effects of carbonic acid dips and packaging films on the shelflife of fresh fish fillets. J. Food Sci. 51: 929-931.

19. Dheeragool, P. 1989. Modified atmosphere packaging of pink prawns (*Pandalus platyceros*). M.Sc. thesis. University of British Columbia, Vancouver, British Columbia.

20. Eklund, M. W. 1982. Significance of *Clostridium botulinum* in fishery products preserved short of sterilization. Food Technol. 36(12): 107-112, 115.

21. Eyles, M. J. 1986. Microbiological hazards associated with fishery products. CSIRO Food Res. Quart. 46: 8-16.

22. Eyles, M. J. and Warth, A. O. 1981. Assessment of the risk of botulism from vacuum-packaged raw fish: A review. Food Technol. Aust. 33: 574-580.

23. Fey, M. S. and Regenstein, J. M. 1982. Extending shelf-life of fresh red hake and salmon using CO_2-O_2 modified atmosphere and potassium sorbate at 1°C. J. Food Sci. 47: 1048-1054.

24. Fletcher, G. C. and Statham, J. A. 1988. Shelf-life of sterile yellow-eyed mullet (*Aldrichetta forsteri*) at 4°C. J. Food Sci. 53: 1030-1035.

25. Fletcher, G. C., Murrell, W. G., Statham, J. A., Stewart, B. J. and Bremner, H. A. 1988a. Packaging of scallops with sorbate: An assessment of the hazard from *Clostridium botulinum*. J. Food Sci. 53: 349-352, 358.

26. Fletcher, G. C., Wong, R. J., Charles, J. C., Hogg-Stec, M. G. and Temple, S. M. 1988b. The storage of cold-smoked New Zealand mackerel. Fish Proc. Bull. No. 11. DSIR, Auckland, New Zealand.

27. Fujii, T., Hirayama, M., Okuzumi, M., Yasuda, M., Nishino, H. and Yokoyama, M. 1989. Shelf-life studies on fresh sardine packaged with carbon-dioxide gas mixture. Bull. Jap. Soc. Sci. Fish. 55: 1971-1975.

28. Fujii, T., Hirayama, M., Okuzumi, M., Nishino, H. and Yokayama, M. 1990. The effect of storage in carbon dioxide - nitrogen gas mixture on the microbial flora of sardines. Nippon Suisan Gakaishi 56: 837.

29. Garcia, C. W. and Genigeorgis, C. 1987. Quantitative evaluation of *Clostridium botulinum* nonproteolytic types B, E and F growth in fresh salmon tissue homogenates stored under modified atmospheres. J. Food Prot. 50: 390-397, 400.

30. Garcia, G. W., Genigeorgis, C. and Lindroth, S. 1987. Risk of growth and toxin production by *Clostridium botulinum* nonproteolytic types B, E and F in salmon fillets stored under modified atmospheres at low and abused temperatures. J. Food Prot. 50: 330-336.

31. Genigeorgis, C. 1985. Microbial and safety implications of the use of modified atmospheres to extend the storage life of fresh meat and fish. Int. J. Food Microbiol. 1: 237-251.

32. Gerdes, D. L., Hoffstein, J. J., Finerty, M. W. and Grodner, R. M. 1989. The effects of elevated CO_2 atmospheres on the shelf-life of freshwater crawfish (*Procambaris clarkii*) tail meat. Lebensm. Wiss. U. Technol. 22: 315-318.

33. Gibson, D. M. 1985. Predicting the shelflife of packaged fish from conductance measurements. J. Appl. Bacteriol. 58: 465-470.

34. Gill, C. O. and Penny, N. 1982. The occurrence of *Clostridium botulinum* at aquatic sites in and around Auckland and other urban areas of the North Island. N. Z. Vet. J. 30: 110-112.

35. Gola, S., Rossi, M., Ghisi, M. and Pirazzoli, P. 1986. Produzione di tossina botulinica in pesce fresco (Merluzzo) confezionato sotto vuoto in materiali plastici a diversa permeabilità all'ossigeno. Ind. Conserve 61: 260-264.

36. Gray, R. J. H., Hoover, D. G. and Muir, A. M. 1983. Attenuation of microbial growth on modified atmosphere-packaged fish. J. Food Prot. 46: 610-613.

37. Haard, N. F., Martins, I., Newbury, R. and Botta, R. 1979. Hypobaric storage of Atlantic herring and cod. Can. Inst. Food Sci. Technol. J. 12: 84-87.

38. Hauschild, A. H. W. 1989. *Clostridium botulinum*, pp. 111-189. In "Foodborne Bacterial Pathogens". (Doyle, M. P. ed), Marcel Dekker, Inc., New York.

39. Hauschild, A. H. W. and Gauvreau, L. 1985. Food-borne botulism in Canada, 1971 - 84. Can. Med. Assoc. J. 133: 1141-1146.

40. Hintlian, C. B. and Hotchkiss, J. H. 1986. The safety of modified atmosphere packaging: A review. Food Technol. 40(12): 70 -76.

41. Hobbs, G. 1976. *Clostridium botulinum* and its importance in fishery products. Adv. Food Res. 22: 135-185.

42. Huss, H. H. 1979. *Clostridium botulinum* in fish. Nord. Vet. Med. 31: 214-221.

43. Huss, H.H. 1980. Distribution of *Clostridium botulinum*. Appl. Environ. Microbiol. 39: 764-769.

44. Hussain, A. M., Ehlermann, D. and Diehl, J. 1977. Comparison of toxin production by *Clostridium botulinum* type E in irradiated and unirradiated vacuum-packed trout (*Salmo gairdneri*). Archiv für Lebensmittelhygiene 28: 23-27.

45. Lannelongue, M., Finne, G., Hanna, M. O., Nickelson, R, and Vanderzant, G. 1982. Storage characteristics of brown shrimp (*Penaeus aztecus*) stored in retail-packages containing CO_2-enriched atmospheres. J. Food Sci. 47: 911-913, 923.

46. Layrisse, M. E. and Matches, J. R. 1984. Microbiological and chemical changes of spotted shrimp (*Pandalus platyceros*) stored under modified atmospheres. J. Food Prot. 47: 453-457.

47. Licciardello, J. J., Ravesi, E. M., Tuhkunen, B. E. and Racicot, L. D. 1984. Effect of some potentially synergistic treatments in combination with 100 Krad irradiation on the iced shelf life of cod fillets. J. Food Sci. 49: 1341 - 1346, 1375.

48. Lilly. T. and Kautter, D. A. 1990. Outgrowth of naturally occurring *Clostridium botulinum* in vacuum-packaged fresh fish. J. Assoc. Off. Anal. Chem. 73: 211-212.

49. Lindroth, S. E. and Genigeorgis, C. A. 1986. Probability of growth and toxin production by nonproteolytic *Clostridium botulinum*

in rockfish stored under modified atmospheres. Int. J. Food Microbiol. 3: 167-181.

50. MacDonnell, M. J. and Colwell, R. R. 1985. Phylogeny of the Vibrionaceae and recommendations for two new genera, *Listonella* and *Shewanella*. Syst. Appl. Microbiol. 6: 171-182.

51. Matches, J. R. and Layrisse, M. E. 1985. Controlled atmosphere storage of spotted shrimp (*Pandalus platyceros*). J. Food Prot. 48: 709-711.

52. McMeekin, T. A., Hulse, L. and Bremner, H. A. 1982. Spoilage association of vacuum packed sand flathead (*Platycephalus bassensis*) fillets. Food Technol. Aust. 34: 278-282.

53. Mills, A. 1985. Fish: New guidelines for CAP. Food Manuf. 60(11): 45.

54. Mitsuda, H., Nakajima, K, Mizuno, H. and Kawai, F. 1980. Use of sodium chloride solution and carbon dioxide for extending shelf life of fish fillets. J. Food Sci. 45: 661-666.

55. Molin, G, Stenstrom, M. and Ternstrom, A. 1983. The microbial flora of herring fillets after storage in carbon dioxide, nitrogen or air at 2°C. J. Appl. Bacteriol. 55: 49-56.

56. Morey, K. S., Satterlee, L. D. and Brown, W. D. 1982. Protein quality of fish in modified atmospheres as predicted by the C-PER assay. J. Food Sci. 47: 1399-1400, 1409.

57. Oberlander, V., Hanna, M. O., Miget, R., Vanderzant, C. and Finne, G. 1983. Storage characteristics of fresh swordfish steaks stored in carbon dioxide-enriched controlled (flow-through) atmospheres. J. Food Prot. 46: 434-440.

58. Ogrydziak, D. M. and Brown, W. D. 1982. Temperature effects in modified-atmosphere storage of seafoods. Food Technol. 36(5): 86-96.

59. Parkin, K. L. and Brown, W. D. 1982. Preservation of seafood with modified atmospheres, p. 453-465. *In* "Chemistry and Biochemistry of Marine Food Products", Martin, R. L., Flick, G. J., Hebard, C. E. and Ward, D. R. (eds). AVI Publ. Co., Westport, CT.

60. Parkin, K. L. and Brown, W. D. 1983. Modified atmosphere storage of dungeness crab (*Cancer magister*). J. Food Sci. 48: 370-374.

61. Partmann, W. 1981. Untersuchungen zur lagerung von verpackten Regenbogenforellen (*Salmo gairdneri*) in luft und kohlendioxid. Fleischtwirtsch. 61: 625-629.

62. Passy, N., Mannheim, C. H. and Cohen, D. 1983. Effect of a modified atmosphere and pretreatments on quality of chilled fresh water prawns (*Macrobachium rosenbergii*). Lebensm. Wiss. U. Technol. 16: 224-229.

63. Post, L. S., Lee, D. A., Solberg, M., Furgang, D., Specchio, J. and Graham, C. 1985. Development of botulinal toxin and sensory deterioration during storage of vacuum and modified atmosphere packaged fish fillets. J. Food. Sci. 50: 990-996.

64. Potter, N. N. 1986. Food Science, p. 209. 4th ed. AVi Publ. Co., Westport, CT.

65. Powrie, W. D., Skura, B. J. and Wu, C. H. 1987. Energy conservation by storage of muscle and plant products at latent zone and modulated subfreezing temperatures. Final Report, ERDAF File 0145B.01916-EP25, Agriculture Canada.

66. Przybylski, L. A., Finerty, M. W., Grodner, R. M. and Gerdes, D. L. 1989. Extension of shelf-life of fresh channel catfish fillets using modified atmosphere packaging and low dose irradiation. J. Food Sci. 54: 269-273.

67. Santos, E. E. M. and Regenstein, J. M. 1990. Effects of vacuum packaging, glazing and erythorbic acid on the shelf-life of frozen white hake and mackerel. J. Food Sci. 55: 64-70.

68. Schvester, P. 1989. Safety and Seafood. Can. Packag. (12): 32-33.

69. Schvester, P. 1989. MAP of seafood: a new and critical analysis to evaluate its potential. Conference Proceedings, Pack Alimentaire '90, Innovative Expositions, Inc., Princeton, NJ, Session A-3.

70. Scott, D. N., Fletcher, G. C. and Summers, G. 1984. Modified atmosphere and vacuum packing of snapper fillets. Food Technol. Aust. 36: 330-332.

71. Scott, D. N., Fletcher, G. C. and Hogg, M. G. 1986. Storage of snapper fillets in modified atmospheres at -1°C. Food Technol. Aust. 38: 234-238.

72. Sharp, W. F., Norback, J. P. and Stuiber, D. A. 1986. Using a new measure to define shelf life of fresh whitefish. J. Food Sci. 51: 936-939, 959.

73. Skura, B. J. 1988. Modified atmosphere packaging: Its potential and dangers, pp. 191-197. *In* "Aquaculture International Congress Proceedings", Aquaculture International Congress, Vancouver, BC.

74. Smith, C. E. 1987. Feasibility of modified atmosphere packaging of fish: A review. Saskatchewan Research Council, Pub. No. I-4202-1-E-87, Saskatoon, SK.

75. Smith, L. D. 1977. Botulism: the Organism, its Toxins, the Disease, p. 91. Charles C. Thomas, Publ. Springfield, IL.

76. Smith, L. Ds., and Sugiyama, H. 1988. Botulism: the organism, its toxins, the disease. Charles C. Thomas, Publisher, Springfield, IL, pp. 88-89.

77. Statham, J.A. 1984. Modified atmosphere storage of fisheries products: The state of the art. Food Technol. Aust. 36: 233-239.

78. Statham, J. A. and Bremner, H. A. 1985. Acceptability of trevalla (*Hyperoglyphe porosa* Richardson) after storage in carbon dioxide. Food Technol. Aust. 37: 212-215.

79. Statham, J. A. and Bremner, H. A. 1989. Shelf-life extension of packaged seafoods - a summary of a research approach. Food Aust. 41: 614-620.

80. Statham, J. A., Bremner, H. A. and Quarmby, A. R. 1985. Storage of morwong (*Nemadactylus macropterus* Bloch and Schreider) in combinations of polyphosphate, potassium sorbate and carbon dioxide at 4°C. J. Food Sci. 50: 1580-1584, 1587.

81. Stenström, I. 1985. Microbial flora of cod fillets (*Gadus morhua*) stored at 2°C in different mixtures of carbon dioxide and nitrogen/oxygen. J. Food Prot. 48: 585-589.

82. Stier, R. F., Bell, L., Ito, K. A., Shafer, B. D., Brown, L. A., Seeger, M. L., Allen, B. H., Porcuna, M. N. and Lerke, P. A. 1981. Effect of modified atmosphere storage on *C. botulinum* toxigenesis and the spoilage microflora of salmon fillets. J. Food Sci. 46: 1639-1642.

83. Taylor, L. Y., Cann. D. D. and Welch, B. J. 1990. Antibotulinal properties of nisin in fresh fish packaged in an atmosphere of carbon dioxide. J. Food Prot. 53: 953-957.

84. Todd, E. C. D. 1985. Economic loss from foodborne disease and non-illness related recalls because of mishandling by food processors. J. Food Prot. 48: 621-633.

85. Villemure, G., Simard, R. E. and Picard, G. 1986. Bulk storage of cod fillets and gutted cod (*Gadus morhua*) under carbon dioxide atmosphere. J. Food Sci. 51: 317-320.

86. Wang, M. Y. and Brown, W. D. 1983. Effects of elevated CO_2 atmosphere on storage of freshwater crayfish (*Pacifastacus leniusculus*). J. Food Sci. 48: 158-162.

87. Wang, M. Y. and Ogrydziak, D. M. 1986. Residual effect of storage in an elevated carbon dioxide atmosphere on the microbial flora of rock cod (*Sebastes* spp.). Appl. Environ. Microbiol. 52:727-732.

88. Wei, C. I., Chen, C. M., Koburger, J. A., Otwell, W. S. and Marshall, M. R. 1990. Bacterial growth and histamine production on vacuum packaged tuna. J. Food Sci. 55: 59-63.

89. Wilhelm, K. A. 1982. Extended fresh storage of fishery products with modified atmospheres: a survey. Mar. Fish. Rev. 44(2): 17-20.

90. Woyewoda, A. D., Bligh, E. G. and Shaw, S. J. 1984. Controlled and modified atmosphere storage of cod fillets. Can. Inst. Food Sci. Technol. J. 17(1): 24-27.

91. Wyatt, C. J. 1978. Concerns, experiences, attitudes and practices of food market managers regarding sanitation and safe food handling procedures. J. Food Prot. 42: 555-560.

Chapter 7

MODIFIED ATMOSPHERE PACKAGING OF FRUITS AND VEGETABLES

William D. Powrie and Brent J. Skura
Department of Food Science, The University of British
Columbia

7.1 Introduction

7.1.1 Food trends

Consumer appeal for fresh, low-calorie, healthy, nutritious and high quality foods has grown steadily in the last 10 years. Such food trends may be attributed to consumer attitudes on lifestyle, nutrition, fitness, health and food quality [7, 103, 128, 239]. The 1990 Grocery Attitudes of Canadians study [104] indicated that 68% of grocery shoppers considered nutrition as extremely or very important. North Americans are concerned about foods in their diet in relation to weight control and chronic disease risks. Government agencies and health-promoting organizations are recommending a reduction of fat, cholesterol and salt in the diet and a greater consumption of fruits, vegetables and cereals with the view that such dietary changes may reduce the risk of heart disease and cancer incidence [7, 285]. Fresh commodities are now considered by consumers to be more nutritious than canned products and more flavourful [128].

Fresh refrigerated prepared fruits and vegetables are becoming popular at the consumer and food service levels as convenience items because the fresh products have been washed, peeled, cut and packaged (sometimes under modified atmospheres). Such minimal processing of produce eliminates inedible portions as waste and brings about ready-to-use products [135]. It has been estimated that by 1995, more than 50% of food dollars spent in American stores will be for fresh, ready-to-eat items [72].

Fruits and vegetables are important components in the diets of humans worldwide to supply nutrients, to satisfy hunger and appetite, and to provide culinary enjoyment. In North America, fruits and vegetables are regarded as one of four major food groups which are required to satisfy the recommended daily nutrient intake. The culinary enjoyment of fresh fruits and vegetables is dependent on their quality attributes which are presented in Table 7.1 [116, 151, 184, 300].

Table 7.1.   Summary of the quality attributes of fresh fruits and vegetables

Category	Sensory quality attributes
Flavour	sweetness sourness bitterness astringency odour
Texture	firmness crunchiness crispness succulence juiciness mealiness smoothness
Appearance	colour shape size glossiness
Nutritional value	

7.1.2 Terminology for fruits and vegetables

As pointed out by Duckworth [73], the common usage of the term, fruit, is not in accordance with the precise botanical nomenclature. For example, a tomato in the botanical sense is a fruit yet it is considered throughout the world as a vegetable, possibly because the tomato has a low sugar content (about 2 to 3%). From the standpoint of food science, a fruit is regarded as a mature fleshy ovary (ovaries) with or without adhering floral parts, and generally hav-

ing a high moisture content, a distinctive ester-like odour, a sufficient sugar content of between about 5 and 22% (Table 7.2) to render
it sweet and pH levels below 4.5 (Table 7.3).   Fruits are considered to
be dessert foods which may have sugar added to them [73, 301].   Fruits
can be classified as simple (grape, cherry), aggregate (raspberry),
multiple (pineapple) and accessory (apples, strawberries).   On the
other hand, a vegetable may be any part of a plant and may be classified as a seed (peas), a bud (lettuce), a leaf (spinach), a stem
(asparagus, potato), a root (carrot) or a fruit (cucumber, green bean,
tomato).   A vegetable can be defined in food science terms as an edible plant part which generally has a low sugar level (Table 7.2) and
a pH of about 5 or higher (Table 7.3), is frequently salted, cooked and
served with a meat dish or sometimes eaten raw as a salad or snack
[301].

Table 7.2.   Sugars in fruits and vegetables (on a percentage fresh
basis)

	Glucose (%)	Fructose (%)	Sucrose (%)
Fruit:			
Apple	1.17	6.04	3.78
Apricot	1.73	1.28	5.84
Blueberry	3.76	3.82	0.19
Cherry (sweet)	6.49	7.38	0.22
Melon, Honeydew	2.56	2.62	5.86
Peach	0.91	1.18	6.92
Pear	0.95	6.77	1.61
Raspberry (red)	2.40	1.58	3.68
Strawberry	2.09	2.40	1.03
Vegetables:			
Asparagus	0.92	1.30	0.28
Broccoli	0.73	0.67	0.42
Brussels sprout	0.66	0.75	0.41
Cabbage	1.58	1.20	0.15
Carrot	0.85	0.85	4.24
Cauliflower	0.83	0.74	0.67
Celery	0.49	0.43	0.31
Cucumber	0.86	0.86	0.06
Lettuce	0.25	0.46	0.10
Onion	2.07	1.09	0.89
Pepper	0.90	0.87	0.11
Potato	0.15	0.09	0.14
Radish	1.34	0.74	0.22
Spinach	0.09	0.04	0.06
Squash, summer	0.77	0.82	0.09
Sweet corn	0.34	0.31	3.03
Sweet potato	0.33	0.30	3.37
Tomato	1.12	1.34	0.01

From Schallenberger and Birch [251]

Table 7.3.   pH values of fruits and vegetables

Fruits		Vegetables	
Lemons	2.2-2.8	Tomatoes	4.0-4.7
Plums	2.8-4.0	Carrot	4.9-5.4
Raspberries	2.8-3.6	Beets	4.9-5.4
Grapefruit	3.0-3.7	Squash	5.0-5.4
Strawberries	3.0-3.3	Beans, green	5.0-6.0
Grapes	3.0-3.4	Spinach	5.1-5.9
Oranges	3.0-4.0	Cabbage	5.2-5.4
Apples	3.1-4.0	Sweet potatoes	5.3-5.6
Cherries	3.2-4.2	Potatoes	5.4-6.0
Peaches	3.4-4.0	Peas	5.8-6.5
Pears	3.6-4.4	Sweet corn	6.0-6.8
Papaya	4.6-5.5		

7.1.3 Market perspective of fresh fruits and vegetables

Fruits and vegetables in the fresh state are appealing to consumers since they are available year-round, have highly desirable flavour and texture, have a low fat content, are a good source of dietary fibre and are nutritious with respect to high levels of vitamin C and β-carotene as well as a variety of essential minerals [160, 214, 256, 283, 284, 310, 312]. In the U.S.A., consumption of fresh fruits and vegetables has increased since 1970 by about 22 and 21%, respectively [229]. On the other hand, consumption of canned fruit and vegetables has dropped by 35 and 6%, respectively. Results of a 1989 fresh produce survey indicated that around 31% of U.S. consumers increased their yearly consumption of fresh vegetables and about 41% increased their consumption of fresh fruit [313].

Consumers consider freshly harvested ripe fruits and appropriately mature vegetables as the ultimate in acceptability from the standpoint of flavour, texture and colour. About 96% of the participants in a 1989 fresh produce survey indicated that they selected fresh fruits and vegetables on the basis of ripeness, freshness, taste and appearance [313]. In the 1988 Grocery Attitudes of Canadians survey, the most important consideration of respondents (shoppers) in choosing a grocery store was freshness of produce [103]. In addition, these shoppers placed "freshness" and "taste" as top priorities when planning meals.

7.2 Quality considerations

7.2.1 Developmental stages of horticultural commodities

The developmental stages of a plant are growth, maturation and senescence [110, 233, 296, 301]. During plant organ growth, cells multiply and become enlarged. Once the increase in size of an organ ceases, the maturation stage commences. Most horticultural commodities are harvested at a specific level of maturity. Some (e.g. snow peas, asparagus) are picked in the immature state. Maturity from a quality standpoint may be defined as "that stage at which a commodity has reached a sufficient stage of development that after harvesting and postharvest handling (including ripening, where required), its quality will be at least the minimum acceptable to the ultimate consumers" [233]. Senescence is generally regarded as the stage when extensive catabolic reactions occur with the consequence of membrane and middle lamella degradation and disorganization of organelles. At this stage, plant organs lose their natural protective barriers and become susceptible to microbial invasion and decomposition [82].

7.2.2 Quality maintenance of fresh fruits and vegetables

The quality attributes (listed in Table 7.1) for fresh fruits and vegetables must be maintained as effectively as possible from the field, through the transportation, storage and distribution systems, and finally to the retail stores to ensure consumer acceptability and buyer appeal [256, 300, 310]. Postharvest chemical, physical and microstructural changes in fresh horticultural produce can lead to decreases in quality attributes and to a greater vulnerability to spoilage microorganisms (spoilogens). Most of these chemical changes are mediated by tissue enzymes. Thus, methods of inhibiting the enzymic activity in fresh produce will be helpful in prolonging the shelf-life. Such methods include pre-cooling (rapid removal of field heat), refrigerated storage, and modified gas atmospheres. The lowering of the temperature of produce reduces the rates of respiration and ripening (by lowering enzyme activity) as well as the microbial decomposition of the plant tissue. However, a low-temperature tissue disorder, called chilling injury, can occur in tropical and subtropical fruits and in a few vegetables when the storage temperature is decreased below 12°C [141, 179, 199]. Presumably this disorder is caused by the physical changes in the lipids and proteins of cell membranes [212]. Lipid peroxidation has been proposed as a critical event in membrane deterioration [191, 257]. Chilling injury symptoms of fruits and vegetables are: (1) brown discoloration; (2) surface pitting; (3) uneven ripening; and, (4) off-flavour development, all of which are linked to enzymic decomposition of naturally occurring compounds [301]. Cellular breakdown on the surfaces of chill-sensitive produce can lead to extensive mould growth [82, 196].

Since fruits and vegetables contain large amounts of water and have high water activities, water is readily lost under low relative humidity conditions. Loss of water from fresh produce can lead to skin wrinkling, loss of crunchiness and crispness, wilting and undesirable colour changes. Proper packaging of fresh commodities can restrict the rate of water loss [155, 210, 301].

7.2.3   Postharvest losses of fresh fruits and vegetables

Consumers may reject fresh horticultural products as physically damaged, decayed or unacceptable sensory quality [17]. In such cases, the products are regarded as postharvest losses [46]. The causes of such losses of perishable produce may be classified as: (1) microbial intrusion and decomposition; (2) metabolic processes; and, (3) physical stresses. Spoilage organisms (spoilogens) of fresh fruits and vegetables are usually fungi, yeasts and bacteria, while viruses are of minor importance [78, 79, 196]. These organisms may originate in the field and indeed invade the growing produce. Subsequently, visible symptoms (e.g., patches of mould mycelia, brown areas) may occur during transport and storage of the harvested produce, particularly in bruised, cut, overripe tissue areas. Abnormal metabolic changes, caused by chilling injury temperatures, oxygen depletion or carbon dioxide stress, may be manifested by tissue discoloration, pitting and extensive tissue softening. Physical stresses occurring during mechanical harvesting, transportation and consumer handling cause bruising, abrasions and cuts in the fruits and vegetables [43]. Some fruit can be harvested at the slightly immature stage to reduce bruising and decrease the softening rate of the flesh, but the expected fullness of the ripe flavour usually is not achieved during postharvest storage.

Studies on the losses of selected fruits and vegetables at the wholesale, retail and consumer levels in New York and Chicago markets have been reviewed by Cappellini and Ceponis [46]. The greatest losses for the fruit category were for strawberries, sweet cherries, blueberries, peaches and nectarines. With strawberries as an example, the wholesale loss was about 6 to 14%, the retail loss was around 5% and the consumer loss was greatest at about 18 to 22%. Microbial spoilage was the principal cause of the losses for the above-mentioned fruits. For vegetables, the greatest losses were noted at the retail level for lettuce, snapbeans, bell peppers and tomatoes. For lettuce, losses between 7 and 14% occurred at the consumer level whereas wholesale and retail losses were about 4 to 6% and 2 to 15%, respectively.

Postharvest losses of fresh horticultural commodities can be obviated by subjecting them to further processing such as canning, freezing and drying. However, such treatments can lead to dramatic alterations in quality attributes. With the canning process, thermal-sensitive compounds in the commodities are decomposed with the result of changes in flavour, colour and texture [131, 225]. Under suitable freezing conditions, horticultural products may retain their flavour and colour features, but the textural quality may be changed dramatically by ice crystal damage to the intact tissue [89].

7.3   An integrated approach to MAP technology of fruits and vegetables

Fresh fruits and vegetables are tissue systems with specific chemical, physical and structural properties which impart quality attributes [143].   Texture is a perceptual manifestation of the physical state of tissue.   Structural and physical characteristics of cell walls, cell turgidity, vacuolar size and fluidity, proportion of fundamental and vascular tissues, and the tenacity of the middle lamellae between cells are all integral contributors to the texture of fruits and vegetables.   The colour of commodities can be attributed to the compartmentalized chlorophylls and carotenoids in plastids, and to anthocyanins in vacuoles of parenchyma cells.   The odour of produce is the result of a release of volatile odoriferous compounds from the vacuoles.

Active enzyme processes such as respiration, ripening and senescence are ongoing in fruits and vegetables.   The development of full-bodied flavours and deep colours of harvested fruits is dependent on anabolism for the synthesis of odoriferous compounds and pigments (part of the ripening process).   The metabolic precursors and the high-energy nicotinamide adenine dinucleotide phosphate-H (NADPH), to be used in these synthetic reactions, may be derived from the pentose phosphate pathway of the respiration process. Respiration also plays an important role in supplying adenosine triphosphate (ATP) for retaining the integrity of membranes, for ethylene production and for the synthesis of *de novo* enzymes such as polygalacturonases which are involved in the ripening process. Sugars and acids are decomposed in the respiration process with the result of changes in sweetness and sourness of commodities.

From the above discussion, it is apparent that the chemical composition, cellular and tissue structural and physical features, and the respiration and ripening processes are intimately interlinked with the quality attributes and the quality maintenance of harvested, stored fruits and vegetables.   Thus, these topics have been reviewed in this chapter as a foundation prelude to the discussion on modified atmosphere packaging (MAP) technology.   The coverage of MAP technology has been divided into the dimensions and benefits of MAP, the complementary factors which have positive implications for the success of MAP technology, the design and functions of package systems, the dynamics of O_2 and CO_2 in MAP commodities and the responses of fruits and vegetables to equilibrium MA (low O_2 and elevated CO_2 levels).   Emphasis has been placed on the pretreatment of cultivar-specific commodities, the establishment of equilibrium MA within package systems having specific permeation rates, and the influence of various levels of O_2 and CO_2 on the respiration and ethylene production rates, and on quality attributes of fruits and vegetables.

For detailed information on postharvest aspects of fruits and vegetables, the following books should be consulted:
- quality [85, 106, 117, 142, 213, 298]
- postharvest biochemistry and physiology [42, 65, 95, 109, 134, 173, 180, 185, 203, 247, 298, 311].
- postharvest process technology [45, 85, 134, 139, 155, 266, 301]
- postharvest handling and packaging [111, 210, 215, 245, 246]

7.4 Chemical composition and structure of fruits and vegetables

7.4.1 Chemical composition

The major components of fresh, harvested fruits and vegetables are water, carbohydrates, proteins and lipids [283, 284]. Water, the continuous phase of plant tissue, represents about 70 to 95% of the fresh weight of horticultural commodities. For most fruits, the protein content ranges from 0.5 to 2% whereas fresh common vegetables have protein levels varying from 1% (lettuce) to 8% (lima beans). The lipid contents of most fruits and vegetables are between about 0.1 and 1%.

Mono-, di-, oligo- and polysaccharides are synthesized by the growing plant as future sources of energy. The amounts of the various carbohydrates in fruits and vegetables are dependent on the stage of plant development when they were harvested and on the cultivar. Although a considerable variation in the carbohydrate content of harvested fruits and vegetables exists in the reported literature, approximate content values are advantageous, particularly from the standpoint of sugars and their contribution to the sweetness of fresh commodities. In addition to the sugar levels, the types of sugars present in a commodity play an important role in the sweetness level of a commodity. On an equimolar basis, fructose is sweeter than sucrose which in turn is sweeter than glucose and maltose [251]. The sugar types and contents of fruits and vegetables are presented in Table 7.2.

Generally, fresh fruits and vegetables possess organic acids and their salts at levels much above those amounts involved in the Krebs cycle of the respiration process [109, 281, 301]. The surplus acids (and acid salts) are usually stored in the vacuoles of the parenchyma cells and exist as buffer solutions [186, 259, 281]. The degree of sourness of a fruit is related to the pH. The pH of a fresh commodity is dependent on the pK values of acids and the ratio of the concentration of free acids and the concentration of their salts [259]. The pH of lemons is low (about 2.2-2.8) due to presence of only about 3% of the total acidity as salts whereas oranges have a much higher pH (about 3.0-4.0) because the percentage of the total acidity as salts is around 15. pH ranges for a variety of fruits and vegetables are presented in Table 7.3. The major acids in fruits and vegetables are citric and malic acids which have similar pK_1 values (3.1 and 3.4, respectively). Between pH levels of 3.5 and 5.5, citric and malic acid buffer solutions (0.01 N acid) have buffer indices of 2.5 and 3.3, respectively [99]. Thus a malic acid buffer solution has a greater resistance to pH change (better buffering action) than a citric acid buffer. Since the pH of saliva is between 6.3 and 6.6, acid fruits with high buffer indices should have the capacity to resist changes in pH and in sourness.

7.4.2   Parenchyma   cell

The chemical components of fruits and vegetables are compartmentalized in organized cellular moieties [83, 86, 247, 299]. Parenchyma cells are the most common units in edible horticultural commodities and are responsible generally for their succulence, crispness, colouration, textural firmness and flavour.   The sizes of parenchyma cells range from around 100 to 600 μm in diameter [195, 224, 299].   Chemical and physical changes in the cells can lead to quality deterioration and microbial spoilage of fruits and vegetables.

A parenchyma cell consists of a cell wall and a protoplast which includes a cytoplasm, a nucleus, vacuoles and ergastic substances such as crystals [86, 247].   Most parenchyma cells in fruits and vegetables have only thin primary walls with thicknesses of 1 to 3 μm [224, 247, 299].   A cell wall possess numerous pores with diameters between about 3.5 to 5.2 nm which are large enough to allow water, sugar and gas molecules to pass freely in and out of the protoplast [47].   The composition of primary walls on a dry weight basis has been found to be about 23% cellulose, 24%, hemicellulose, 34% pectic polysaccharides (pectic substances) and 19% glycoproteins [144].   Cellulose exists as microfibrils which are arranged in a parallel fashion and at right angles to the long axis of the cell [271].   The microfibrils provide high tensile strength and plasticity but not much elasticity to the cell walls [203, 247, 271].   Thus the cells resist stretching as water enters by osmosis to produce an osmotic pressure.   If the cell loses water, the cell walls, not being sufficiently elastic, collapse with the loss of turgidity.   In such a situation, the tissue becomes wilted.

Microfibrils are embedded in a matrix which is an aqueous dispersion of the hemicelluloses, pectic polysaccharides and glycoproteins [130, 144, 165].   The hemicellulose fraction consists of xyloglucans (-1,4-D-glucosyl residues as a backbone with branches of D-xylosyl residues) and glucuronoarabinoxylan.   Pectin polysaccharides have been separated as rhamnogalacturonan I, homogalacturonan, rhamnogalacturonan II, arabinans, galactans and arabinogalactan.   Proteins in primary walls generally exist in the glycosylated form and contain around 20% hydroxyproline. Experimental evidence indicates that an interconnective pattern of the macromolecular components of the primary walls exists [144, 165] and several models have been proposed.   For example, xyloglucan is presumably strongly bonded to the surfaces of the cellulose microfibrils by hydrogen bonding [5].   Arabinogalactan and rhamnogalacturonan may be covalently bonded to the xyloglucan. Non-covalent bonding between pectin polysaccharides in the cell wall involves calcium ions as bridges [119].   The position of glycoproteins in the primary cell is not known.   Some parenchyma cells which are intended to support plant tissue may secrete a secondary wall between the primary wall and the plasma membrane (plasmalemma).   Secondary walls (about 4 μm thick) are made up of about 41 to 45% cellulose, 30% hemicelluloses and 22 to 28% lignin [247].

Walls of adjacent cells are cemented together by a gel-like layer (middle lamella) composed of a protopectin complex of pectic polysaccharides with cross-bridging calcium and magnesium ions [19, 21, 130, 146, 165, 216, 271].   Thin tubular structures, called plasmodesmata, pass through the cemented adjacent walls in the paired pit-field areas of the primary walls [21].   Plasmodesmata act as connectors between the adjacent cells for the rapid transfer of solutes [114].

Since cell walls are not in perfect contact with each other, intercellular gas spaces are present in plant tissues.   In mature apple tissue, about 20 to 36% of the total volume is made up of intercellular spaces whereas the percentage volume of intercellular spaces in mature peaches and potatoes is about 15 and 1 to 2, respectively [36, 40, 41, 194, 195, 232].   Because of the interconnecting network of all of the intercellular spaces in parenchyma tissue of fruits and vegetables, oxygen in the surrounding air can diffuse rapidly into spaces of the tissue and pass into the cells through exposed cell walls [41].   Burton [41] found that a potato had no consistent oxygen gradient from the periphery to the centre, thus showing that the oxygen distribution system was highly efficient for adequate aerobic respiration in cells.   On the other hand, vascular regions in contrast to parenchyma tissue apparently act as barriers to gas diffusion.

The cytoplasm consists of the plasma membrane (plasmalemma), tonoplast, microtubules, microfilaments, organelles (such as endoplasmic reticulum, mitochondria, ribosomes and Golgi bodies), plastids, microbodies and spherosomes [83, 84, 86, 88, 247]. The term cytosol is used to designate the aqueous medium in which organelles and other relatively large particulate bodies are suspended [52].   The plasmalemma, the membrane next to the cell wall, regulates the flow of solutes into and out of the cells [185].   The membrane around the vacuole is called the tonoplast which is selectively permeable to water and solutes with the consequence of cell turgidity [185, 247].   The endoplasmic reticulum (ER), an assembly of tiny membrane tubes called tubules attached to balloon-like bodies called cisternae, is responsible for the transport of solutes within a cell and between cells through the plasmodesmata [114].   Enzymes and proteins are transported from the ER to the cell wall for specific functions.   Ribosomes are associated with ER for the synthesis of proteins. Ribosomes are also present in the free form in the cytosol and are also present in chloroplasts and mitochondria.   Some ribosomes are attached to form a long chain which is held together by a m-RNA strand.   This ribosome chain is called a polyribosome or polysome [247].

The mitochondria contain enzymes which are responsible for the respiration in the plant cell and for the production of ATP [109, 247].   The mitochondrion is a sausage-shaped or oval-shaped organelle with dimensions of about 4 µm in length and 0.5 µm in diameter [86, 88, 109, 247].   A plant cell contains numerous mitochondria, generally in the vicinity of 200, but sometimes up to 2000.   The outer membrane of a mitochondrion is smooth whereas the inner membrane is folded into a series of tubular protrusions called cristae. The mitochondrion consists of two compartments: 1. the intermembrane space between the outer and inner membranes and 2. the matrix within the inner membrane system. Small particles on the inner surface of the inner membrane contain several enzymes such as adenosine triphosphatase (ATPase) and succinic dehydrogenase. The granular matrix, with a protein content as high as 500 mg/mL, contains most of the enzymes in the Krebs cycle [225, 247]. Glycolytic products for the Krebs cycle and oxidative phosphorylation system are transported from the cytosol through the highly permeable outer membrane of the mitochondrion into the intermembrane space.

Plastids, surrounded by a double membrane system, include chloroplasts containing chlorophylls and carotenoids and chromoplasts containing carotenoids and leucoplasts which are colourless [86, 247]. Chloroplasts are responsible for the green colour of leaf and stem vegetables such as spinach, lettuce, broccoli and asparagus and some fruits such as a green apple. Chloroplasts have active enzyme systems which may break down chlorophylls during the storage of the produce. Chromoplasts are present in roots, stems, leaves and fruits which may be coloured yellow, orange or red. β-carotene is the most common carotenoid in fruits and vegetables, and is valuable as provitamin A. During the postharvest ripening of tomatoes, the red-coloured carotenoid, called lycopene, is synthesized in the chromoplasts of cells.

Vacuoles are cellular bodies (solutions) which are enclosed by membranes called tonoplasts [186, 287]. A young, dividing cell contains numerous small vacuoles which are formed from the ER. During the plant development stage, they coalesce to form one large vacuole which may occupy up to 95% of the total volume of a mature cell in a harvested fruit or vegetable. Vacuoles contain inorganic salts, anthocyanins, acids, sugars, phenols and enzymes [30, 185, 287]. In fruits, the pH of the vacuole generally ranges from about 3 to 4.5 whereas the pH of the surrounding cytosol is around 6 to 7.5. In lettuce, pH values of the vacuole and cytoplasm have been estimated to be 5.7 and 6.7, respectively [262]. The vacuolar enzymes may breakdown compounds in plant organs during postharvest storage. If the tonoplasts are broken during storage of fruits and vegetables, undesirable chemical reactions leading to quality deterioration may occur. For example, phenolic compounds in vacuoles may come in contact with phenolases in the cytosol upon tonoplast rupture with the consequence of phenolic oxidation to brown pigments [272].

The plasmalemma, tonoplast, endoplasmic reticulum, chloroplast, nucleus, mitochondrium [109], and microbodies have similar structural features [33, 109, 120, 173, 225, 247]. As viewed under the electron microscope, each of these membranes appear as two electron-dense (dark) layers separated by an electron-transparent (light) layer and has a thickness of between 7.5 to 10 nm. About one-half to two-thirds of the dry weight of a membrane is protein and the remainder is lipid. The proportions of proteins and lipids vary from membrane to membrane in the parenchyma cell.

The principal lipids of plant cell membranes are phospholipids, glycolipids and sterols [109, 120, 173]. The most abundant phospholipids are phosphatidyl choline, phosphatidyl ethanolamine, phosphatidyl glycerol and phosphatidyl inositol. The phospholipids are organized with their hydrophobic tails interacting with each other through van der Waals forces to form a lipid bilayer. The hydrophilic portions of the phospholipids are on the two surfaces of the bilayer and interact with cell water. Sterols (about one-fifth of the weight of phospholipids) are scattered amongst the phospholipids in a parallel fashion, and function as stabilizers of the hydrophobic interior. Three different types of proteins existing in the membranes of parenchyma cells are enzymes, carriers and structure proteins.

The structural integrity of membranes is dependent on the presence of calcium ions which act as bridges between phosphate portions of phospholipids. A deficiency of calcium in membranes can lead to higher solute diffusion rates, an increase in respiration rate and elevation of ethylene production rate [33, 87, 176]. Stabilization of the plasmalemma with calcium may inhibit the conversion of 1-aminocyclopropane-1-carboxylic acid (ACC) to ethylene by the ethylene-forming enzyme [176]. Bangerth *et al.* [14] have suggested that calcium in the membrane structure of tonoplasts limit substrate diffusion to the cytoplasm.

7.4.3 Plant tissues

The tissues of fruits and vegetables are classified as dermal, fundamental and vascular. These tissues are responsible for specific colour, flavour and textural characteristics of horticultural commodities [83, 86]. The dermal tissue provides an outer protective covering for underlying tissue. In the primary stage, the plant organs (e.g. asparagus stem) possess dermal tissue called epidermis having small, thin-walled cells as structural units. During secondary plant growth, the epidermis may be replaced by a periderm which is composed of cork cells (e.g. skin of potato tubers).

Fundamental tissues have been categorized as parenchyma, collenchyma and sclerenchyma [83, 86, 115, 299]. Parenchyma is regarded as ground tissue in which vascular and other types of tissue are imbedded. Some parenchyma cells may be modified to form thicker primary cell walls and are the units of supporting tissue called collenchyma. Other modifications of parenchyma cells bring about the formation of thick, hard secondary walls which contain up to about 28% lignin. These cells are grouped together in the form of long fibres and sclereids. This tissue called sclerenchyma is resistant to physical stress. Fibres of the sclerenchyma may be responsible for the stringy texture of vegetables [115].

Vascular tissue has been classified as phloem and xylem [82, 86, 115, 247, 299]. Phloem tissue consists of long sieve tubes and phloem fibres surrounded by metabolically-active companion and phloem parenchyma cells. Sieve tubes are made up of elongated living cells, called sieve elements, which are interconnected with each other at pore-containing junctures (sieve plates). The thick-walled phloem fibres provide structural stability for the sieve tubes. Phloem sieve tubes are responsible for the active transport of solutes such as amino acids, sugars and minerals from one point in a plant to another. When a sieve element is cut, the loss of pressure in the sieve tube results as a consequence of nutrient solution outflow [247]. Such loss initiates the blockage of the sieve pores by a phloem protein (P-protein). In general, around 90% of the solutes trans-ported in the phloem consists of non-reducing sugars, particularly sucrose. Xylem tissue, the water transporting vascular system, con-sists of four types of cells: vessel elements, fibres, xylem parenchyma cells and tracheids. The xylem parenchyma cells are the only ones in the living state. The xylem sap (very low solute content) is transported in the elongated tracheids and vessel ele-ments.

Roots, stems, shoots, leaves and fruits can be distinguished on the basis of the distribution and relative amounts of the vascular and fundamental tissues [83, 84, 115, 299]. The stem generally is made up of a centrally located pith (made up of parenchyma cells) sur-rounded by vascular tissue and a parenchymatous cortex located between the epidermis and the vascular tissue. On the other hand, the root may consist of only vascular and epidermal tissue and have no pith or cortex, particularly in plants with secondary growth. The predominant cells in the root are parenchyma. In the leaf, the vas-cular system is a network of vascular bundles (veins) which are made up of xylem and phloem. The bundles are embedded in photo-synthetic parenchyma called the mesophyll, which consists of chloroplast-containing cells. The epidermal tissue of the leaf is composed of guard cells of the stomata and cutin-coated cells. Again the majority of cells in the leaf are parenchyma. The edible portions of fruits are composed largely of parenchyma cells with small vascu-lar bundles spread throughout the fundamental tissue [299]. The wall of the ovary in mature fruits and vegetables is called the peri-carp which consists of the outer wall (exocarp, usually one cell thick), middle wall (mesocarp containing conducting tissues) and the inner wall (endocarp) surrounding locules with seeds [224].

7.5 Respiration in fruits and vegetables

7.5.1 Overview

Respiration in the cells of fruits and vegetables is the metabolic process involving the breakdown (catabolism) of complex organic compounds such as sugars, organic acids, amino acids and fatty acids into lower molecular weight molecules with the accompa-nying production of energy (ATP and heat) through oxidation-re-duction enzymic reactions [10, 65, 109, 116, 247, 301]. The principal carrier of free energy is ATP which is used to: (1) sustain metabolic reactions in cells; (2) transport metabolites, and (3) maintain cellular organization and membrane permeability. Moreover, compounds formed in respiration pathways can be utilized in the biosynthesis (anabolism) of amino acids, fatty acids, aromatic compounds and pigments which may be important in quality acceptance of fruits and vegetables. Aerobic respiration includes stepwise energy-releasing reactions in which organic compounds are oxidized to CO_2 and the absorbed oxygen is reduced to H_2O [42, 109, 247, 274, 312]. The summation of all the reactions for the aerobic respiratory breakdown of the most common substrate, glucose, is:

$$C_6H_{12}O_6 + 6 O_2 \rightarrow 6 CO_2 + 6 H_2O + \text{energy}. \tag{1}$$

When one mole of glucose is oxidized in the respiration process to CO_2 and H_2O, 36 moles of ATP are formed. Each mole of ATP possesses 31.8 kJ or 7.6 kcal of useful energy which is stored in the terminal phosphates of ATP and is released during hydrolysis. The total energy for the 36 moles would be 1140 kJ or 274 kcal. Theoretically, the total free energy change for the oxidation of one mole of glucose is 2870 kJ or 686 kcal. Thus the energy transformed into chemical energy in the form of ATP represents about 40% of the total energy output and about 60% is converted into heat. Certainly, the production of thermal energy during the respiration of stored fruits and vegetables is particularly important in quality deterioration which increases with an increase in temperature.

In the absence of microatmospheric oxygen, anaerobic respiration (fermentation) occurs in fruits and vegetables [64]. Under anaerobic conditions, much lower amounts of energy (2 moles of ATP = 64 kJ or 15.2 kcal) and CO_2 are formed from 1 mole of glucose than that under aerobic conditions. Further the metabolic end-products are lactic acid, acetaldehyde and ethanol from the degradation of pyruvic acid. Off-flavours sometimes are evident in fruits and vegetables stored in atmospheres with very low levels of O_2 (below 2%).

Aerobic respiration reactions in stored fruits and vegetables can be grouped into four reaction schemes, namely, glycolysis, pentose phosphate pathway, Krebs or citric acid cycle, and oxidative phosphorylation. On the other hand, the respiratory reactions in harvested plant organs held under anaerobic conditions are limited to the glycolysis and pentose phosphate pathway systems.

7.5.2 Glycolysis

Glycolysis (sometimes referred to as the Emden-Meyerhof-Parnas or EMP pathway) is the reaction scheme for the sequential enzymic breakdown of glucose or fructose to pyruvic acid with no oxygen entering the pathway and no CO_2 being released [109, 225, 247, 274, 311]. These reactions take place in the cytosol of plant cells. The overall reaction scheme can be written as:

$$C_6H_{12}O_6 + 2\ ADP^{2-} + 2H_2PO_4^- + 2\ NAD^+ \rightarrow 2\ \text{pyruvate} + 2\ ATP^{3-}$$

$$+ 2 NADH + 2H^+ + 2H_2O \qquad\qquad\qquad (2)$$

As such, the low energy adenosine diphosphate (ADP) is converted to a high-energy adenosine triphosphate (ATP) by phosphorylation with inorganic phosphate. The cofactor, nicotinamide adenine dinucleotide (NAD^+), is a hydrogen acceptor and is reduced to NADH by the addition of two electrons and a H^+. Such reduction occurs during the enzymic transformation of glyceraldehyde-3-phosphate to 1,3-bisphosphoglycerate.

For each hexose molecule converted into two molecules of pyruvic acid, two molecules of NAD^+ are reduced to two NADH which may pass into the mitochondria of a plant cell to be oxidized by O_2 or remain in the cytosol to provide electrons for synthesis of amino acids, fatty acids and other complex components of the cell. The oxidation of a cytosolic NADH molecule in a mitochondrion will produce only two molecules of ATP.

In glycolysis, two ATP molecules are decomposed to ADP in the phosphorylation of hexoses whereas four ATP molecules are produced per hexose molecule. Thus a net gain of two ATP molecules is evident in the conversion of hexose to pyruvate.

The glycolysis pathway for glucose is presented in Figure 7.1. The glucose is phosphorylated at position 6 with the transfer of a phosphate group from ATP by the enzyme, hexokinase, and then isomerized to fructose-6-phosphate. Thereafter, the fructose-6-phosphate is phosphorylated by phosphofructokinase to fructose-1-6-bisphosphate which is cleaved to form dihydroxyacetone phosphate and glyceraldehyde-3-phosphate in the presence of aldolase. Dihydroxyacetone phosphate is isomerized by triose phosphate isomerase to glyceraldehyde-3-phosphate. The catalytic action of glyceraldehyde-3-phosphate dehydrogenase along with NAD^+ as an oxidizing agent brings about the oxidation and phosphorylation of glyceraldehyde-3-phosphate to 1,3-bisphosphoglycerate. 3-phosphoglycerate is formed by the dephosphorylation of 1,3-bisphosphoglycerate in the presence of phosphoglycerate kinase with the creation of ATP. 3-phosphoglycerate is isomerized by phosphoglyceromutase to 2-phosphoglycerate which is converted to phosphoenolpyruvate along with a loss of H_2O. Finally, pyruvate is formed by the action of pyruvate kinase on phosphoenolpyruvate with the formation of another ATP.

7.5.3   Pentose phosphate pathway

An alternative reaction route for the conversion of glucose to glyceraldehyde-3-phosphate is the pentose phosphate pathway (in the cytosol) for the generation of reducing power by the reduction of nicotinamide adenine dinucleotide phosphate ($NADP^+$) to NADPH [109, 247, 274]. Only NADPH not NADH acts as an electron donor in reductive biosynthesis of complex compounds whereas NADPH and NADH can be oxidized in the mitochondrion to ATP. In the pentose phosphate pathway, NADPH is formed when glucose is oxidized through a series of reactions to ribulose-6-phosphate.

$$\text{Glucose-6-phosphate} + 2\ \text{NADP}^+ + H_2O \rightarrow \text{ribulose-5-phosphate} + 2$$

$$\text{NADPH} + 2H^+ + CO_2 \tag{3}$$

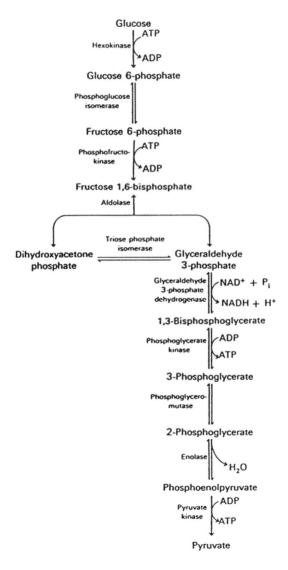

Figure 7.1.   Pathway of glycolysis [274].

The pentose phosphate pathway involves the dehydrogenation of glucose-6-phosphate to 6-phosphoglucono-δ-lactone with the production of a molecule of NADPH. The next reactions include the hydrolysis of the 6-phosphoglucono-δ-lactone to 6-phosphogluconate which in turn is oxidatively decarboxylated to ribulose-5-phosphate with another NADPH molecule formed and CO_2 released. The ribulose-5-phosphate can be isomerized by phosphopentose isomerase to ribose-5-phosphate and epimerized by phosphopentose epimerase to xylulose-5-phosphate. Ribose-5-phosphate and xylulose-5-phosphate interact through transketolase to form glyceraldehyde-3-phosphate and sedoheptulose-7-phosphate, and both reacting together (catalysed by transaldolase) to form erythrose-4-phosphate and fructose-6-phosphate. Erythrose-4-phosphate, is an essential starting compound for the synthesis of phenolic compounds such as anthocyanins [247]. Ribose-5-phosphate is the precursor of ribose and deoxyribose, to be used for the synthesis of RNA and DNA, respectively. Erythrose-4-phosphate and xylulose-5-phosphate can react in the presence of transketolase to form glyceraldehyde-3-phosphate and fructose-6-phosphate.

The importance of the glycolysis pathway and the pentose phosphate pathway varies from organ to organ, from species to species and with the stage of development of an organ. The pentose phosphate cycle is dominant in some mature fruits but as the ripening process progresses, the pentose phosphate cycle is replaced by the glycolysis pathway (e.g. banana).

7.5.4 Anaerobic respiration

Pyruvic acid, under anaerobic conditions, can be reduced to lactic acid by lactate dehydrogenase with NADH or can be decarboxylated to acetaldehyde by carboxylase with the formation of CO_2 [64, 247]. Acetaldehyde can be reduced by alcohol dehydrogenase in the presence of NADH to form ethanol and NAD^+. This regenerated NAD^+ can re-enter the glycolysis pathway as a oxidizing agent. Thus, it is clear why glycolysis can proceed under anaerobic conditions.

The microatmospheric oxygen concentration at which anaerobic respiration of a fruit or vegetable commences is known as the extinction point. The extinction point has been defined as the lowest O_2 concentration at which alcohol production ceases. Above this extinction point, sufficient oxygen is transported to the mitochondria in plant cells for aerobic respiration. Boersig *et al.* [26] has suggested the term, anaerobic compensation point, as an indicator of anaerobic respiratory transition, since alcohol is formed in some commodities under aerobic conditions. It is defined as the O_2 concentration at which CO_2 evolution is minimum.

7.5.5 Krebs cycle

When pyruvate migrates into the mitochondrial matrix from the cytosol of a cell, it is oxidatively decarboxylated with pyruvic acid dehydrogenase in an irreversible reaction to an acetyl unit which combines with coenzyme A (CoA) to form acetyl CoA [109, 225, 247, 274, 311].   The acetyl unit enters into the Krebs cycle (Figure 7.2) by condensing with oxaloacetate from the cycle to form citrate. Two carbons enter the Krebs cycle as an acetyl and two carbons leave as two CO_2 molecules.

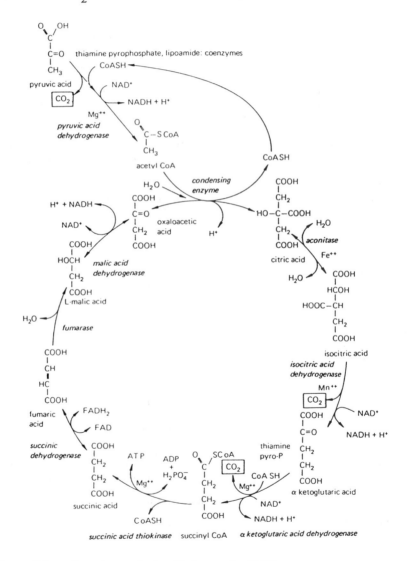

Figure 7.2.   Reactions of the Krebs cycle [247].

Oxidation-reduction reactions take place in the Krebs cycle with six electrons being transferred to three NAD^+ molecules and two electrons transferred to a flavin adenine dinucleotide (FAD) molecule. Although O_2 is not involved directly in the reactions of the Krebs cycle, the cycle can operate only under aerobic conditions in a commodity. The reason is that NAD^+ and FAD in the mitochondrion can be regenerated only by the transfer of electrons from NADH and $FADH_2$ to O_2 (in the oxidative phosphorylation pathway) with the formation of CO_2.

The overall reaction of the Krebs cycle is:

$$2 \text{ pyruvate} + 8NAD^+ + 2FAD + 2ADP^{2-} + 2H_2PO_4^- + 4H_2O \rightarrow 6CO_2$$
$$+ 8NADH + 2FADH_2 + 2ATP^{3-} + 8H^+ \qquad (4)$$

NADH and $FADH_2$ are energy-rich molecules, each of which possess two electrons with high transfer potential. For each of the eight NADH molecules, three ATP molecules are formed by oxidative phosphorylation in the mitochondrion, whereas two ATP molecules are produced for each of the two $FADH_2$ molecules. On top of this, two high-energy phosphate bonds in the form of ATP are generated by cleavage of the thioester bonds of succinyl coenzyme A. Thus a total of thirty high-energy phosphate bonds are created from the two pyruvates entering the Krebs cycle.

7.5.6 Oxidative phosphorylation

NADH and $FADH_2$, formed in glycolysis and the Krebs cycle, are high-energy molecules, each having a pair of electrons with a high transfer potential. When the electron pair in NADH and $FADH_2$ is transferred through a chain of electron carriers in the inner mitochondrial membrane to O_2, free energy is liberated and used to produce ATP from ADP and a phosphate ion [108, 247, 274]. The electrons can flow through the electron carrier chain because NADH and $FADH_2$ have a negative reduction potential (strong reductants) and O_2 has a positive reduction potential (strong oxidant). The reduced O_2 interacts with H^+ ions to form H_2O. This process of oxidation of NADH and $FADH_2$, electron flow through the electron-carrier chain and the phosphorylation of ADP to ATP is called oxidative phosphorylation.

The electron-carrier chain is made up of three enzyme complexes called NADH-Q reductase, cytochrome reductase and cytochrome oxidase and two interlinking mobile carriers called ubiquinone (Q) and cytochrome c. These are present in the inner mitochondrial membrane.   The enzymes contain electron-carrying prosthetic groups, namely, flavins, iron-sulfur (Fe-S) clusters, hemes and copper ions.   The Fe-S proteins (in NADH-Q reductase and cytochrome reductase) and cytochromes can receive or transfer only one electron at a time whereas flavoproteins and ubiquinone can receive or transfer two electrons.   A flavoprotein accepts electrons from a glycolytic NADH and passes them to ubiquinone.   The reactions in the electron-carrier chain are presented in Figure 7.3.

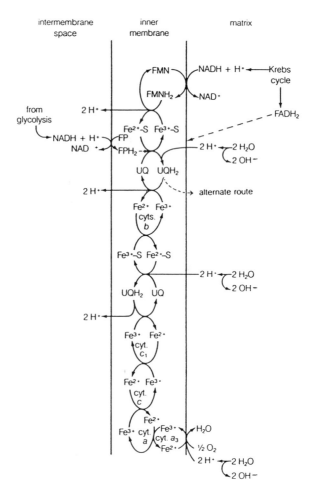

Figure 7.3.   Pathway   of   the   mitochondrial   electron-transport system [247].

The electrons from NADH of the matrix enter the chain at the NADH-Q reductase (NADH dehydrogenase) with the transfer of two electrons and two H^+ to the flavin mononucleotide (FMN) prosthetic group to produce $FMNH_2$ (see Figure 7.3). Next, the electrons of $FMNH_2$ are transferred to the Fe-S clusters which represent the other type of prosthetic group in the NADH-Q reductase with the ferric ion (Fe^{3+}) of the Fe-S cluster being reduced to a ferrous ion (Fe^{2+}) and the two H^+ ions being transported to the intermembrane space.

The NADH-Q reductase is the first of three proton pump assemblies that transfer protons (H^+) through the inner membrane to the intermembrane space (cytosolic side of the inner membrane) with the creation of a proton-motive force made up of a pH gradient and transmembrane electric potential. ATP is formed when two protons ($2H^+$) flow to the matrix through the ATPase complex. The electron of Fe^{2+} of a Fe-S cluster in NADH-Q reductase is transferred to ubiquinone along with H^+ to form semiquinone which then is reduced by a second electron and H^+ to ubiquinol (QH_2). The two electrons from $FADH_2$ formed in the Krebs cycle are transferred to Fe-S clusters of NADH-Q reductase and subsequently to ubiquinone (Q) along with $2H^+$ to form QH_2 for entry into the electron-carrier chain.

The second proton pump assembly in the chain is cytochrome reductase which catalyses the transfer of electrons from QH_2 to cytochrome c (this water-soluble protein is not attached to the membrane) and brings about the transfer, across the inner mitochondrial membrane, of H^+ to the intermembrane space. Another ATP molecule will be formed when two H^+ pass to the matrix through the ATPase complex. The cytochrome reductase contains a Fe-S protein, and cytochromes b and c_1, which are responsible for electron transport from QH_2 to cytochrome c.

The third and last proton pump assembly is the cytochrome oxidase which possesses cytochrome a and a_3, and copper ions. This enzyme catalyses the transfer of electrons from reduced cytochrome c to O_2 and the transport of the protons (H^+) to the intermembrane space.

As shown in Figure 7.3, three pairs of H^+ are pumped into the intermembrane space when the electrons pass through the electron-transfer system from a NADH molecule of the Krebs cycle to O_2

[247].   When these three pairs of H^+ are transferred back to the matrix through the ATPase complex, three ATP molecule are formed. With the oxidation of $FADH_2$ and glycolytic NADH, only two pairs of H^+ are pumped into the intermembrane space and thus only two ATP molecules are formed from these high-energy molecules.

Electrons are not usually transported through the electron-transfer chain to O_2 unless ADP is available for phosphorylation to ATP.   Thus the rate of oxidative phosphorylation is determined by the concentration of the matrix ADP.   Further, the reaction rate in the Krebs cycle will be influenced by the ADP level in the matrix because of the requirement of NAD^+ and FAD as incoming products from the electron-transfer chain.   ADP coming into the inner mitochondrial membrane is exchanged with ATP from the matrix in an antiport system.   ATP readily diffuses through the outer mitochondrial membrane to the cytosol.

7.5.7   Heat of respiration

Mitochondria have an alternate route (cyanide-resistant or cyanide-insensitive respiration) in the electron-transfer pathway with the branch-off at ubiquinone (Figure 7.3).   In this alternate route, electrons are transferred through a flavoprotein and a terminal oxidase to O_2 [170].   The free energy is converted into heat energy but is not used for the formation of ATP.   Thus, uncoupling of oxidative phosphorylation is a means of generating heat.   Toledo *et al.* [278] calculated the heat of respiration of shelled peas, sweet corn and apples in air and optimum MA.   He found that low O_2/high CO_2 levels lowered the respiration heat values of the commodities to about 30% of those in air.

7.5.8   Respiration rate

The rate of respiration of a fruit or vegetable is an indication of the rate of: 1. catabolic changes of metabolites of ATP production; and, 2. quality deterioration.   The respiration rate can be roughly related to the potential storage life of a commodity [115, 245]. The aerobic respiration rate of fresh fruits and vegetables is highest at the immature stage (cell division and enlargement) and gradually decreases as the maturation process proceeds.   However, with some fruits at the beginning of the ripening stage, the rates of respiration and associated metabolic processes (e.g. ethylene production) increase dramatically and subsequently decrease with storage time [10].   This rise in respiration rate is known as climacteric.   Some fruits and all vegetables except tomatoes do not display a climacteric at any time during ripening [10].   On the basis of respiration patterns, fruits have been classified as climacteric and non-climacteric as shown in Table 7.4 [10, 115, 217, 299, 301].   During ripening, the climacteric fruits synthesize larger amounts of ethylene than do the non-climacteric fruits.

Table 7.4.   Climacteric   and   non-climacteric   fruits   and   vegetables

Climacteric commodity	Non-climacteric commodity
Apple	Blueberry
Apricot	Cherry
Avocado	Cucumber
Banana	Grape
Mango	Grapefruit
Papaya	Lemon
Passion fruit	Melon
Peach	Orange
Pear	Pineapple
Plum	Strawberry
Tomato	

If carbohydrates such as glucose and fructose are completely oxidized in the aerobic respiration process of fresh commodities, the volume of O_2 uptake per unit time is about equal to the volume of CO_2 released per unit time.   However, if organic acids are oxidized in fruits and vegetables, more CO_2 is generated than the volume of O_2 consumed.   The respiratory quotient (RQ), the ratio of the CO_2 released per unit time to the O_2 consumed per unit time, is useful for assessing the types of substrates used in the respiration process of a fruit or vegetable.   The RQ for sugars, long-chain fatty acids and short-chain organic acids are about 1.0, 0.7 and 1.3, respectively. However, if several substrates are involved in respiration, the RQ would be a combined value.   During the storage of a fresh commodity, the change in the types of respiratory substrates can be assessed by the alteration in the RQ.

The rate of respiratory enzymic reactions increases with the increase in temperature of a fruit or vegetable from about -1 to 35°C [245, 301].   The temperature quotient (Q_{10}) of a fruit or vegetable has been used to predict the rate of respiration (R_2) at a temperature T_1 (°C) in relation to the respiration rate (R_1) at a temperature 10°C below T_1 $(T_1 - 10°C)$ [245].   The equation for Q_{10} is:

$$Q_{10} = R_2/R_1$$

For temperature intervals other than 10°C, the following equation for Q_{10} can be used:

$$Q_{10} = (R_2/R_1)^{10/(T_2 - T_1)}$$

where T_2 (higher) and T_1 (lower) are any temperatures in degrees Celsius and R_2 and R_1 are the respective rates of respiration. Generally, the difference between T_2 and T_1 should not be greater than 10°C to ensure realistic values [245].

The Q_{10} values for fruits and vegetables are temperature dependent and vary from about 1 to 7 in the temperature range of 0 to 30°C. Many fresh commodities have Q_{10} values higher in the 0 to 10°C storage temperature range than those in the 10 to 25°C range. Some Q_{10} values for vegetables at various storage temperatures are presented in Table 7.5 [245].

The rates of respiration in mg CO_2/kg/h for several fruits and vegetables at 5°C are presented in Table 7.6. It is of interest to note that the respiration rate varies widely with different cultivars within a commodity. Robbins et al. [236] found that with four cultivars of raspberries, the respiration rate ranged from 162 to 264 mL CO_2/kg/h and was positively correlated with ethylene evolution. Raspberry cultivars with low respiration rates should be preferable for the fresh market since the rapidly respiring cultivars had a higher incidence of rot and were less firm.

Table 7.5.   Temperature quotients (Q_{10}) for vegetables

Commodity	Temperature, °C			
	0-5	5-10	10-15	15-20
Asparagus	3.3	4.2	1.2	2.3
Broccoli	5.2	4.6	3.9	2.7
Brussels sprouts	4.9	2.7	1.5	---
Carrots	2.0	2.3	1.5	2.8
Cauliflower	1.8	2.4	1.7	2.7
Celery	3.5	4.0	1.7	---
Lettuce, leaf	2.0	1.7	2.3	2.3
Lettuce, Romaine	---	2.7	1.5	2.2
Melon, cantaloupe	5.2	2.5	4.6	1.7
Radishes	2.2	3.4	2.5	2.1
Spinach, winter	6.8	2.8	2.8	---
Sweet corn on cob	6.0	1.9	2.0	2.8
Tomatoes, ripening	---	---	---	2.1

From Ryall and Liston [245]

Table 7.6.   Respiration rates of fruits and vegetables

Respiration rate at 5°C (mg CO_2/kg/h)	Commodities
5-10	Apple, grape, potato, onion
10-20	Apricot, carrot, cabbage, cherry, lettuce, peach, plum, pepper, tomato
20-40	Blackberry, cauliflower, lima bean, raspberry, strawberry
40-60	Brussels sprouts, green onion, snap bean
>60	Asparagus, broccoli, mushroom, pea, spinach, sweet corn

From Kader [149]

When tissue is damaged by peeling and slicing, the respiration rates may increase appreciably.   According to McLachlan and Stark [190], broccoli and carrots held at 10 and 20°C had higher respiration rates when peeled and cut (Table 7.7).   Cutting the whole heads of broccoli into individual florets brought about an increase of about 40% in respiration rate at both 10 and 20°C.   The respiration rate of carrots was increased more than 7 times when they were julienne sliced.

Table 7.7.   Respiration rates of cut and uncut vegetables at 10 and 20°C

Vegetable	Respiration rate (mL CO_2/kg/h)	
	10°C	20°C
Broccoli:		
cut florets	78	147
uncut heads	59	104
Carrots:		
julienne	65	145
peeled, whole	12	26
unpeeled, whole	9	26

From McLachlan and Stark [190]

7.6   Ripening   of   fruits

Ripening, a term confined to fruit and fruits consumed as vegetables (e.g. tomato, peppers, egg-plant), occurs within the maturity stage and sometimes at the beginning of the senescence stage [110, 173, 233, 301]. Ripening represents a series of catabolic and anabolic processes which transform the mature fleshy ovaries into edible commodities with acceptable colour, flavour and texture [9, 10, 173]. Some of the metabolic processes in ripening fruit are presented in Table 7.8.

Many fruits such as cherries, raspberries and oranges are not harvested until they have reached a level of ripeness for optimal quality acceptability whereas other fruits such as bananas and apples may be harvested in a pre-ripe condition to prolong storage time and obviate physical tissue damage (e.g., bruising) during harvesting, handling and shipping [155]. At the onset of ripening, some fruits undergo a rapid rise (climacteric) in respiration rate with the concomitant formation of the plant hormone, ethylene, and ATP to be used as an energy source in the ripening enzyme synthesis and other anabolic and catabolic reactions [9, 10, 37, 164].

Table 7.8.   Changes   occurring   during   fruit   ripening

Degradative	Synthetic
Degradation of chloroplast	Maintenance of mitochondrial structure
Breakdown of chlorophyll	Formation of carotenoids and anthocyanins
Starch hydrolysis	Interconversion of sugars
Destruction of organic acids	Increased TCA cycle activity
Inactivation by phenolic compounds	Synthesis of flavour volatiles
Hydrolysis of pectins	Increased amino acid incorporation
Activation of hydro-lytic enzymes	Increased transcription and translation
Initiation of membrane leakage	Preservation of selective membranes
Ethylene-induced wall softening	Ethylene production

From Baile and Young [10]

7.6.1 Ethylene synthesis and action

Ethylene is produced from methionine by a pathway which involves: (1) the formation of S-adenosyl-methionine (SAM) in the presence of ATP; (2) the conversion of SAM to 1-aminocyclopropane-1-carboxylic acid (ACC) catalysed by ACC synthase; and, (3) ACC oxidation (O_2 required) and deamination to ethylene. An ethylene-forming enzyme (EFE) is regarded as essential for conversion of ACC to ethylene [10, 15, 33, 289, 307]. Lieberman [175] has suggested that the enzymes for ethylene synthesis are located in the plasmalemma. Kende *et al.* [159] found that ethylene can be synthesized in vacuoles by an ethylene-forming enzyme system. Since ACC is not synthesized in the vacuole, a transport system must be involved for efficient ACC transfer to the vacuole.

The maximum ethylene production rates for various fruits is presented in Table 7.9. As the respiration rate rises in climacteric fruits and fruit vegetables at the onset of ripening, the rate of ethylene synthesis increases from around 0.1 µL/kg/h to 10 to 100 µL/kg/h and has a profound influence on the ripening process. Non-climacteric fruits and vegetables synthesize ethylene at a rate of 0.1 to 1 µL/kg/h. The relationship between ethylene production and the onset of respiratory rise in climacteric fruits is variable with the respiration rise preceding ethylene production in some fruits (e.g. banana, honeydew melon, muskmelon), with ethylene production coinciding with the rise (e.g. apricot) or in other fruits (e.g. mango, tomato), with ethylene production preceding the respiratory rise [10]. Baile and Young [10] suggested that the endogenous ethylene produced within the climacteric fruits is not the initiating factor for the induction of the respiratory climacteric.

Table 7.9.   Maximum ethylene production rates by fruits and vegetables at 20°C

Range (µL/kg/h)	Commodity
0.01-0.1	Cherry, citrus, grape, strawberry, asparagus, cauliflower
0.1-1.0	Blueberry, cucumber, pineapple, raspberry, pepper
1.0-10.0	Banana, honeydew melon, mango, tomato
10.0-100.0	Apple, avocado, cantaloupe, nectarine, papaya, peach, pear, plum

From Kader [149]

Non-climacteric fruits can be divided into those which respond to exogenous ethylene and those which are unresponsive [164]. The responsive fruits are the citrus and cucurbit families (e.g. oranges, cucumbers). Unresponsive fruits include the strawberry, pineapple, cherry, grape and bell pepper. The treatment of pre-climacteric fruits with low levels of exogenous ethylene (1 to 10 ppm) accelerates the onset of the climacteric and is only effective before the influence of endogenous ethylene [164]. However, the shape of the curve of the climacteric cycle is not altered [10]. Ethylene removal from pre-climacteric fruits can delay ripening. Non-climacteric fruits are responsive to exogenous ethylene at any point in the ripening period when the ripening rate increases as the concentration of exogenous ethylene increases to a maximum level [9, 10].

Ethylene can bring about undesirable changes in some vegetables from the standpoint of loss of chlorophyll in leafy vegetables, formation of a bitter compound (isocoumarin) in carrots, lignification in asparagus (stringiness) and russet spotting on lettuce [115, 116, 295]. The increased activity of phenylalanine ammonia lyase and the enhanced synthesis of phenolic compounds may be implicated in the ethylene-induced quality deterioration of some of these commodities [1]. Phenylalanine ammonia lyase catalyses the deamination of phenylalanine to cinnamic and coumaric acids, both of which serve as intermediates for the formation of bitter compounds and lignin. Hicks et al. [129] noted that, at a 5 ppm level of ethylene in the microatmosphere of stored cabbage induced a significant loss in green colour during a 1-month period. When packaged spinach was held in air with 10 ppm of ethylene at 25°C, chlorophyll a and b decreased rapidly within a 4-day storage period [295]. At day 4 of storage, the chlorophyll a and b contents of the ethylene-treated spinach were about one-half those with no ethylene treatment (air only). About 4 ppm of ethylene can cause considerable yellowing in Brussels sprouts, broccoli, cauliflower and cabbage [295]. The mechanisms for chlorophyll degradation in produce exposed to ethylene may involve chlorophyllase and peroxidase-hydrogen peroxide system [295].

Oxygen in plant cells is required for the biosynthesis of ethylene and for ethylene action, such as ripening [303, 304]. When O_2 is absent in tissue, ethylene biosynthesis does not occur [305]. Oxygen-dependent reactions are involved the conversion of ACC to ethylene and the recycling of 5-methylthioadenosine to methionine [289]. Thus, the rate of ethylene synthesis has been found to be directly related to O_2 concentration [163]. The ethylene-forming enzyme has a half-maximal activity at about 2% O_2 [164]. Banks et al. [16] found that when apple slices were stored in 6% or lower O_2, ACC accumulated and ethylene production was reduced. Apples, stored in 3% O_2, synthesized two to three times less ethylene per unit time compared to those held in air [289]. Moreover, O_2 is also required for ethylene action, and the sensitivity of fruit to ethylene drops as the O_2 level declines below 8% [38, 39]. Presumably O_2 must be bound at or near the ethylene receptor site before ethylene can bring about ripening. When O_2 decreases to 3%, the binding of ethylene is reduced to about 50% of that in air [38].

In leaf tissue, CO_2 can increase ethylene production at levels below 1%, whereas for fruits and tubers the saturation levels would be somewhat higher.   Moderately elevated CO_2 levels inhibit, promote or have no effect on the synthesis of ethylene [1].   However, at about 10% CO_2, the hormonal activity of ethylene in most fruits can be prevented at around 1 μL/L of ethylene and thus ripening can be retarded.   Burg and Burg [38, 39] have suggested that CO_2 acts as a competitive inhibitor with the ability to displace ethylene from the binding site of a metalloenzyme.   On the other hand, Beyer [23, 24] theorized that since ethylene at the receptor site can be oxidized by cellular enzymes to ethylene oxide, CO_2 may act as a feedback or mass-action inhibitor.   Chaves and Tomas [50] found that a short exposure of apples to 20% CO_2 decreased ethylene production and increased ACC and thus suggested that high CO_2 levels can inhibit the ethylene-forming   enzyme.

7.6.2   Softening of fruit tissue

The softening of fruit tissue during ripening is important from the standpoint of textural acceptability [19, 33, 113, 119, 144]. The alteration of polymeric carbohydrates in the middle lamellae and primary cell walls of ripening fruits is responsible for tissue softening [21, 33, 132, 144, 146, 224].   The main components in the middle lamella are pectic polysaccharides whereas the primary cell wall contains about 34% pectic polysaccharides [5, 19 144 271].   The pectic polysaccharides in the middle lamellae of some fruits presumably undergo polygalacturonase hydrolysis to form water-soluble pectic substances during ripening [19, 21, 33, 60].   A correlation between the appearance of polygalacturonase isoenzymes in ripening tomatoes and a dissolution of the middle lamellae has been reported [60].   Ultrastructural studies of apple and pear tissue indicated the dissolution of the middle lamellae as ripening progressed [21].

Cell wall degradation of some ripening fruit occurs in concert with the formation of hydrolases such as endopolygalacturonase, exopolygalacturonase and cellulase [33, 112, 132, 216, 280].   In ripe apple tissue, only exopolygalacturonase is present whereas in ripe strawberries, polygalacturonase activity has not been detected [133]. The above-mentioned hydrolases in the cell walls of ripe fruit are usually synthesized at the beginning of the ripening period [33, 144].   Ethylene presumably is involved in the triggering of transcription and translation in the formation of proteins, particularly enzymes, in climacteric fruit [112, 144].   Other types of enzymes are synthesized in cells of some ripening fruits such as strawberries for the formation of soluble polyuronides which may be transferred to the cell walls and thereby reduce the structural integrity [133]. Such infusion of soluble polyuronides into the cell walls may be responsible for the softening of tissue in ripening strawberries without enzymic breakdown of pectic polysaccharides.

Membrane structures of the plasmodesmata, tonoplast, mitochondrium, endoplasmic reticulum and chromoplasts remain essentially intact in ripening tomatoes, apples and pears [21, 60, 189, 219].

7.7 Strategy for prolonging the shelf life of fresh fruits and vegetables

As discussed in a previous section, the edible tissue of fresh fruits and vegetables is composed of interconnected, turgid parenchyma cells with compartmentalized chemical constituents. The pectic components in the protopectin of the middle lamella are required to provide cell-cell adhesion which contributes to the textural characteristics of the produce.   The cell wall constituents are organized in a systematic pattern to provide a firm, cohesive, pliable protective coating for the protoplast of each cell and to endure high osmotic pressures.   The turgid cell walls contribute to the firmness, crunchiness and crispness (popping sound) of tissue.   The pleasant colour of fruits and vegetables can be attributed to pigments such as chlorophyll in the chloroplast, carotenoids in the chromoplasts and anthocyanins in the vacuoles.   The typical odours of commodities have been attributed to numerous volatile aromatic compounds present in the vacuole.   Acids and sugars, which are responsible for the taste attributes of produce, are located in the vacuole.   The complex carbohydrate, starch, exists in mature cells as starch granules which are commonly decomposed in fruit tissue to sugars during the ripening period.   In some vegetables such as sweet corn, the sugars are condensed into starch during storage with the loss of acceptable texture and sweetness.

In the course of ripening and early stages of senescence, fruits and vegetables gradually lose their quality attributes and eventually unacceptable levels are reached.   The tissue becomes soft due to enzymic changes of the pectic components in the middle lamellae and in the cell walls.   As such, the tissue no longer has a desirable rigidness for resistance-to-bite and crunchiness.   With a loss of tensile strength, cell walls may balloon outward and the tissue may lose its turgidity and crispness.   Membranes such as the tonoplast may weaken and break with the consequence of weeping of fluid from the tissue (e.g. raspberries).   As respiration proceeds, sugars and acids will be metabolized to the extent that sweetness and sourness are diminished.

To reduce the rate of quality deterioration of stored fruits and vegetables, physical and chemical treatments of tissue are required to impede catabolic and anabolic processes such as respiration, ripening and senescence so that the native intact tissue structure and the just-picked chemical composition can be maintained as much as possible.   Further, it should be noted that intact tissue is resistant to the invasion of spoilogens and thus microbial decomposition of tissue can be limited by preserving the structural integrity of produce.   Some aerobic respiration may be required to provide sufficient ATP and NADPH for the synthesis of pigments and flavour compounds and for the maintenance of membrane structural integrity.

With the knowledge that modified atmospheres (MA) consisting of specific levels of CO_2 and O_2 can inhibit respiration rate, ripening and microbial growth in fresh fruits and vegetables, package systems may be designed and constructed to accommodate commodities (which may have been minimally processed previously) for prolonged storage.   The gas and moisture permeabilities as well as the physical properties of the packaging material must meet the specifications for: (1) maintaining definitive levels of CO_2 and O_2; (2) obviating gas pressure build-up; (3) preventing tissue bruising and crushing; (4) restricting tissue transpiration; and, (5) inhibiting the condensation of water vapour on the interior surfaces.   The above design model is called a modified atmosphere packaging (MAP) system.

The MAP system can be useful for preserving the freshness of: (1) whole large fruits and vegetables such as oranges, apples, peppers and head lettuce; (2) small soft fruits such as raspberries, strawberries and cherries; (3) vegetable pieces such as shredded lettuce, and broccoli and cauliflower florets; and, (4) table-ready cut fruits such as pineapple rings, apple slices and orange segments.

7.8   Dimensions of modified atmosphere packaging

From the standpoint of fresh fruits and vegetables, a modified atmosphere (MA) is regarded as a gas mixture which has a composition different from that of air, and which surrounds the produce to bring about beneficial effects, for extending the shelf-life of the commodities [150, 152, 154, 178, 309].   The mixture is made up primarily of O_2, CO_2 and N_2 with perhaps small amounts of noble gases. Controlled atmosphere (CA) storage is considered to be a process of introducing a specific MA around fresh produce and maintaining the specific composition within a sealed storage chamber during the designated storage period [70, 150, 178, 245].

The first scientific study on the effect of MA on the storage life of horticultural products was carried out by the French chemist, J.E. Berard, in the 1800s [62, 70].   He found that fruits in an atmosphere lacking O_2 did not ripen.   In the 1920s, Kidd and West in England initiated research on the influence of O_2 and CO_2 levels in the MA of refrigerated apples, pears and berries on the respiratory activity, ripening and storability [70].   Their results indicated that low O_2 and moderately high CO_2 levels in the MA reduced the quality deteriorative reaction rates.   R.M. Smock at Cornell University, with knowledge on postharvest MA physiology, helped to perfect the technology of CA fruit storage in the 1940s [62, 178, 266].   Thousands of reports have been published on the effectiveness of MA on the lengthening of the storage life of fruits and vegetables [70, 178].

From such studies, recommended MA conditions for a variety of commodities has been assembled [150, 168]. Such MA knowledge along with the principles of postharvest respiration, transpiration, ripening, senescence and natural defences against microbial attack has been used to provide a sound scientific basis for MAP technology of fresh fruits and vegetables. The physiological response of a fruit or vegetable to a recommended equilibrium MA in a suitable package system should be similar to the response of the commodity in CA storage with the same recommended MA composition and temperature. However, in the case of an MA packaged commodity, the establishment of the recommended equilibrium MA may take several days, particularly if a gas flush into the package headspace is not employed.

In the 1950s and early 1960s, several reports were published on the effectiveness of polymeric film as packages for extending the storage life of fresh commodities [77, 121, 227, 279]. Polymeric films such as polythene, polystyrene, pliofilm, cellulose and cellulose acetate were used in the intact and perforated forms with the goal of reducing O_2 and increasing CO_2 levels in the microatmosphere of the commodity-containing packages [227, 279]. In 1962, Tomkins [279] presented the results of a study on the gas compositional changes in polymeric film-covered trays containing respiring apples during storage at various temperatures. During the storage of the apples in polythene film-covered trays at 15°C, the CO_2 content in the microatmosphere rose rapidly over the first 50 h and gradually levelled off over the next 25 h at which point the rate of CO_2 produced by respiration was equal to the rate of CO_2 diffusing through the polymeric film to the environmental air. The O_2 level in the microatmosphere fell continuously over a period of about 160 h before levelling off to an equilibrium value. Tomkins [279] also noted that, with more apples per unit package volume, the equilibrium CO_2 level increased and O_2 concentration dropped as low as 0%. When the temperature of the apples in the polythene film-covered trays was increased from 0 to 20°C, the equilibrium CO_2 levels rose up to four times greater. The effect of various microatmospheres on the quality maintenance of fruits and vegetables has been examined by Hardenberg [122].

Modified atmosphere packaging (MAP) of fresh fruits and vegetables is a process involving: 1. the introduction of whole, separated, segmented or cut produce into a package system with either air or an input gas mixture (gas flush) having a specific MA composition, as the initial surrounding microatmosphere; 2. the closure (sealing) of the package system; and, 3. the refrigerated storage of the packaged produce at -1 to 12°C. During the first few days of storage, the produce with an oxygen-containing microatmosphere progressively converts the O_2 to CO_2 by means of respiration reactions [154, 227, 309, 310]. The objective in the design of a MA package system is to achieve and maintain a suitable equilibrium MA (low O_2 and elevated CO_2 contents) around the produce during storage for optimum retention of quality attributes and inhibition of microbial growth. Reduced O_2 and elevated CO_2 in the microatmosphere around a fruit or vegetable reduce respiration rate, ripening and microbial growth; the effects of low O_2 and high CO_2 contents are additive for both respiration and ripening control [154, 309]. Unsuitable MA may be responsible for inducing physiological damage, preventing wound healing, enhancing the senescence process and producing off-flavour compounds. Generally O_2 levels lower than about 1% in the microatmosphere will bring about anaerobic respiration [26]. CO_2 levels of 10% and above can inhibit the growth of spoilogens, but may cause physiological damage to CO_2-sensitive commodities [82, 150, 152, 153].

The term, commodity-generated or passive MAP, has been used in the literature to designate the matching of commodity respiratory characteristics with the gas permeabilities of package system so that a suitable equilibrium MA can passively evolve through the consumption of O_2 and the evolution of CO_2 in the respiration process [154, 168, 309].

An active modified atmosphere, on the other hand, can be established by withdrawing air from the package system with vacuumization and by back flushing with a selected gas mixture. The advantage of active modification of the microatmosphere is the rapid establishment of a desired gas mixture. Adsorbents and absorbents may be included in the package system to reduce the contents of O_2, CO_2, ethylene and water vapour [20, 154, 168].

With the advent of engineered plastic polymeric films with a wide range of gas permeabilities, beneficial modified atmospheres can be attained with fresh fruits and vegetables in flexible film package systems [210]. Further, the development of microporous polypropylene breathable membranes for attachment to package films (over small punch openings or windows) has been helpful in the design of package systems for high respiring commodities such as broccoli and cauliflower. Microperforated plastic film for over-wrapping whole fruits and vegetables has been used to provide relatively high air flow into the microatmosphere and thus ensure adequate O_2 for aerobic respiration even under temperature abuse (15°C or higher) conditions [100, 205].

When produce is present in a package system within which an optimum equilibrium MA exists, the following benefits would be expected:

    (1)   reduction of respiration rate
    (2)   lowering of ethylene production

(3) inhibition of the initiation of ripening
(4) decrease in rate of ripening and senescence
(5) reduction of tissue water loss and maintenance of cell turgidity
(6) minimization of nutrient decomposition
(7) inhibition of microbial growth and spoilage
(8) reduction of specific physiological disorders such as chilling injury
(9) maintenance of membrane and cell wall integrity and natural barriers against microbial invasion
(10) encouragement of wound healing

From the above-mentioned benefits of MAP for fruits and vegetables, several commercial advantages can be expected [2, 177]. These include:

(1) extension of shelf-life with retention of quality attributes
(2) lowering spoilage and waste
(3) retaining the freshness of produce without preservatives or irradiation
(4) harvesting fresh produce at consumer-acceptable ripeness levels
(5) reducing transit costs by providing additional storage time to transport produce by truck, train or ship rather than by air
(6) providing ready-to-eat cut product
(7) conserving energy by excluding energy-intensive processes (e.g. freezing, thermal sterilization).

Are the benefits of MAP fruits and vegetables apparent to consumers? A study by Agriculture Canada [4] was initiated to study the consumer perceptions of MAP food products. The respondents' initial spontaneous reaction to MAP was that it was a good idea (about 47% of respondents), and was valuable for longer storage of products (18% of respondents). Three distinct clusters of respondents were identified: convinced budgeters (36%), leaners (34%) and concerned traditionalists (30%). The convinced budgeters, being cost-conscious of their food budget, agreed on the advantages of MAP. The leaners, although positive towards MAP, did not believe that MAP offered any significant advantage over current packaging. The perceptions of the concerned traditionalists towards MAP were the harmful effects of MAP on the environment and themselves. The results of the study implied that a significant segment of the Canadian population would pay premium prices for MAP products, particularly fresh produce and single-serving products. Some MAP opportunity targets for horticultural products were identified as lettuce, strawberries, peaches, cauliflower, mushrooms, blueberries, raspberries, broccoli and pears.

7.9 Complementary factors to modified atmosphere packaging

To successfully achieve the goals of MAP of fruits and vegetables, several factors as well as suitable MA in the headspaces of package systems must be considered. These factors include: (1) cultivar and maturity selection for required quality attributes and physical properties; (2) harvest and postharvest handling; (3) spoilogen contamination; (4) minimal pretreatment processing; and, (5) storage temperature effects.

7.9.1 Cultivar and maturity selection

For prolonging the shelf life of fresh commodities under MAP, high-quality standards must be established for the input produce. Colour, texture, flavour and tissue integrity are some quality factors that must be considered [255, 256]. These attributes are governed by cultivar, growing conditions, and maturity at harvest [85, 213]. Considering a commodity, the cultivar is the major factor (as compared to agronomic practices and weather) in quality variability since the genetic make-up of the plant determines the structural and chemical features of importance as they relate to quality attributes of fresh-market produce. Wasserman [294] has pointed out that the correct molecular targets (enzymes), which are related to commodity quality and storability, must be identified prior to plant genetic manipulation. It appears that genetic engineering has promise as a means of improving the quality of fruits and vegetables. With respect to new disease-resistant crop plants developed through genetic engineering, the safety of the crops must be investigated thoroughly [90].

Although selection of cultivars on the basis of harvest quality is essential, product stability during handling, storage and distribution are important selection criteria [241]. Cultivars with low respiration rates should be considered for MAP. Robbins et al. [236] found that low respiring raspberry cultivars produced less endogenous ethylene and had a lower degree of rot than high respirers during storage at 0°C for periods up to 36 days.

Certainly, if fruits and vegetables with optimum eating quality could be stored without any appreciable deterioration, then these commodities should be harvested at maturity levels where they are at the peak-of-perfection [213]. MAP may be helpful in retarding the deterioration of commodities harvested at these maturity levels to prolong the storage life. However, if MAP produce is expected to survive as acceptable-in-quality under temperature-abuse conditions, rough handling treatment and long storage periods, then slightly immature pre-climacteric fruits (e.g. apples) and fruit vegetables (e.g. tomatoes) should be considered as input commodities. When slightly immature raspberries and strawberries are packed under MAP, the firmness is retained and mould growth is not apparent during several weeks of storage at 0°C. However, the fresh fruit flavour does not develop to the fullest during the storage period.

Climacteric fruits, to be cut into pieces prior to MAP, should be ripened so that they have optimum flavour prior to packaging.

Suitable cultivars for MAP must be resistant to mechanical injury and crushing. The cells in the injury-resistant tissue are turgid, tightly held together by pectic substances and slightly elastic in order to absorb stress. Fruits which are harvested mechanically may be unsuitable for MAP since a slight bruise or split in the abused tissue may lead to microbial growth and spoilage.

7.9.2   Harvest and postharvest handling

With regard to fresh fruits and vegetables, mechanical harvesting has the advantages of reduced labour input, relatively low cost, and swift, efficient harvesting at a desired maturity; however, it can cause physical damage to the edible, fresh tissues [148, 155, 158, 200, 207, 254, 275]. Mechanical harvesting has been used for recovery of green beans, carrots, sweet corn, celery, radishes and tomatoes for the fresh market without appreciable damage [43, 155]. These vegetables have structural features that resist, to some degree, physical stress. On the other hand, many fruits have delicate tissue with thin, primary cell walls which are susceptible to small stresses with the result of bruises [228]. According to Shewfelt [254], Peleg [215] and Thai et al. [277], physically damaged tissue of fruits and vegetables may not be evident immediately after mechanical harvesting but may arise as latent damage in the form of bruises. Harvested commodities to be stored for prolonged periods must possess a minimum of tissue damage for retention of firmness and resistance to microbial growth [34]. Miller and Smittle [192] found that blueberries harvested by machine were softer, and had greater decay and higher moisture loss than hand-picked fruit after 1, 2 and 3 weeks of storage at 3°C. Machine-harvested berries from the cultivar, Climax, were firmer and less decayed than those from another cultivar, Woodward. Suitable cultivars of fruit must be selected on the basis of resistance of mechanical injury and crushing [43]. For MAP, fruits and vegetables must have no fissures in the epidermal tissue and only minor if any bruises. Hand-picked fruits and some vegetables are preferred as input produce for MAP to ensure quality retention over prolonged storage.

Rolle and Chism [240] pointed out that tissue injury due to physical impact brings about stress ethylene production, increased respiration at the site of injury, accumulation of secondary metabolites and membrane breakage resulting in the decompartmentalization of both substrates and enzymes to initiate deteriorative enzyme reactions. Tissue injury can bring about physical changes in cell membranes with the accompaniment of enzyme degradation of lipid components [96, 97, 156]. Free fatty acid formed by lipid hydrolysis can cause the disruption of organelles [96]. Free radicals formed during the autoxidation of polyunsaturated lipids in cells of commodities can damage the membranes and bring about an increase in permeability and a release of vacuolar components [191, 209].

Wound-induced alterations in cell membranes can increase the rate of ethylene synthesis [187], presumably by a pathway similar to that in ripening fruit [306]. Following tissue injury, ACC synthase located in membranes increases in activity with the consequence that ACC accumulates and ethylene production increases [31, 137, 308]. Since ethylene promotes ripening and senescence, it is obvious that commodities with significant physical injuries (wounds) should not be used in the MAP process [154].

Respiration can be increased by wound induction in fruits and vegetables along with accelerated ATP generation [43, 240]. With tissue wounding, respiratory inhibitor mechanisms in the Krebs cycle and the electron-transfer chain are disrupted, whereupon accelerated catabolism of sugars ensues with additional ATP available for wound healing [156, 157, 169]. Polyphenol oxidase, present in a wide range of fruits, may interact with decompartmentalized phenolic compounds from the vacuoles of disrupted cells to form melanin pigments [156, 188, 235, 272].

From the above discussion, physical injury during harvesting and handling of commodities must be obviated to ensure that MAP produce can be stored for lengthy periods without appreciable quality loss.

7.9.3   Minimal   pretreatment   processing

Minimal processing is the process conversion of raw harvested fresh fruits and vegetables into sound, clean, safe, convenient, nutritious, appealing and storable commodities which have retained, to a high degree, their fresh-produce (fresh-like) quality attributes. The operations in minimum processing include: (1) washing to remove soil, insects, pesticides and other extraneous matter, to apply an antimicrobial agent (e.g. fungicide) and to cool the produce; (2) trimming to remove unsound tissue (e.g. bruised, fissured, mouldy); (3) shearing inedible portions (e.g. peeling, coring) from desirable edible segments; (4) cutting edible tissue into suitable shapes and sizes; (5) cooling and temperature conditioning; and, (6) food additive diffusion and infusion for pH adjustment, microbial control, oxidative reaction control (prevent enzymatic browning), and texture modification [135, 136, 162, 254]. These minimal process operations should be valuable adjuncts to MAP for successfully prolonging shelf-life of commodities.

7.9.3.1   Washing

Washing of fruits and vegetables may be carried out by water immersion or spray techniques, the choice being dependent on the purpose of washing and the delicacy of the tissue [193]. The wash water should have a low microbial count and a suitable temperature for effective soil removal. Bolin *et al.* [29] found that thorough washing of shredded lettuce to remove free cellular contents was necessary for prolonged storage. Some produce (e.g. strawberries, raspberries) should not be washed because of the rapid water imbibition by the tissue to create a water-logged product and/or because residual wash water enhances microbial spoilage. The application of fungicide in a wash water will be discussed in section 7.9.3.4, 'Spoilogen control'.

7.9.3.2   Tissue   shearing

Trimming, pitting, peeling, coring, slicing and dicing are operations which shear the tissue by mechanical rupture of tissue. Such shearing causes decompartmentalization of cellular components and the bruising of tissue near the shear faces [240]. Both decompartmentalization and bruising of tissue lead to oxidative reactions such as the enzymatic browning reaction with the consequence of product darkening and off-flavour development. Bolin *et al.* [190] observed that the slicing of lettuce with a dull knife reduced considerably the quality of shredded lettuce over a 21-day refrigerated storage period, compared to the quality of lettuce prepared with a sharp knife. The sharp-edge shearing action on tissue brings about structural and metabolic changes similar to those of mechanically injured and wounded tissue. The respiration rate of cut produce is increased markedly [240]. McLachlan and Stark [190] reported that the cutting of whole heads of broccoli into individual florets was responsible for an increase in respiration rate of about 40% at both 10 and 20°C. Further, when carrots were julienne-sliced, the respiration rate increased more than of seven times. A detailed discussion the enzymic changes in bruised and wounded fruit and vegetable tissue has been presented in section 7.9.2, 'Harvest and postharvest handling'.

7.9.3.3   Cooling   and   temperature   conditioning

The process operations of removing field and respiratory heat energy are important for reducing the rates of respiration and ripening, for lowering transpiration and for decreasing microbial growth [118, 193, 254]. The Q_{10} values presented in Table 7.5 for vegetables are indicators of temperature-drop effectiveness in promoting the reduction of respiration rates. Commodities with high respiration rates should be cooled immediately and rapidly after harvest to preserve the just-harvested quality attributes by the reduction of metabolic deteriorative reactions. These temperature-dependent reactions include glycolysis, pentose phosphate pathway, Krebs cycle, oxidative phosphorylation and ethylene synthesis. Fungal spoilogen growth is retarded as the temperature drops toward 0°C. However, since some decay fungi cannot be controlled on commodities held at 5°C and below [270], additional complementary antimicrobial techniques should be employed (see section 7.9.3.4).

Cooling methods for removing thermal energy from produce prior to storage and shipment include forced-air cooling, hydrocooling, hydraircooling, ice contact cooling and vacuum cooling [193, 245, 246, 254, 301]. The cooling method must be selected on the basis of the commodity characteristics (size, shape, impact resistance, respiration rate), desired cooling rate, handling system (container types and sizes, stacking design), flow-through rate, capital investment and operating cost.

Tropical and subtropical fruits and some fruit vegetables are susceptible to a low-temperature tissue disorder called chilling injury [56, 74, 75, 76, 141, 179, 180, 199, 212, 240, 282]. Symptoms of chilling injury include surface and internal browning, surface pitting, failure to degreen, development of textural defects (mealy and woolly), uneven ripening or failure to ripen, off-flavour development and increased microbial susceptibility [301]. Generally, the lowest injury-safe temperatures (critical threshold temperatures) for fruits and vegetables are in the range of 5 to 13°C [155, 199, 301]. With a decrease in produce temperature below the critical threshold value, the rate of tissue injury decreases but the severity of tissue damage increases [286]. The proposed mechanisms of chilling injury have been reviewed recently [92, 98, 141, 212, 257]. Chilling commodities at low temperatures may cause an increase in respiration [74, 75, 76, 174] and induce ethylene production [252, 290].

Reduction of chilling injury in some commodities may be accomplished by temperature conditioning, intermittent warming and humidity control [49, 138, 240, 254, 288]. Conditioning of watermelon and grapefruit at moderate temperatures for specific time periods can reduce chilling injury when they are subjected to low temperature storage [124, 125, 218]. Intermittent warming of some produce (citrus, cucumbers, peppers) above and below the critical threshold temperatures can decrease chilling injury [66, 291].

Low temperatures for cut lettuce and citrus salads are essential for prolonged storage [28, 29, 136, 244]. Cut lettuce held at 10°C lost appreciable quality but was still acceptable after 7 days whereas that held at 2°C was still highly acceptable after 14 days [29]. At the 20-day storage time, the quality was about the same as the lettuce after 7 days of storage at 10°C [29]. Rushing and Senn [136] found that the shelf life of citrus salads was less than 7 days at 10°C and 12 to 16 weeks at -1°C.

7.9.3.4   Spoilogen control

Microbial spoilage of postharvest fruits and vegetables must be minimized prior to MA packaging. Spoilogens are microorganisms which cause spoilage of harvested fruits and vegetables whereas pathogens are organisms which are responsible for diseases in the growing plant. During the growth of horticultural crops in the field, plant pathogens may invade the organs to be used for food and cause irreversible damage [54, 69, 79, 268]. Application of antimicrobial agents to the crops at specific stages in development is a common practice [155, 196, 197]. Preharvest application of growth hormones has been found to be helpful in maintaining quality attributes by delaying the onset of ripening and senescence processes [94, 254]. Such quality maintenance is reflected in the resistance of produce to microbial growth.

Harvested fruits and vegetables must be inspected with the aim of removing cut, bruised, punctured, overripe and spoiled (decayed) commodities [155, 213, 254] and further, they must be held under strict spoilage control conditions. Control of spoilogen growth on commodities during bulk storage may include: application of antimicrobial agents to reduce spoilogen population; rapid air and water cooling; removal of excess surface water; maintenance of suitable storage temperatures and relative humidities, and sanitation practices for cold storage rooms [80, 155, 196].

Vegetables are decomposed primarily by bacteria which may cause soft rot, blemishes or surface defects [78, 181, 182]. *Erwinia carotovora, Flavobacterium* and *Pseudomonas marginalis* can cause a considerable amount of spoilage in vegetables both in the field and after harvest [32, 181, 245]. These organisms attack the vegetable tissue by secreting pectinesterase and polygalacturonases [272]. Chlorine in water (hypochlorite) at 10 ppm available chlorine is a very effective agent for rapidly inactivating vegetative bacteria. Control of spoilage bacteria without chemicals is possible by subjecting vegetables to low temperature (0 to 5°C). Fortunately, vegetables (other than fruit vegetables) are not chill-injured and thus can be stored close to the freezing point for maximum spoilogen inhibition.

The microflora of fruits with pH values below 4.5 are fungi and lactic acid bacteria, with fungi as the major contributors to fruit spoilage [32, 78, 197, 269]. Fungi adhere strongly to the surfaces of the invaded commodities and thus resist removal by postharvest washing [272]. The cuticle, a thin layer of complex lipids and phenolic acids on fruits, acts as a physical barrier to fungi. However, fungi can break down the cuticle by secreting cutinase, which catalyses the hydrolysis of cutin, the major component of the cuticle [272]. Cells near the surfaces of fruits contain relatively high levels of phenols and phenolases which are involved in the mechanism of the resistance-to-invasion to fungi [272]. Phenols oxidized by phenolases in the vicinity of invading spoilogens are considered to be natural antifungal agents. The rate of fungal growth is related to temperature, with minimal growth occurring at temperatures close to the freezing points of fruits. However, some fruits are susceptible to chilling injury and must be held at about 10°C or above.

Several synthetic fungicides are available for use in postharvest sprays and dips [80, 81]. Broad spectrum fungicides such as sodium orthophenylphenate and benomyl have been successful as a prestorage dip for apples to prevent black and white rot, and *Alternaria* rot. These postharvest fungicides have also been used for peaches, oranges, cherries, mangoes, pineapples, strawberries and kiwi [196]. Other fungicides used for postharvest treatment of produce include dehydroacetic acid, imazalel, biphenyl and benzimidazole [208].

Chlorine in wash water is effective at levels of 10 to 200 ppm to inactivate moulds, yeasts and bacteria [78]. For effective bacterial disinfection of commodities such as cut green salad products and other vegetables, the chlorinated water must have a pH of 6 or lower, at which point the highly effective undissociated hypochlorous acid is the dominant species [297]. Although 10 ppm available chlorine in the wash water will inactivate vegetative organisms, higher levels are required in commercial operations since available chlorine is lost through the interaction with amino acids, proteins and other amine compounds in the tissue and tissue juices of the commodities. Beuchart and Brackett [22] found that chlorine treatment of shredded lettuce did not inhibit the growth of *Listeria monocytogenes*.

Hot-water treatment of fruits and vegetables can inactivate fungi in the temperature ranges of 43-60°C for specific periods of time between 1 and 20 min [78, 270]. For example, fungal organisms on blueberries can be controlled with a hot-water dip at 49 to 51°C for 1.5 to 2.5 min. Papaya, being more resistant to heat treatment damage, can be dipped in hot water at 43 to 49°C for 20 min to inactivate fungi such as *Colletotrichum* and *Rhizopus* [270]. The hot-water dip treatment, sometimes in combination with fungicide application, has been successful in reducing microbial surface contamination on cantaloupes, nectarines, plums, papaya, oranges, lemons, tomatoes, and mangoes [48, 270, 273]. With commodities such as raspberries and strawberries which are injured by a hot-water treatment, a hot-air treatment may be effective. By exposing strawberries and raspberries to humid air at 44°C for 30 to 60 min, *Botyris* and *Rhizopus* have been controlled [57, 264, 302]. Thermal treatments are valuable for tropical and subtropical commodities which are susceptible to low-temperature injury.

7.9.3.5   Food additive diffusion and infusion

The diffusion or infusion of food additives into fruit and vegetable tissue may be of value in reducing the pH, modifying the textural attributes, inhibiting microbial growth and preventing discoloration.

Some fruits such as papaya and cantaloupe, and fruit vegetables such as tomatoes and peppers, may have pH values above pH 4.6 (see Table 7.3). When these fresh fruits are cut for salads and desserts, and packaged under MAP conditions, spoilogen and possibly human pathogen growth may occur [55, 242]. Generally, the lower the pH level of a cut produce, the less the growth of spoilogens and human pathogens. At pH 4.6 and below, human pathogens are unable to grow and form toxins in fruits, under either aerobic or anaerobic conditions [276]. pH adjustment to 4.6 or below would be advantageous from the standpoint of MAP, cut fruit being safe from human pathogens and being long-term storable. Lemon juice (about pH 2.5) or acidulants such as citric and malic acids [99] can be added to cut fruits for pH adjustment prior to MA packaging. O'Beirne [205] pointed out that dipping potatoes in an ascorbic acid/citric acid solution prior to packaging inhibited the growth of aerobic and anaerobic organisms, and restricted the growth and toxin production of *Clostridium botulinum*.

Calcium in plant tissues is involved in the delaying of senescence, reducing respiration, decreasing ethylene production, increasing tissue firmness and preventing enzymatic browning [220, 253]. The reasons for calcium involvement in these physiological changes are: (1) the interaction of calcium ions with phosphate groups of the phospholipid molecules in membranes to provide membrane structural integrity; and (2) the interaction of calcium ions with carboxyl groups of pectic polysaccharides to form calcium-bridge polymer complexes in the middle lamellae and cell walls [13, 14, 176, 221, 240]. Reduced respiration rates and enzymatic browning in cells with an increased calcium content may be attributed to the calcium alteration of fluidity of the tonoplast structure to the degree that substrate diffusion from the vacuole to the cytoplasm is limited [14, 211, 221]. The increase in tissue firmness with the elevation of tissue calcium is caused by the interaction of the calcium ions with pectic polysaccharides in both the middle lamellae and parenchyma cell walls. Ethylene production is influenced by calcium interaction in the plasmalemma where ethylene-forming enzyme system pre-sumably resides.

Postharvest calcium infusion (infiltration) into fruits and vegetable fruits may be advantageous from the standpoint of adding further storage days onto MAP produce. Poovaiah [221] reported that calcium infiltration into apples increased the ascorbic acid content, reduced the respiration rate and decreased ethylene production rate. According to Bolin and Huxsoll [27], calcium diffusion (2 min dip of 2% $CaCl_2$) into fresh-peeled peach halves was responsible for a firmer texture compared to the control untreated samples held for a storage period of between 2 and 5 weeks at 2°C. Fresh apricot halves dipped in 3% $CaCl_2$ (2 min) attained a tissue calcium level of 200 ppm and had higher textural value than the untreated halves stored at 2°C for 4 to 5 weeks. The taste (e.g., bitterness, saltiness) of the $CaCl_2$-treated apricot halves was not different from that of the un-treated halves.

Chemical preservatives such as antimicrobial agents can be useful for extending the shelf-life of cut fruits, particularly those with pH values below 4.5. Studies have shown that the effectiveness of antimicrobial agents such as benzoate and sorbate improves as the pH level drops below 5 [34, 267]. When fresh citrus salads were treated with either sodium benzoate or potassium sorbate, Rushing and Fenn [244] found that their antimicrobial activity was effective at a storage temperature of about 0°C in maintaining better quality of the salads compared to the untreated controls. Shelf-life exten-sion of the treated salads did not occur at storage temperatures of about 5 and 10°C. According to Heaton et al. [126], sodium benzoate at a level of 0.06% was effective in preserving the freshness of peach slices at 0°C.

Discoloration of cut fruit occurs in the presence of air when phenolic compounds are released from the vacuoles and come in contact with polyphenol oxidase in the cytoplasm. Enzymatic browning in fresh fruits and vegetables can be inhibited or pre-vented by: (1) reducing agent treatment (ascorbic acid, erythorbic acid, ascorbyl palmitate, bisulfite and cysteine); (2) enzyme inhibition or inactivation (chelating agents); (3) oxygen removal from tissue and microatmosphere; and, (4) metallic ion treatment (calcium) to stabilize membranes and cell walls [205, 243, 248, 249]. Bolin and Huxsoll [27] found that a 2% calcium chloride dip (2 min) retarded the browning of fresh peach halves more effectively than an ascorbic acid dip. However, this effect was not observed with fresh apricot halves. Sapers *et al.* [250] found that cut apples and potatoes, treated with a sodium ascorbate-$CaCl_2$ as a dip or infiltrated by either vacuum or pressure, inhibited browning for storage periods up to 20 days at 4°C. Waterlogging of tissue was experienced with some infiltration treatments. Ponting *et al.* [222] found that the storage life of apple slices could be increased by dipping the slices in a 1% ascorbic acid 0.1% calcium salt solution at pH 7 for 3 min. Pretreatment of raw potato strips with 10% ascorbic acid chip in combination with MAP (5% O_2, 10% CO_2, 85% N_2 flush) inhibited enzymatic browning for 1 week at 5°C [204, 205, 206].

Ethylene may be used as a supplemental gas in a ripening room to initiate ripening, increase the ripening process and ensure uniform ripening of pre-climacteric, slightly immature fruit such as melons, tomatoes, bananas and mangoes. The threshold concentrations of ethylene in fruit tissue for ripening varies from about 0.1 to 1 ppm [234]. The recommended concentrations of ethylene in a ripening room are 10 to 100 ppm [234]. Levels of ethylene higher than 100 ppm will not increase the ripening rate and indeed may lead to an explosive air-ethylene mixture. The optimum temperature for ripening with ethylene is between 18 and 25°C with a relative humidity of 90 to 95% [234, 245]. Periodic air exchanges in the ripening room are essential to remove respiratory CO_2 which reduces the effectiveness of the ethylene gas. The duration of ethylene treatment should be about 24 to 72 h depending on the type of fruit, cultivar, stage of maturity and the desired level of ripeness. Generally, moderately ripe to full ripe fruits are required for MAP in the whole and peeled, cut forms.

Treatment of apples at -1°C with a high level of environmental CO_2 (20%) for 10 days before conventional storage at -1°C in 2.5% O_2 and 1% of CO_2 reduced the rate of softening and acidity loss over a 7-month storage period [58]. The dissolved CO_2 in the tissue presumably altered the structural integrity of enzyme-containing membranes in the apples and thus disrupted the deteriorative enzymic pathways.

7.10 Packaging systems

For MAP of fruits and vegetables to be successful in prolonging the shelf life, suitable package systems are essential, particularly for adapting to abusive conditions such as temperature fluctuations and physical impacts during storage and distribution. A package system must be designed around (1) the properties of a food;

(2) the at-site utility; (3) the distribution conditions; (4) the necessity for film transparency; and (5) the gas barrier requirements. The package system for a fresh fruit or vegetable may be as simple as a heat-sealed pouch constructed from a polypropylene flexible film. On the other hand, the package system may be more complex with a heat-sealed plastic pouch within a larger sealed gas-containing pouch which resides in a sealed cardboard box. The complexity of a package system is dependent on the number and types of functions required to ensure quality retention of fruits or vegetables during storage and in the distribution and retail sectors [18, 102, 111, 202].

The major functions of a MAP package system are: (1) protection against microbial and insect invasion into the food; (2) ease of handling food at distribution, retail and consumer levels; (3) resistance to physical forces such as shock, vibrations and compression impacting on food; (4) regulation of gas and water vapour migration into and out of the package interior; and (5) ease of cooling produce.

The selection of materials for the construction of MAP package systems depend on the following criteria [123, 154, 210]:

 (1) specific gas and water vapour permeabilities
 (2) compatibility with food
 (3) reliable sealability
 (4) suitable heat transfer characteristics
 (5) anti-fog characteristics
 (6) physical abuse resistance
 (7) ease of opening package
 (8) machinability
 (9) recyclability
 (10) cost-effectiveness

Several types of package systems have been used for MAP [202, 210]:

 (1) tray or double tray (rigid or semi-rigid) lidded with a plastic film
 (2) flexible plastic pouch (bag)
 (3) plastic film overwrap of a tray
 (4) bag(s)-in-a-bag
 (5) bag(s)-in-a-carton (paperboard)
 (6) bag(s)-in-a-box (paperboard, styrofoam)
 (7) box(es)-in-a-bag

Trays for package systems may be constructed of styrofoam, moulded plastic film, coinjected plastic, plastic-coated pre-moulded pulp and paperboard-plastic film laminates [210].

Plastic resins have been developed for the manufacture of films, sheets and coatings to be used as components of food package systems [210]. Flexible plastic films are available as a monolayer and as coextruded or laminated multi-layer material with O_2 transmission rates (CO_2 transmission rates of a film are about 2 to 10 times greater than the O_2 rates) between 0.1 (for ethylenevinyl alcohol) and 13,000 (for low-density polyethylene) $mL/m^2/24$ h/1 atm/25°C/25 µm [154]. High-barrier films restrict almost totally the transfer of O_2 and CO_2 through the plastic walls, whereas low-barrier (permeable) films such as polyvinyl chloride have very high gas transmission rates. The high-barrier flexible films include ethylenevinyl alcohol (EVOH), nylon and polyvinylidine chloride (PVDC), the intermediate-barrier film would be polyethylene terephthalate (PET) and low-barrier (permeable) films would be represented by high-density polyethylene (HDPE), low-density polyethylene (LDPE), polypropylene (PP), ethylenevinyl acetate and polyvinyl chloride (PVC). Laminated and coextruded multilayer flexible films and sheets are generally combinations of the above-mentioned plastic resins.

For MAP of fruits and vegetables, package systems may consist of either low-, medium- or high-barrier films. Whole fruits and vegetables, and cut vegetables under MAP require sufficient O_2 (generally above 1%) and adequate CO_2 (up to the CO_2 tolerance level) in the microatmosphere of a package system to provide near-optimum preservation of fresh commodities without anaerobic respiration. For ongoing oxygen flow into the microatmosphere of a commodity-containing package system, a low-barrier flexible film with a specific O_2 transmission rate is required. Respiratory CO_2 also must diffuse out of the package system through the plastic wall. The influx of O_2 and the efflux of CO_2 are governed by the permeability of the plastic film, thickness and surface area of the film, and the partial pressure gradients of the gases inside and outside of the package system [127, 154, 168, 172]. In an ideal package system, the rate of O_2 consumption in the aerobic respiration of a commodity is equal to the influx rate of O_2. Moreover, the rate of CO_2 production from aerobic respiration is equal to the efflux rate of CO_2. Thus, the O_2 and CO_2 transmission rate requirements for a plastic film with known film area and thickness can be estimated for each commodity.

High-barrier flexible polymeric films and high-barrier plastic sheets in the form of trays may be used as components of a MAP package system for aerobic-respiring produce when perforated areas or windows, covered by either a porous diaphragm or a micro-porous membrane, are included to allow exchange of internal and external gases. Myers [201] invented package systems consisting of either a sealed, gas-filled, perforated high-barrier film bag containing a tray of produce or, alternatively, a thermoformed high-barrier gas-filled tray containing produce with the mouth lidded with a perforated high-barrier film. According to Myers [202], such perforated package systems have the advantage of using one film type for all fresh commodities, and the quality of the film and process can be easily controlled. Magnen [183] described a flexible high-barrier bag with a window closed by a gas-permeable silicon elastomer diaphragm for prolonging the storage life of respiring fruits and vegetables. Anderson [6] developed a package system composed of a high-barrier container (sealed bag or lidded tray) with a microporous membrane over a window for controlling the O_2 and CO_2 microatmosphere around a fruit or vegetable. This system is called Fresh Hold'.

Cut fruits can be held in sealed high-barrier containers under specific conditions. Powrie *et al.* [223] have developed a technique for preserving the freshness of cut fruits such as pineapple, mango, grapefruit and papaya by flushing a high-barrier container with an oxygen-containing gas prior to package sealing, and subsequently cold-shocking the fruit pieces. Shelf-life times at 1°C for the fruit pieces ranged from 8 to 16 weeks.

MAP equipment for the manufacture of containers, the filling of containers, and the gasification and sealing of package systems are readily available [2, 3, 210]. This equipment can be categorized as:

(1)   horizontal   form-fill-seal
(2)   vertical   form-fill-seal
(3)   tray-stretch   overwrap
(4)   pre-made   bag-fill-seal

MA can be introduced into a package system by either gas flushing or gas compensated vacuum [3, 205]. The flushing technique involves the purging of air from the package system with a continuous stream of a preselected gas mixture prior to sealing. This technique can be used with any of the above-mentioned packaging equipment. In the gas compensated technique, a vacuum is created in the package system to remove the air (1% or less residual) and thereafter, a preselected gas mixture is flushed in as the MA prior to sealing. The gas compensated technique involves more costly equipment and a slower production rate compared to the flushing technique.

The flushing of a commodity-containing package system with a specific MA gas mixture can have many rapid-response advantages such as: (1) reducing oxidative deteriorative reactions; (2) inhibiting microbial growth; and, (3) producing a pillow package to decrease physical damage to a delicate commodity. With some fruits and vegetables having high respiration rates, rapid quality losses occur at ambient temperatures when they are stored in a package system having air in the headspace. By flushing the headspace with a low O_2/high CO_2 gas mixture, the respiration rates and other oxidative reactions would be reduced, particularly during the slow cooling of the commodities. Cut fruits (e.g. apples, peaches) and vegetables (e.g. potatoes, mushrooms, lettuce) in a sealed package system with air can attain surface browning through the oxidative enzymatic browning reactions. Such reactions can be inhibited by flushing package systems with low O_2 gas mixtures such as 5% O_2, 5% CO_2 and

90% N_2 for cut lettuce, 5% O_2, 10% CO_2, 85% N_2 or 3% O_2 and 97% N_2 for potato strips and 100% N_2 or 50% CO_2 and 50% N_2 for apple slices [11, 12, 204, 205, 206].

Oxygen absorbents (powdered FeO) in packets may be placed in package systems to complement the gas mixture flush for reducing respiration and ripening rates [154, 168]. The amount of FeO required to lower the O_2 level in the package system to a beneficial value can be calculated [154]. As the package system is cooled, the rate of O_2 absorption by the absorbent decreases.

An important objective of a producer of MAP fresh fruits and vegetables is to properly seal package systems. The seal strength is dependent on the type of film, sealer temperature, jaw pressure, thickness of plastic film and presence of foreign matter. A slight wrinkle in the seal area may be responsible for gas leakage. The simplest method to determine seal integrity is by placing the package system under water and observing bubble formation with hand pressure on the package or with vacuumization of the test water chamber [68]. A small amount of surfactant in the water provides easier bubble detection. A sensitive, reliable method for the detection of seal leaks in MAP package systems involves the estimation of CO_2 loss from the package under vacuum by an IR detector system [68].

7.11 Dynamics of gases within fruits and vegetables under MAP

When enzyme-active fruits and vegetables are held in a gaseous environment within a closed gas-permeable package system, the gaseous phase is in a state of flux [154, 265]. The dynamic state of the gases within the package system is due to the on-going utilization of O_2 in the respiration pathways for the oxidation of hexoses and other energy sources and to the evolution of respiratory CO_2. During the respiration of an MAP commodity, concentration gradients are established between the gases in the tissue and the microatmosphere gases, and between the microatmosphere gases and the outside air. With a depleting O_2 environment in the tissue, O_2 is transferred from the outside air through the packaging material into the microatmosphere and finally into the tissue. The rate of O_2 transfer to the cells will depend on the partial pressure differences along the path, the diffusion coefficient of the packaging material, resistance of gas diffusion in the tissue, and the temperatures of the environmental air, the microatmosphere, the commodity and the package system. The accumulating respiratory CO_2 in the tissue will travel in the opposite direction with the rate of transfer to the outside air being governed by same factors mentioned above for O_2 transfer rate. Eventually, a gas equilibrium will be established in the microatmosphere with specific levels of O_2 and CO_2 [101, 127]. Many researchers have developed predictive equations for selecting polymeric films with gas permeabilities required to achieve O_2 and CO_2 levels in the microatmosphere for MA packages [51, 127, 154, 168, 309].

7.11.1 Diffusion of O_2 into cell

As the partial pressure of O_2 (pO_2) declines in the microatmosphere, a level is reached in a fruit or vegetable when insufficient O_2 is available for maximum rate of respiration. As the pO_2 is reduced further, the respiration rate declines as a function of pO_2. The reduction of the respiration rate may be attributed to the suppression of low O_2 affinity enzymes such as ascorbic acid oxidase and glycolic acid oxidase [41]. At O_2 levels between 1% (pO_2 = 0.01 atm) and 7% (pO_2 = 0.07 atm) in the microatmosphere of a commodity, the respiration rate is approximately halved [36, 309]. For example, an apple respiration rate was half-inhibited at pO_2 = 0.04 atm in the microatmosphere, when, at the centre of the apple, the pO_2 was 0.016 atm [36]. In broccoli, the respiration rate is cut in half by a drop in O_2 level from about 21% to 5% [309]. The O_2 gradient between the microatmosphere and the centre of the apple represents a 60% drop in pO_2. This pO_2 drop can be attributed to the restriction of O_2 diffusion through the various tissue barriers.

The skin, the intercellular spaces, the cell walls and the cytosol are the barriers of O_2 movement to the mitochondria in the cells. Before O_2 passes into the skin or epidermal tissue of a commodity, it must diffuse through a thin boundary gas layer at a rate which depends on the O_2 gradient, the effective surface area available for gas molecule reception, microatmospheric gas velocity and temperature [36]. The skin and epidermal tissue may possess pores called stomata and lenticels which are the passages for the O_2. The surface areas of leaves have high frequency distributions of stomata. The apple leaf has about 40,000 stomata per cm^2 [61], whereas apple skin has only 1.8 lenticels per cm^2 and green banana skin possesses about 480 stomata per cm^2 [53, 145]. Cranberry, tomato and green pepper skins contain no pores, and all of the gases diffuse through holes in the pedicel scar. Resistance (r) of gas diffusion through the boundary gas layer and skin of most fruits and vegetables with pores varies from 5000 to 50,000 s/cm with O_2, CO_2 and ethylene [36].

After passage through the skin, O_2 is transported through interconnected intercellular gas spaces (microconduits) by diffusion within an O_2 gradient between just under the skin and commodity centre [36, 154]. If the tissue of a ripe fruit is cut, the released cellular fluid tends to flow into the gas spaces to create waterlogging. The diffusion of gas through cell sap in the microconduits would be much slower than the gas-filled intercellular spaces [36].

The O_2 is then transported through the cell walls and cytosol to the mitochondria under an O_2 gradient. Diffusion of O_2 into the cytoplasm is dependent on the O_2 gradient, the cell wall thickness, diffusion coefficient of the O_2 in water and permeability coefficient of the plasmalemma [36]. Finally, the dissolved O_2 diffuses into the mitochondrial outer membrane, and into the matrix.

7.11.2 Solubility characteristics of O_2, CO_2 and N_2

The transport of respiratory gases in tissue is not only in the gaseous state (in intercellular spaces) but also in the dissolved form (in the cytoplasm and vacuole). The amount of a gas dissolved in water is dependent on the chemical nature of the gas (molecular weight, polarity), the water temperature and the gas pressure above the water [167]. The solubility of a gas can be expressed by an absorption coefficient which is the volume of gas, reduced to 0°C and 1 standard atmosphere (101.3 kPa), dissolved in a unit volume of water at a specific temperature. The absorption coefficients for O_2, CO_2 and N_2 are presented in Table 7.10. CO_2 is roughly 30 times more water-soluble than O_2 which is about twice as soluble as N_2. With a temperature rise of water from 0 to 20°C the solubility of each of the three gases is reduced by about one half. The solubility of gases is also determined by the solute content and gas pressure. Fruits and some vegetables contain appreciably amounts of sugars (Table 7.2). As the sugar concentration increases in water, the solubility of a gas is reduced under a constant pressure. However, as pointed out in a previous section, most of the sugars in parenchyma cells are located in the vacuoles, whereas the cytoplasm possesses a relatively low concentration of sugars. All of the three gases (O_2, CO_2, N_2) obey Henry's Law at moderate temperatures and low pressures. Normally, the pressure within an MA package system would be at a value similar to the external air pressure (altitude-dependent), particularly if the package system is composed of flexible components.

CO_2 is an important chemical compound in MAP fruits and vegetables since it lowers the rates of respiration, ethylene production and ripening and inhibits microbial growth. To gain an insight into the mechanisms of the above-mentioned inhibitions, the chemistry of CO_2 in water should be known. When CO_2 is dissolved in water, a small amount is hydrated to H_2CO_3. The K_{hyd} value has been determined to be 2.6×10^{-3}. Between pH 1 and 5.5, a CO_2 solution contains about 2% of the CO_2 as H_2CO_3 and the remainder exists as dissolved CO_2 [63]. When the pH values of a CO_2 solution rises from 5.5 to 8.0, the H_2CO_3 dissociates to H^+ and HCO_3^- ($K_a = 4.3 \times 10^{-7}$) with:

(1) a gradual increase in HCO_3^- to about 75% of the total CO_2 in water; (2) a decrease of dissolved CO_2 to around 23%; and (3) H_2CO_3 remaining about 2% of the total CO_2 [59]. With the cytosol of parenchyma cells having pH values in the vicinity of 6.5, the CO_2 species in the cytoplasm would probably be at levels of about 2% CO_2 as H_2CO_3, 8% CO_2 as HCO_3^- and 90% dissolved CO_2. The addition of CO_2 to vegetable tissue will probably bring the pH of the aqueous phase down to the value of pK_a (6.3) depending on the buffer index [44].

Table 7.10.   The   effect   of   temperature   on   the   absorption coefficients of N_2, O_2 and CO_2 in water

Water temperature, °C	N_2	O_2	CO_2
0°	0.024	0.049	1.71
10°	---	0.038	1.19
20°	0.015	0.028	0.88
30°	0.013	0.026	0.67

7.12 Responses of fruits and vegetables to low O_2 and high CO_2 levels in modified atmospheres

7.12.1 General

The responses of fruits and vegetables, in whole and cut forms, to low O_2 (1 to 5%) and elevated CO_2 (3-20%) levels in microatmospheres surrounding the commodities have been reviewed extensively [25, 35, 45, 70, 139, 147, 150, 152, 154, 205, 227, 289, 309, 310]. In general, reduced O_2 and elevated CO_2 levels can be beneficial by decreasing the respiration rate, ripening process, ethylene synthesis rate, ethylene sensitivity to ripening and tissue softening [152, 154, 309, 310]. Kader [149] has assigned recommended MA levels of O_2 and CO_2 for the maintenance of quality attributes of a variety of fruits and vegetables and these are presented in Tables 7.11 and 7.12.

MA may play an important role on preventing or inhibiting microbial invasion and growth on fruits and vegetables. Fruits and vegetables with natural defence structures can be preserved by MAP. The reduction of ripening of fruit by MA is beneficial for retaining the integrity of epidermal tissue and thus maintaining resistance to spoilogens [82]. CO$_2$ levels of 10% and higher can inhibit fungal growth and some bacteria can be inactivated [71].

Table 7.11.   Recommended MA conditions for vegetables

Commodity	Temperature range, °C	Modified atmosphere O$_2$	CO$_2$
Asparagus	0-5	air	5-10
Beans, snap	5-10	2-3	5-10
Broccoli	0-5	1-2	5-10
Brussels sprouts	0-5	1-2	5-7
Cabbage	0-5	3-5	5-7
Cantaloupe	3-7	3-5	10-15
Cauliflower	0-5	2-5	2-5
Corn, sweet	0-5	2-4	10-20
Cucumber	8-12	3-5	0
Honeydew, melon	10-12	3-5	0
Lettuce	0-5	2-5	0
Mushrooms	0-5	air	10-15
Peppers, bell	8-12	3-5	0
Spinach	0-5	air	10-20
Tomatoes,			
mature-green	12-20	3-5	0
partly ripe	8-12	3-5	0

From Kader [150]

Table 7.12.   Recommended MA conditions for fruits

Commodity	Temperature range, °C	Modified atmosphere O$_2$	CO$_2$
Apple	0-5	2-3	1-2
Apricot	0-5	2-3	2-3
Avocado	5-13	2-5	3-10
Banana	12-15	2-5	2-5
Cherry (sweet)	0-5	3-10	10-12
Grapefruit	10-15	3-10	5-10
Kiwi fruit	0-5	2	5
Mango	10-15	5	5
Papaya	10-15	5	10
Peach	0-5	1-2	5
Pear	0-5	2-3	0-1
Pineapple	10-15	5	10
Strawberry	0-5	10	15-20

From Kader [150]

Table 7.13.   Tolerance of selected fruits and vegetables to low O_2 levels

Minimum O_2 level tolerated (%)	Commodity
1	apple, broccoli, mushroom, pear, most sliced (cut) fruit and vegetables
2	apple, apricot, Brussels sprouts, cabbage, cantaloupe, cauliflower, celery, cherry, corn (sweet), green bean, kiwi, nectarine, papaya, peach, pear, pineapple, plum, strawberry
3	artichoke, avocado, cucumber, pepper, tomato
5	asparagus, citrus fruits, green pea, potato, sweet potato

From Kader and Morris [153]

Table 7.14.   Tolerance of selected fruits and vegetables to elevated CO_2 levels

Maximum CO_2 level tolerated (%)	Commodity
2	apple (Golden Delicious), apricot, artichoke, celery, Chinese cabbage, grape, lettuce, pear, pepper (sweet), sweet potato, tomato
5	apples (most cultivars), avocado, banana, Brussels sprouts, cabbage, carrot, cauliflower, cranberry, kiwi, mango, nectarine, orange, papaya, pea, peach, pepper (chilli), plum, radish
10	asparagus, broccoli, cucumber, grapefruit, green bean, lemon, parsley, pineapple, potato
15	blackberry, blueberry, cantaloupe, cherry, corn (sweet), mushroom, raspberry, spinach, strawberry

From Kader and Morris [153]

Kader and Morris [153] have assembled data on the relative tolerance limits of fruits and vegetables to reduced O$_2$ and elevated CO$_2$ levels in MA. These tolerance limits of O$_2$ and CO$_2$ for a variety of fruits and vegetables are presented in Tables 7.13 and 7.14. When the O$_2$ level drops below the O$_2$ tolerance value for a commodity, specific physiological injuries and quality deterioration may occur. With a CO$_2$ level above the CO$_2$ tolerance level for the same commodity, different physiological injuries and quality deterioration may occur. Kader and Morris [153] pointed out that the tolerance limits can vary as a function of type of commodity, cultivar, storage temperature, physiological age, storage time and added supplemental gases. For example, various types of lettuce are influenced differently under high CO$_2$ levels. Crisphead lettuce responded to 1% CO$_2$ in an MA by developing brown stain during refrigerated storage whereas Romaine lettuce tolerated CO$_2$ levels up to 12% [35]. When Brecht *et al.* [35] exposed several cultivars of crisphead lettuce to elevated CO$_2$, they observed different degrees of brown stain injury. Low temperature as an adjunct to MA is not always beneficial for vegetables. It has been found that lettuce stored at 0°C under MA attained brown stain whereas at a storage temperature of 10°C, no brown stain was evident.

Below O$_2$ levels of 1 to 2% in the microatmosphere of a commodity, the aerobic respiration shifts over to the anaerobic respiration which involves only glycolysis and the decarboxylation of pyruvic acid to acetaldehyde, which is then converted to ethanol. Anaerobic volatile compounds are formed in some fruits and vegetables to the extent that the commodities are no longer acceptable due to intense off-flavour [150, 154].

7.12.2 Metabolic responses

7.12.2.1 Respiration

The lowering of the respiration rate of a commodity by an MA has the following benefits: (1) reduces the loss of sugars and thus the sweetness is retained; (2) reduces the decomposition of organic acids with the retention of sourness and acceptable pH; (3) decreases the amount of heat produced in the alternate route of electron-transfer chain; (4) lowers the rate of ATP synthesis in the oxidative phosphorylation reactions with the consequence of a decrease in ethylene production and synthesis of ripening enzymes; and (5) reduction in CO$_2$ production to obviate the gas-ballooning of MAP packages. From the above considerations, it is obvious that the respiration rate decline can lengthen the shelf-life of fresh fruits and vegetables [310].

The principles of respiration enzymology have been presented in section 7.5 (Respiration of fruits and vegetables) and these will serve as a basis for understanding the effect of low O_2 and high CO_2 levels on the respiratory metabolic pathways. Beneficial respiratory responses of fruits and vegetables to lowering the O_2 level do not take place until the level reaches about 12% or less [25, 154, 310]. For apples and broccoli, the respiration rates are reduced by about one-half when the O_2 levels are dropped to about 4 to 5% [25, 309]. Robinson *et al.* [238] found that the respiratory response of 30 fruits and vegetables to 3% O_2 at 0, 10 and 20°C varied widely. The reduction of respiration rate in a 3% O_2 microatmosphere at 0°C in comparison to an air (21% O_2) microatmosphere ranged from 10 to 46%, whereas at 10 and 20°C, the reduction was between 20 and 60%. When whole fruits and vegetables are sliced, the respiration rate increases and the heat production rises [12, 140, 169, 243, 295]. Rosen and Kader [243] found that the lowering of O_2 in the microatmosphere to 2% or below reduced the evolution rate of CO_2 by pear slices at 2.5°C over a 7-day period.

The reduction of respiration rates has been attributed to the inhibition of the activity of oxidases such as a polyphenol oxidase, ascorbic acid oxidase and glycolic acid oxidase [41]. Since potato polyphenol oxidase has a Michaelis constant, K_m, of about 5×10^{-4} M, which is an indication of a low affinity for O_2, this oxidase would not be active with an internal tissue O_2 content of 1 to 2% or with a microatmosphere O_2 level of around 3%. When the levels of internal tissue O_2 decline to around 0.1%, the terminal oxidase (cytochrome oxidase) in the oxidative phosphorylation pathway is presumably not active or functional since the K_m of this oxidase is around 7×10^{-8} M. At this point, the aerobic respiration has shifted to the anaerobic respiration exclusively under the glycolysis and pentose phosphate pathways and the pyruvic decarboxylation pathway [35, 152]. Off-odour volatile compounds are generated in the anaerobic respiratory process, but the types and amounts of these compounds and biosynthetic pathways are not known.

With MA CO_2 levels of 5% and higher, commodities have reduced respiration rates [35, 152, 154, 309, 310]. As shown in Figure 7.4, as the CO_2 content of the microatmosphere around broccoli at 0°C increased from 0 to 12% in either air or 5% O_2 gas mixture, the respiration rates declined progressively. With a 2% O_2 level in the microatmosphere of broccoli, the respiration rate decreased when from 5% CO_2 was added. Kader [152] pointed out that 10% CO_2 added to air reduced the respiratory rate of commodity to about the same extent as a 2% O_2 drop. When the CO_2 content in the microatmosphere of black currants and citrus fruits reached around 10%, anaerobic respiration was evident with the formation of acetaldehyde and ethanol [67, 263].

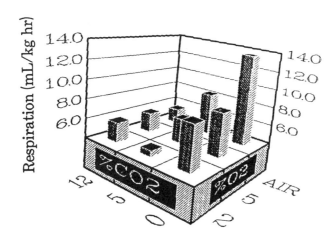

Figure 7.4.   Effect of O_2 and CO_2 levels on the respiration rates of broccoli at 0° for 7 days [309].

Elevated CO_2 contents in microatmosphere around fruits and vegetables have in inhibitory effect on enzymes in the Krebs cycle and glycolysis pathways.   Succinate has been found to accumulate in apples, apricots, carrots, peaches, pears, cherries and lettuce when the microatmospheric CO_2 was elevated up to 20% [93, 198].   Such a situation implies that succinic dehydrogenase activity was inhibited. Ranson *et al.* [230, 231] found that CO_2 inhibited succinic acid oxidase in isolated castor bean mitochondria.   Shipway and Bramlage [258] reported that, with isolated apple mitochondria exposed to a gas with 6% CO_2, malate oxidation was enhanced whereas the oxidation of citrate, α-ketoglutarate, succinate, fumarate and pyruvate was reduced.   They attributed these results to the structural and conformational changes in the mitochondria.   According to Monning [198], apple tissue responded to a 10% CO_2 gas mixture by an inhibition of glycolysis and succinic dehydrogenase activity along with a reduction in the formation of citrate/isocitrate, and α-ketoglutarate. Malic dehydrogenase activity was not altered by the CO_2 treatment. Kerbel *et al.* [161] noted that a 10% CO_2-air treatment of apples reduced the activities of ATP:phosphofructokinase and PPi:phosphofructokinase in the glycolytic pathway.   Further they found an increase in the concentrations of fructose-6-phosphate and fructose-2,6-bisphosphate and a reduction of fructose-1,6-bis-phosphate in the apples. Presumably the inhibitory influence of the elevated CO_2 on the respiration rate can be attributed, in part, to the reduction of activities of the phosphofructokinases.

The respiratory quotient (RQ), as indicated previously is de-pendent on the types of energy-source substrates used in the glyco-lysis pathway and the possible utilization of CO_2 for carboxylic acid synthesis in the Krebs cycle. Under MA conditions, the RQ of pro-duce may change gradually [154, 289]. According to Fidler and North [91], the RQ for apples decreased markedly when the MA had low O_2/high CO_2 levels or a high CO_2 level.

7.12.2.2   Ethylene synthesis and sensitivity

When fruits and vegetables are exposed to microatmospheres with levels less than 8% O_2, the ethylene production decreases and the sensitivity of the commodities to ethylene is reduced [38, 39, 289]. At an O_2 level of 2.5%, ethylene production was reduced to one-half the rate in comparison with air. Burg and Burg [38] observed that at a 3% O_2 level in the microatmosphere, ethylene binding was reduced to around 50% of that in air. The reason for this O_2 effect is because the conversion of 1-aminocyclopropane-1-carboxylic acid to ethy-lene requires O_2.

When the CO_2 level in fruits is elevated, the rate of ethylene synthesis may be reduced or promoted or may have no effect, depending on the CO_2 content and the commodity type [289]. In some commodities, the increase in ethylene production rate occurs only when the CO_2 is sufficiently high to cause physiological injury [152]. Arpaia et al. [8] found that elevated CO_2 levels enhanced the ethylene-induced physiological disorder in kiwi.

7.12.3 Ultrastructural   changes

When mature green Bartlett pears were exposed to elevated CO_2 levels in the microatmosphere, Frenkel and Patterson [93] ob-served that tonoplast was altered.

7.12.4 Quality   changes

Since quality attributes such as taste, odour, texture and colour are dependent on the chemical composition and structural features of fruits and vegetables, alteration of chemical components in tis-sues through degradative reactions could lead to consumer unac-ceptability. Retention of the chemical composition and the intact structure of freshly harvested produce during storage may be at-tained by low O_2 and high CO_2 microatmospheres in MAP.

7.12.4.1   Flavour

Flavour is the human perception which includes taste and odour sensations. During storage of fruits and vegetables, sugars may be converted into starch (peas, sweet corn) and bitter components (isocoumarin) may be synthesized under ethylene inducement. Thus the taste profile of a fruit or vegetable changes with time of storage, the rate being dependent on the temperature and the MA. Goodenough [105] reported that tomatoes at 12.5°C in an MA of 5% O_2 and 5% CO_2 had an increase in glucose, fructose and citric acid levels, and a decrease in starch and malic acid contents. In apples exposed to a 2.5% CO_2 microatmosphere, the acid decomposition rate was reduced [171]. When carrots were stored under MA conditions to reduce ethylene production, the formation of the bitter component, isocoumarin, was prevented [40]. The synthesis of volatile odour compounds is influenced by low O_2 levels. Knee and Hatfield [166] reported that low levels of esters were synthesized in apples under a 2% O_2, because the precursor alcohols were synthesized at a slow rate. After moving the stored apples to air, the ester synthesis was increased. Off-flavours in fruits and vegetables can be produced by subjecting them to low O_2 (lower than 2%) and high levels of CO_2 (15% and above). Anaerobic off-flavour volatiles may be decomposed by submitting the produce to air storage. When broccoli is held under MA at 0.5% or less O_2 and more than 15% CO_2, a mercaptan/H_2S odour is evident in a few days at 5°C.

7.12.4.2   Colour

The retention of pigments in stored fruits and vegetables is essential for visual acceptability by consumers. The degradation of pigments in commodities can be reduced considerably by MA [139, 266, 292, 293]. According to Singh et al. [260, 261], chlorophyll loss can be retarded in crisphead lettuce by reduced O_2 (2 to 5%) and elevated CO_2 (2.5%) levels in the microatmosphere. With elevated CO_2 contents in microatmospheres, pigments in green beans, apricots and peaches have been reduced [293]. Knee [163] found that the rate of chlorophyll degradation in the peel of apples was reduced by one-half by MA with 2.5 to 4% O_2 levels. Goodenough and Thomas [107], upon storage of tomatoes at 12°C in 2.5 to 4% O_2 and 4% CO_2 microatmosphere, found that the chlorophyll degradation and lycopene synthesis were reduced.

7.12.4.3   Texture

The tenderness of broccoli and asparagus can be retained during storage by increasing the CO_2 in the microatmosphere. When Lipton [178] stored asparagus spears at 4°C under MA of about

12% CO_2, textural toughening was retarded considerably. Broccoli stored in a 10% CO_2 microatmosphere at 5°C retained the tenderness over a 2-week period. It is of interest to note that the lowering of O_2 had no influence on tenderness retention of asparagus and broccoli.

The storage of harvested commodities under MA conditions can inhibit the synthesis of polygalacturonases and thus reduce the softening of tissue. Goodenough *et al.* [108] found that when mature green tomatoes were stored at 12.5°C in an MA of 5% O_2 and 5% CO_2 for up to 8 weeks, the synthesis of polygalacturonase was inhibited. Upon exposure of these stored tomatoes to air, the polygalacturonase was synthesized, the tissue was softened and lycopene was synthesized. Knee [163] reported that rate of tissue softening of apples was reduced by one-half when exposed to 2.5 to 4% O_2 levels.

7.13 The future of MAP

The vision of food scientists is to prolong for months the retention of fresh quality attributes of fruits and vegetables, as whole and cut forms, in convenient, low-cost packages. Obstacles to long-term storage of these commodities involve degradative enzyme reactions, membrane disruption, and microbial invasion and spoilage, as well as package systems which do not have adequate adaptability to wide temperature fluctuations. Process manipulation of the enzyme pathways of plant tissue and spoilogens by physical and chemical pretreatments and by modification of O_2 and CO_2 levels in microatmospheres of package systems has been successful, to some degree, in controlling the degradative alterations in commodities. Considerable fundamental knowledge has been published on the mechanisms involved in the suppression of degradative reactions in fresh commodities under low O_2 and elevated CO_2 levels in microatmospheres. However, much more basic knowledge on chemical, physical and microstructural changes in stored commodities under process stress conditions is needed to support technological advances in MAP. In particular, more research should be directed towards:

(1)   the assessment of the network structure of the intercellular spaces (microconduits), and their disruption during ripening, bruising and cutting of fruits and vegetables;

(2)   the determination of the transport phenomena of exogenous and endogenous gases in whole and cut commodities under MAP at temperatures between 0 and 15°C;

(3)   the accumulation of data on the rates of respiration and ethylene production and quality deterioration of whole and cut commodities under MAP conditions with various concentrations of O_2 and CO_2 and at temperatures from 0 to 15°C;

(4)   the elucidation of mechanisms for the reduction of the rate of respiration and ethylene production, and for quality changes in fruits and vegetables under MAP;

(5)   the assessment of the modes of action by ethylene in com-
        modities under MAP towards alterations of quality attributes;

(6)   the evaluation of combined pretreatment and MAP method-
        ologies for altering the enzymic pathways to reduce quality
        deterioration;

(7)   the assessment of the natural defence (chemical and mi-
        crostructural) mechanisms in MAP commodities for the re-
        striction of spoilogen growth and the promotion of the syn-
        thesis of phytoalexins;

(8)   the determination of the benefits of low O_2/elevated CO_2 gas
        flushes in MAP fruits and vegetables for reducing respiration
        and ripening rates, increasing resistance to spoilogen growth
        and improving quality retention;

(9)   the examination of the mechanisms for the synthesis of
        anaerobic off-flavour compounds and the development of
        preventative off-flavour techniques for MAP commodities
        having low O_2 microatmospheric levels.

The design of effective MA package systems is dependent on
the chemical and physical properties of the package components
(e.g. polymeric film, microporous membranes, styrofoam and paperboard trays) as well as information on the gas dynamics in the commodity-containing package systems. Unfortunately, detailed data on
the physical properties of package components are not available.
For example, gas permeabilities of many polymeric films with and
without antifog surfactants, at various temperatures and relative humidities, have not been reported. The on-going commercial suc-
cesses in MAP technology for fruits and vegetables will rely on
commercial availability of new functional types of barrier and permeable polymeric films. Further, the design of suitable package
systems to cope with temperature abuse situations will be invaluable.

References

1. Abeles, F.B. 1973. Ethylene in Plant Biology.   Academic Press, New York.
2. Agriculture Canada. 1990.   Modified Atmosphere Packaging.   A. An Extended Shelf-Life Packaging Technology.   Food Development Division, Agriculture Canada, Ottawa.
3. Agriculture Canada. 1990.   Modified Atmosphere Packaging. B. Investment Decisions.   Food Development Division, Agriculture Canada, Ottawa.
4. Agriculture Canada.   1990.   Modified Atmosphere Packaging. C. The Consumer Perspective.   Food Development Division, Agriculture Canada, Ottawa.
5. Albersheim, P. 1978.   Concerning the structure and biosynthesis of primary cell walls of plants.   Int. Rev. Bioch. 16: 127-150.
6. Anderson, H.S. 1989.   Controlled atmosphere package.   U.S. Patent   No. 4,842,875.
7. Anon. 1990.   Action towards healthy eating.   The Report of the Communications/Implementation Committee.   Health and Welfare Canada, Ottawa.
8. Arpaia, M.L., Mitchell, F.G., Kader, A.A. and Mayer, G. 1985. Effect of 2% O_2 and varying concentrations of CO_2 with or without C_2H_4 on the storage performance of kiwi. J. Am. Soc. Hort. Sci. 110: 200-203.
9. Baile, J.B. 1975.   Synthetic and degradative processes in fruit ripening, pp. 5-19.   *In* Postharvest Biology and Handling of Fruits and Vegetables.   Haard, N.F. and Salunkhe, D.K. (eds.). AVI Publ. Co., Westport, CT.
10. Baile, J.B. and Young, R.E. 1981.   Respiration and ripening in fruits - retrospect and prospect, pp. 1-39.   *In* Recent Advances in the Biochemistry of Fruits and Vegetables.   Friend, J. and Rhodes, M.J.C. (eds.).   Academic Press, London.
11. Ballantyne, A. 1987.   Modified atmosphere packaging of selected prepared vegetables.   Technical Memo. No. 464.   Campden Food Preservation Research Association, Chipping Campden, Gloucestershire.
12. Ballantyne, A., Stark, R. and Selman, J.D. 1988.   Modified atmosphere packaging of shredded lettuce.   Int. J. Food Sci. Technol. 23: 267-274.
13. Bangerth, F. 1979.   Calcium related to physiological disorders of plants.   Ann. Rev. Phytopathol. 17: 97-122.
14. Bangerth, F., Dilley, D.R. and Dewey, D.H. 1972.   Effect of postharvest calcium treatment on internal breakdown and respiration of apple fruits. J. App. Soc. Hort. Sci. 97: 679-682.
15. Banks, N.H. 1985.   The oxygen affinity of 1-amino-cyclopropane-1-carboxylic acid oxidation of sliced banana fruit tissue, pp. 29-36.   *In* Ethylene and Plant Development.   Roberts, J.A. and Tucker, G.A. (eds.).   Butterworths, Oxford.
16. Banks, N.H., Elyatem, S.M. and Hammat, M.T. 1985.   The oxygen affinity of ethylene production by slices of apple fruit tissue. Acta Hort. 157: 257-260.

17. Baritelle, J.L. and Gardner, P.D. 1984. Economic losses in the food and fiber system: from the perspective of an economist. *In* Postharvest Pathology of Fruits and Vegetables, pp. 4-10. Moline, H.E. (ed.). Publ. NE-87. Agric. Exp. Sta. University of California, Davis, CA.

18. Barmore, C.R. 1987. Packaging technology for fresh and minimally processed fruits and vegetables. J. Food Qual. 10: 207-217.

19. Bartley, I.M. and Knee, M. 1982. The chemistry of textural changes in fruit during storage. Food Chem. 9: 47-58.

20. Bedrosian, K. and Schiffman, R.F. Controlled atmosphere produce package. U.S. Patent No. 4,423,080.

21. Ben-Arie, R., Kislev, N. and Frenkel, C. 1979. Ultrastructural changes in the cell walls of ripening apple and pear fruit. Plant Physiol. 64: 197-202.

22. Beuchat, L.R. and Brackett, R.E. 1990. Survival and growth of *Listeria monocytogenes* on lettuce as influenced by shredding, chlorine treatment, modified atmosphere packaging and temperature. J. Food Sci. 55: 755-758.

23. Beyer, E.M. 1981. Ethylene action and metabolism, pp. 107-121. *In* Recent Advances in the Biochemistry of Fruits and Vegetables. Friend, J. and Rhodes, M.J.C. (eds.). Academic Press, London.

24. Beyer, E.M. 1985. Ethylene metabolism, pp. 125-137. *In* Ethylene and Plant Development. Roberts, J.A. and Tucker, G.A. (eds.). Butterworths, Oxford.

25. Blanpied, G.D. 1990. Controlled atmosphere storage of apples and pears, pp. 266-299. *In* Food Preservation by Modified Atmospheres. Calderon M. and Barkai-Golan, R. (eds.). CRC Press, Boca Raton, FL.

26. Boersig, M.R., Kader, A.A. and Romani, R.J. 1988. Aerobic-anaerobic respiratory transition in pear fruit and cultured pear fruit cells. J. Am. Soc. Hort. Sci. 113: 869-873.

27. Bolin, H.R. and Huxsoll, C.C. 1989. Storage stability of minimally processed fruit. J. Food Proc. Preserve. 13: 281-292.

28. Bolin, H.R. and Huxsoll, C.C. 1991. Effect of preparation procedures and storage parameters on quality retention of salad-cut lettuce. J. Food Sci. 56: 60-67.

29. Bolin, H.R., Stafford, A.E., King, A.D. and Huxsoll, C.C. 1977. Factors affecting the storage stability of shredded lettuce. J. Food Sci. 42: 1319-1321.

30. Boller, T. and Kende, H. 1979. Hydrolytic enzymes in the central vacuoles of plant cells. Plant Physiol. 63: 1123-1132.

31. Boller, T. and Kende, H. 1980. Regulation of wound ethylene synthesis in plants. Nature 286: 259-260.

32. Brackett, R.E. 1987. Microbiological consequences of minimally processed fruits and vegetables. J. Food Qual. 10: 195-206.

33. Brady, C.J. 1987. Fruit ripening. Ann. Rev. Plant Physiol.38: 155-178.

34. Branen, L. and Davidson, P.M. 1983. Antimicrobials in Foods. Marcel Dekker, New York.

35. Brecht, P.E. 1980. Use of controlled atmospheres to retard deterioration of produce. Food Technol. 34(3): 45-50.

36. Burg, S.P. 1990. Theory and practice of hypobaric storage, pp. 353-372. *In* Food Preservation by Modified Atmospheres.

Calderon, M. and Barkai-Golan, R. (eds.). CRC Press, Boca Raton, FL.

37. Burg, S.P. and Burg, E.A. 1962. Role of ethylene in ripening. Plant Physiol. 36: 179-189.

38. Burg, S.P. and Burg, E.A. 1967. Molecular requirements for the biological activity of ethylene. Plant Physiol. 42: 144-152.

39. Burg, S.P. and Burg, E.A. 1969. Interaction of ethylene, oxygen and carbon dioxide in the control of fruit ripening. Qual. Plant Water. Veg. 19: 185-200.

40. Burton, W.G. 1974. Some biophysical principles underlying the controlled atmosphere storage of plant material. Ann. Appl. Biol. 78: 149-168.

41. Burton, W.G. 1978. Biochemical and physiological effects of modified atmospheres and their role in quality maintenance, pp. 97-110. *In* Postharvest Biology and Biotechnology. Hultin, H. and Milner, M. (eds.). Food and Nutrition Press, Westport, CT.

42. Burton, W.G. 1982. Post-Harvest Physiology of Food Crops. Longman, Inc., New York.

43. Burton, C.L. and Brown, G.K. 1984. Quality retention strategies for mechanically harvested fresh-market fruit, pp. 42-49. *In* Postharvest Pathology of Fruits and Vegetables. Moline, H.E. (ed.). Publ. NE-87. Agric. Exp. Sta., University of California, Davis, CA.

44. Butler, J.N. 1982. CO_2 Equilibria and their Applications. Addison-Wesley, London.

45. Calderon, M. and Barkai-Bolin, R. 1990. Food Preservation by Modified Atmospheres. CRC Press, Boca Raton, FL.

46. Cappellini, R.A. and Ceponis, M.J. 1984. Postharvest losses in fresh fruits and vegetables, pp. 24-30. *In* Postharvest Pathology of Fruits and Vegetables. Moline, H.E. (ed.) Public. NE-87. University of California, Berkeley, CA.

47. Carpita, N.C. 1982. Limiting diameters of pores and the surface structure of plant cell walls. Science 218: 813-814.

48. Carter, W.W. 1981. Reevaluation of heated water dip as a postharvest treatment for controlling surface and decay fungi of muskmelon fruits. Hort. Sci. 16: 334-335.

49. Chalutz, E., Waks, J. and Schiffman-Nadel, M. 1985. Reduced susceptibility of grapefruit to chilling injury. Hort. Sci. 20: 226-228.

50. Chaves, A.R. and Tomas, J.O. 1984. Effect of a brief CO_2 exposure on ethylene production. Plant Physiol. 76: 88-91.

51. Chinnan, M.S. 1989. Modeling gaseous environment and physicochemical changes of fresh fruits and vegetables in modified atmosphere storage, pp. 189-202. *In* Quality Factors of Fruits and Vegetables. Jen, J.J. (ed.). ACS Symp. Series 405. American Chemical Society, Washington, D.C.

52. Clegg, J.S. 1983. What is the cytosol? Trends in Bioch. Sci. 8: 436-437.

53. Clements, H.F. 1935. Morphology and physiology of the pome lenticles in *Pyrus malus*. Bot. Gaz. 97: 101.

54. Conway, W.S. 1984. Preharvest factors effecting postharvest losses from disease, pp. 11-16. *In* Postharvest Pathology of

Fruits and Vegetables. Moline, H.E. (ed.). Public. NE-87. University of California, CA.

55. Corlett, D.A. 1989. Refrigerated foods and use of hazard analysis and critical control point principles. Food Technol. 43(2): 91-94.

56. Couey, H.M. 1982. Chilling injury of crops of tropical and subtropical origin. Hort. Sci. 17: 162-165.

57. Couey, H.M. and Follstad, M.N. 1966. Heat pasteurization for control of postharvest decay in fresh strawberries. Phytopathol. 56: 1346-1347.

58. Couey, H.M. and Olsen, K.L. 1975. Storage response of "Golden Delicious" apples after high carbon dioxide treatment. J. Am. Hort. Sci. 100: 148-151.

59. Covington, A.K. 1985. Potentiometric titrations of aqueous carbonate solutions. Chem. Soc. Rev. 14: 265-281.

60. Crooks, P.R. and Grierson, D. 1983. Ultrastructure of tomato fruit ripening and the role of polygalacturonase isoenzymes in cell wall degradation. Plant Physiol. 72: 1085-1093.

61. Curtis, O.F. and Clark, D.G. 1950. An Introduction to Plant Physiology. McGraw-Hill, New York.

62. Dalrymple, G.D. 1969. The development of an agricultural technology: controlled atmosphere storage of fruits. Technol. Cult. 10(1): 35-48.

63. Daniels, J.A., Krishnamurthi, R. and Rizvi, S.S.H. 1985. A review of effects of carbon dioxide on microbial growth and food quality. J. Food Prot. 48: 532-537.

64. Davies, D.D. 1980. Anaerobic metabolism and the production of organic acids, pp. 581-611. In The Biochemistry of Plants. Vol. 2. Metabolism and Respiration. Davies, D.D. (ed.). Academic Press, New York.

65. Davies, D.D. 1980. The Biochemistry of Plants. Vol. 2. Metabolism and Respiration. Academic Press. New York.

66. Davis, P.L. and Hoffmann, R.C. 1973. Reduction of chilling injury of citrus fruits in cold storage by intermittent warming. J. Food Sci. 38: 871-873.

67. Davis, P.L., Roe, B. and Bruemmer, J.H. 1973. Biochemical changes in citrus fruits during controlled atmosphere storage. J. Food Sci. 38: 225-229.

68. Demorest, R.L. 1988. Non-destructive leak detection of blister packs and other sterile medical packages. J. Packaging Technol. 2: 182-190.

69. Dennis, C. 1983. Postharvest Pathology of Fruits and Vegetables. Academic Press, New York.

70. Dilley, D.R. 1990. Historical aspects and perspectives of controlled atmosphere storage, pp. 187-196. In Food Preservation by Modified Atmospheres. Calderon, M. and Barkai-Golan, R. (eds.). CRC Press, Boca Raton, FL.

71. Dixon, N.M. and Kell, D.B. 1989. The inhibition by CO_2 of the growth and metabolism of micro-organisms. J. Appl. Bacteriol. 67: 109-136.

72. Dougherty, R.H. 1990. Future prospects for processed fruit and vegetable products. Food Technol. 44(5): 124-125.

73. Duckworth, R.B. 1966. Fruit and Vegetables. Pergamon Press, London, England.

74. Eaks, I.L. and Morris, L.L. 1956. Respiration of cucumber fruits associated with physiological injury at chilling temperatures. Plant Physiol. 31: 308-314.
75. Eaks, I.L. 1960. Physiological studies of chilling injury in citrus fruits. Plant Physiol. 35: 632-636.
76. Eaks, I.L. 1980. Effect of chilling in respiration and volatiles of California lemon fruit. J. Amer. Soc. Hort. Sci. 105: 865-869.
77. Eaves, C.A. 1960. A modified atmosphere system for packages of fruit. J. Hort. Sci. 35: 110-117.
78. Eckert, J.W. 1975. Postharvest diseases of fresh fruits and vegetables - etiology and control, pp. 81-117. *In* Postharvest Biology and Handling of Fruits and Vegetables. Haard, N.F. and Salunkhe, D.K. (eds.). AVI Publ. Co., Westport, CT.
79. Eckert, J.W. 1978. Pathological diseases of fresh fruits and vegetables. J. Food Bioch. 2: 243-249.
80. Eckert, J.W. 1983. Control of postharvest diseases with antimicrobial agents, pp. 265-285. *In* Postharvest Physiology and Crop Preservation. Lieberman, M. (ed.). Plenum Press, New York.
81. Eckert, J.W. and Ogawa, J.M. 1988. The chemical control of postharvest diseases: Deciduous fruits, berries, vegetables and root/tuber crops. Ann. Rev. Phytopathol. 26: 433-469.
82. El-Goorani, M.A. and Sommer, N.F. 1981. Effect of modified atmospheres on postharvest pathogens of fruits and vegetables. Hort. Rev. 3: 412-461.
83. Esau, K. 1965. Plant Anatomy. John Wiley and Sons, Inc., New York.
84. Esau, K. 1977. Anatomy of Seed Plants. Second Ed. John Wiley and Sons, Inc., New York.
85. Eskin, N.A.M. 1991. Quality and Preservation of Fruits. CRC Press, Inc., Boca Raton, FL.
86. Fahn, H. 1982. Plant Anatomy. Third Ed. Pergamon Press, Oxford.
87. Faust, M.I. and Shear, C.B. 1972. The effect of calcium on the respiration of apples. J. Am. Soc. Hort. Sci. 97: 437-439.
88. Fawcett, D.W. 1982. The Cell. Second Ed. W.B. Saunders, Inc., Philadelphia, PA.
89. Fennema, O., Powrie, W.D. and Marth, E. H. 1973. Low-Temperature Preservation of Foods and Living Matter. Marcel Dekker, Inc. New York.
90. Fenwick, G.R. Johnson, I.T. and Hedley, C.L. 1990. Toxicity of disease-resistant plant strains. Trends in Food Sci. Technol. 1: 23-25.
91. Fidler, J.C. and North, C.J. 1967. The effect of storage on the respiration of apples. The effects of temperature and concentration of carbon dioxide and oxygen on the production of carbon dioxide and uptake of oxygen. J. Hort. Sci. 42: 189-206.
92. Frenkel, C. 1991. Disruption of macromolecular hydration - a possible origin of chilling destabilization of biopolymers. Trends in Food Sci. Technol. 2: 39-42.
93. Frenkel, C. and Patterson, M.E. 1973. Effect of carbon dioxide on activity of succinic dehydrogenase in "Bartlett" pears during cold storage. Hort. Sci. 8: 395-396.

94. Frenkel, C. Dyck, R. and Haard, N.F. 1975.   Role of auxin in the regulation of fruit ripening,   pp. 19-34.   *In* Postharvest Biology and Handling of Fruits and Vegetables.   Haard, N.F. and Salunkhe, D.K. (eds.).   AVI Publ. Co. Inc., Westport, CT.

95. Friend, J. and Rhodes, M.J.C. 1981.   Recent Advances in the Biochemistry of Fruits and Vegetables.   Academic Press, London.

96. Galliard, T. 1979.   The enzymic degradation of membrane lipids in higher plants, pp. 121-132.   *In* Advances in the Biochemistry and Physiology of Plant Lipids.   Appelquist, L-A. and Liljenberg, C. (eds.).   Elsevier Biomedical Press, Amsterdam.

97. Galliard, T., Matthews, J.A., Fishwick, M.J. and Wright, A.J. 1976. The enzymic degradation of lipids resulting from physical disruption of cucumber fruit.   Phytochemistry 15: 1731-1734.

98. Gardner, H.W. 1980.   Lipid enzymes: Lipids lipoxygenases and "hydroperoxidases",   pp. 447-504.   *In* Autoxidation in Food and Biological System.   Simic, M.G. and Karel, M. (eds.).   Plenum Press, New York.

99. Gardner, W.H. 1966.   Food Acidulants.   Allied Chemical Co., New York.

100. Geeson, J.D. Everson, H. and Browne, M. 1988.   Micro-perforated films for fresh produce.   Grower (April): 33-34.

101. Geeson, J.D., Browne, K.M., Maddison, J., Shepherd, J. and Guaraldi, F. 1985.   Modified atmosphere packaging to extend the shelf life of tomatoes.   J. Food Technol. 20: 339-349.

102. Gilbert, S.G. 1985.   Food/package compatibility.   Food Technol. 39(12): 54-56.

103. GPMC. 1988.   Grocery Attitudes of Canadians 1988.   Grocery Products Manufacturers of Canada,   Don Mills, Ontario.

104. GPMC. 1990.   Grocery Attitudes of Canadians 1990.   Grocery Products Manufacturers of Canada, Don Mills, Ontario.

105. Goodenough, P.W. 1982.   Comparative biochemistry of tomato fruit during ripening on the plant or retarded ripening.   Food Chem. 9: 253-267.

106. Goodenough, P.W. and Atkin, R.K. 1981.   Quality in Stored and Processed Vegetables and Fruit.   Academic Press, New York.

107. Goodenough, P.W. and Thomas, T.H. 1980.   Comparative physiology of field-grown tomatoes during ripening on the plant and retarded ripening in controlled atmospheres.   Ann. Appl. Biol. 94: 445-455.

108. Goodenough, P.W., Tucker, G.A., Grierson, D. and Thomas, T.H. 1982.   Changes in color, polygalacturonase, monosaccharides, and organic acids during storage of tomatoes.   Phytochemistry 21: 281-284.

109. Goodwin, T.W. and Mercer, E.I. 1983.   Introduction to Plant Biochemistry.   Second Ed. Pergamon Press, Oxford.

110. Gortner, W.A., Dull, G.G. and Krauss, B.H. 1967.   Fruit development, maturation, ripening and senescence: A biochemical basis for horticultural terminology.   Hort. Sci. 2: 141-144.

111. Gray, J.I., Harte, B.R. and Miltz, J. 1987.   Food Product-Package Compatibility.   Technomic Publ. Co., Lancaster, PA.

112. Grierson, D., Slater, A., Maunders, M., Crookes, P., Tucker, G.H., Schuch, W. and Edwards, K. 1985.   Regulation of the expression of tomato fruit ripening genes: the involvement of ethylene,

pp. 147-161. *In* Ethylene and Plant Development. Roberts, J.A. and Tucker, G.A. (eds.). Butterworths, London.

113. Grierson, D., Tucker, G.A. and Robertson, N.G. 1981. The molecular biology of ripening, pp. 149-160. *In* Recent Advances in the Biochemistry of Fruits and Vegetables. Friend, J. and Rhodes, M.J.C. (eds.). Academic Press, London.

114. Gunning, B.E.S. and Overall, R.L. 1983. Plasmodesmata and cell-to-cell transport in plants. BioScience. 33: 260-265.

115. Haard, N.F. 1985. Characteristics of Edible Plant Tissues, pp. 857-911. *In* Food Chemistry. Second Ed. O.R. Fennema (ed.). Marcel Dekker, Inc., New York.

116. Haard, N.F. 1984. Postharvest physiology and biochemistry of fruits and vegetables. J. Chem. Ed. 61: 277-283.

117. Haard, N.F. and Salunkhe, D.K. 1975. Postharvest Biology and Handling of Fruits and Vegetables. AVI Publ., Westport, CT.

118. Hall, E.G. 1974. Biological aspects of the cooling and freezing of fruit and vegetables. Part 1, pp. 37-73. *In* Refrigeration Applications to Fish, Fruit and Vegetables in South East Asia. FAO-Int. Inst. Refrig., Paris.

119. Hall, M.A. 1981. Cell wall structure in relation to texture, pp. 53-64. *In* Quality in Stored and Processed Vegetables and Fruit. Goodenough, P.W. and Atkins, P.K. (eds.). Academic Press, New York.

120. Hall, J.L. and Baker, D.A. 1975. Cell membranes, pp. 39-77. *In* Ion Transport in Plant Cells and Tissues. North-Holland Publ. Co., Amsterdam.

121. Hardenburg, R.E. 1956. How to ventilate packaged produce. Prepackage Age. 76: 14.

122. Hardenburg, R.E. 1971. Effect of in-package environment on keeping quality of fruit and vegetables. Hort. Sci. 6: 198-201.

123. Harte, B.R. and Gray, J.I. 1987. The influence of packaging on product quality, pp. 17-29. *In* Food Product-Package Compatibility. Gray, J.I., Harte, B.R. and Miltz, J. (eds.). Technomic Publ. Co., Lancaster, PA.

124. Hatton, T.T. and Cubbedge, R.H. 1982. Conditioning Florida grapefruit to reduce chilling injury during low temperature storage. J. Am. Soc. Hort. Sci. 107: 57-60.

125. Hatton, T.T. and Cubbedge, R.H. 1983. Preferred temperature for prestorage conditioning of 'Marsh' grapefruit to prevent chilling injury at low temperatures. Hort. Sci. 18: 721-722.

126. Heaton, E.K., Boggess, T.S. and Li, K.C. 1969. Processing refrigerated fresh peach slices. Food Technol. 23(7): 96-100.

127. Henig, Y.S. 1975. Storage stability and quality of produce packaged in polymeric films, pp. 144-152. *In* Postharvest Biology and Handling of Fruits and Vegetables. Haard, N.F. and Salunkhe, D.K. (eds.). AVI Publ. Co., Westport, CT.

128. Hicks, R. 1990. Consumer Food Trends for the 1990s. Food Development Division, Agriculture Canada, Ottawa.

129. Hicks, J.R., Ludford, P.M. and Masters, J.F. 1982. Effects of atmosphere and ethylene on cabbage metabolism during storage, p. 309. *In* Controlled Atmosphere for Storage and Transport of Perishable Agricultural Commodities. Richardson, D.G. and Meheriuk, M. (eds.). Timber Press, Beverton, OR.

130. Hobson, G.E. 1981. Enzymes and texture changes during ripening, pp. 123-132. *In* Recent Advances in the Biochemistry of Fruits and Vegetables. Friend, J. and Rhodes, M.J.C. (eds.). Academic Press, London.

131. Hoyem, T. and Kvale, O. 1977. Physical, Chemical and Biological Processing. Applied Science Publ., London.

132. Huber, D.J. 1983. The role of cell wall hydrolases in fruit softening, pp. 169. *In* Horticultural Reviews. Janick, J. (ed.). AVI Publ. Co., Westport, CT.

133. Huber D.J. 1984. Strawberry fruit softening: The potential roles of polyuronides and hemicelluloses. J. Food Sci. 49: 1310-1315.

134. Hultin, H.O. and Milner, M. 1978. Postharvest Biology and Biotechnology. Food and Nutrition Press, Westport, CT.

135. Huxsoll, C.C. and Bolin, H.R. 1989. Processing and distribution alternatives for minimally processed fruits and vegetables. Food Technol. 43(2): 124-128.

136. Huxsoll, C.C., Bolin, H.R. and King, A.D. 1989. Physicochemical changes and treatments for lightly processed fruits and vegetables, pp. 203-215. *In* Quality Factors of Fruits and Vegetables. Jen, J.J. (ed.). ACS Symp. Series 405. American Chemical Society, Washington, D.C.

137. Hyodo, H., Tamaka, K. and Yoshisaka, J. 1985. Induction of 1-aminocyclopropane-1-carboxylic acid synthase in wounded mesocarp tissue of winter squash fruit and the effect of ethylene. Plant and Cell Physiol. 26: 161-167.

138. Ilker, Y. and Morris, L.L. 1975. Alleviation of chilling injury of okra. Hort. Sci. 10: 324-325.

139. Isenberg, M.F.R. 1979. Controlled atmosphere storage of vegetables. Hort. Rev. 1: 337-394.

140. Iversen, E., Wihelmsen, E. and Criddle, R.S. 1989. Calorimetric examination of cut fresh pineapple metabolism. J. Food Sci. 54: 1246-1249.

141. Jackman, R.L. Yada, R.Y., Marangoni, A., Parkin, K.L. and Stanley, D.W. 1988. Chilling injury. A review of quality aspects. J. Food Qual. 11: 253-278.

142. Jen, J.J. 1989. Quality Factors of Fruits and Vegetables. ACS Symp. Series 405. American Chemical Society, Washington, D.C.

143. Jen, J.J. 1989. Chemical basis of quality factors in fruits and vegetables, pp. 1-9. *In* Quality Factors of Fruits and Vegetables. Jen, J.J. (ed.). ACS Symp. Series 405. American Chemical Society, Washington, D.C.

144. John, M.A. and Dey, P.M. 1986. Postharvest changes in fruit cell wall, pp. 139-193. *In* Advances in Food Research. Vol. 30. Chichester, C.O. Mrak, E.M. and Schweigert, B.S. (eds.). Academic Press Inc. New York.

145. Johnson, B.E. and Brun, W.A. 1966. Stomatal density and responsiveness of banana fruit stomates. Plant Physiol. 41: 99-101.

146. Joslyn, M.A. 1962. The chemistry of protopectin: A critical review of historical data and recent developments, pp. 1-107. *In* Advances in Food Research. Vol. 11. Chichester, C.O., Mrak, E.M. and Stewart, G.F. (eds.). Academic Press, New York.

147. Kader, A.A. 1980. Prevention of ripening in fruits by use of controlled atmospheres. Food Technol. 34(3): 51-54.

148. Kader, A.A. 1983.   Influence of harvesting methods on quality of deciduous tree fruits.   Hort. Sci. 18: 409-411.
149. Kader, A.A. 1985.   Postharvest biology and technology: An overview, pp. 3-7.   *In* Postharvest Technology of Horticultural Crops.   Kader, A.A., Kasmire, R.F., Mitchell, F.G., Reid, M.S., Sommer, W.F. and Thompson, J.F. (eds.).   Special Publ. 3311, University of California, Davis, CA.
150. Kader, A.A. 1985.   Modified atmosphere and low-pressure systems during transport and storage, pp. 58-67.   *In* Postharvest Technology of Horticultural Crops.   Kader, A.A., Kasmire, R.F., Mitchell, F.G., Reid, M.S., Sommer, W.F. and Thompson, J.F. (eds.).   Special Publ. 3311. University of California, Davis, CA.
151. Kader, A.A. 1985.   Quality factors: Definition and evaluation for fresh horticultural crops, pp. 118-121.   *In* Postharvest Technology of Horticulture Crops.   Kader, A.A., Kasmire, R.F., Mitchell, F.G., Reid, M.S., Sommer, W.F. and Thompson, J.F. (eds.).   Special Publ. 3311.   University of California, Davis, CA.
152. Kader, A.A. 1986.   Biochemical and physiological basis for effects of controlled and modified atmospheres on fruits and vegetables.   Food Technol. 40(5): 99-104.
153. Kader, A.A. and Morris, L.L. 1977.   Relative tolerance of fruits and vegetables to elevated CO_2 and reduced O_2 levels, pp. 260-265.   *In* Controlled atmospheres for the Storage and Transport of Horticultural Crops.   Dewey, D.H. (ed.).   Department of Horticulture.   Michigan State University, East Lansing, MI.
154. Kader, A.A., Zagory, D. and Kerbel, E.L. 1989.   Modified atmosphere packaging of fruits and vegetables.   CRC Crit. Rev. Food Sci. Nut. 28(1): 1-30.
155. Kader, A.A., Kasmire, R.F., Mitchell, F.G., Reid, M.S., Sommer, N.F. and Thompson, J.F. 1985.   Postharvest Technology of Horticultural Crops, Special Publ. 3311. University of California, Davis, CA.
156. Kahl, G. 1978.   Biochemistry of Wounded Plant Tissues.   Walter de Gruyter and Co., Berlin.
157. Kahl, G. 1983.   Wound repair and tumor induction in higher plants, pp. 193-213.   *In* The New Frontiers in Plant Biochemistry.   Akazawa, T., Ashai, T. and Imaski, H. (eds.).   Japan Scientific Societies Press, Tokyo.
158. Kasmire, R.F. 1983.   Influence of harvesting methods on quality of nonfruit vegetables.   Hort. Sci. 18: 421-423.
159. Kende, H., Acaster, M.A. and Guy, M. 1985.   Studies on the enzymes of ethylene biosynthesis, pp. 23-27.   *In* Ethylene and Plant Development.   Roberts, J.A. and Tucker, G.A. (eds.).   Butterworths, London.
160. Klein, B.P. 1987.   Nutritional consequences of minimal processing of fruits and vegetables.   J. Food Qual. 10: 179-193.
161. Kerbel, E.L., Kader, A.A. and Romani, R.J. 1988.   Effects of elevated CO_2 concentrations on glycolysis in intact "Bartlett" pear fruit.   Plant Physiol. 86: 1205-1209.
162. King, A.D. and Bolin, H.R. 1989.   Physiological and microbiological storage stability of minimally processed fruits and vegetables.   Food Technol. 43(2): 132-135.

163. Knee, M. 1980. Physiological responses of apple fruits to oxygen concentrations. Ann. Appl. Biol. 96: 243-253.
164. Knee, M. 1990. Ethylene effects in controlled atmosphere storage of horticultural crops, pp. 225-235. *In* Food Preservation by Modified Atmospheres. Calderon, M. and Barkai-Golan, R. (eds.). CRC Press, Boca Raton, FL.
165. Knee, M. and Bartley, I.M. 1981. Composition and metabolism of cell wall polysaccharides in ripening fruits, pp. 133-148. *In* Recent Advances in the Biochemistry of Fruits and Vegetables. Friend, J. and Rhodes, M.J.C. (eds.). Academic Press, London.
166. Knee, M. and Hatfield, S.G.S. 1981. The metabolism of alcohols by apple fruit tissue. J. Sci. Food Agric. 32: 593-600.
167. Knoche, W. 1980. Chemical reactions of CO_2 in water, pp. 3-11. *In* Biophysics and Physiology of Carbon Dioxide. Bauer, C., Gros, G. and Bartels, H. (eds.). Springer-Verlag, Berlin.
168. Labuza, T.P. and Breene, W.M. 1989. Applications of "active packaging" for improvements of shelf-life and nutritional quality of fresh and extended shelf-life foods. J. Food Proc. Preserv. 13: 1-69.
169. Laites, G.G. 1978. The development and control of respiratory pathways in slices of plant storage organs, pp. 421-446. *In* Biochemistry of Wounded Tissues. Kahl, G. (ed.). Walter de Gruyter and Co., Berlin.
170. Lance, C. 1981. Cyanide-insensitive respiration in fruits and vegetables, pp. 63-87. *In* Recent Advances in the Biochemistry of Fruits and Vegetables. Friend, J. and Rhodes, M.J.C. (eds.). Academic Press, London.
171. Lau, O.L. and Looney, N.E. 1982. Improvement of fruit firmness and acidity in controlled-atmosphere-stored "Golden Delicious" apples by a rapid O_2 reduction procedure. J. Am. Soc. Hort. Sci. 107: 531-534.
172. Lee, J.J.L. 1987. The design of controlled and modified packaging systems of fresh produce. *In* Food Product-Package Compatibility. Gray, J.I., Harte, B.R. and Miltz, J. (eds.). Technomic Publ. Co., Lancaster, PA.
173. Leshem, Y., Halevy, A.H. and Frenkel, C. 1986. Processes and Control of Plant Senescence. Elsevier Science Publ., Amsterdam.
174. Lewis, D.A. and Morris, L.L. 1956. Effect of chilling storage in respiration and deterioration of several sweet potato varieties. Proc. Amer. Soc. Hort. Sci. 68: 421-428.
175. Lieberman, M. 1979. Biosynthesis and Action of Ethylene. Ann. Rev. Plant Physiol. 30: 533-589.
176. Lieberman, M. and Wang, S.Y. 1982. Influence of calcium and magnesium on ethylene production by apple tissue slices. Plant Physiol. 69: 1150-1155.
177. Lioutas, T.S. 1988. Challenges of controlled and modified atmosphere packaging: A food company perspective. Food Technol. 42(9): 78-86.
178. Lipton, W.J. 1975. Controlled atmospheres for fresh vegetables and fruits - why and when, pp. 130-143. *In* Postharvest Biology and Handling of Fruits and Vegetables. Haard, N.F. and Salunkhe, D.K. (eds.). AVI Publ. Co., Westport, CT.

179. Lyons, J.M. and Breidenbach, R.W. 1987. Chilling injury, p. 305. *In* Postharvest Physiology of Vegetables. Weichmann, J. (ed.). Marcel Dekker, Inc., New York.
180. Lyons, J.M., Graham, D. and Raison, J.K. 1979. Low Temperature Stress in Crop Plants. Academic Press, New York.
181. Lund, B.M. 1981. The effect of bacteria on post-harvest quality of vegetables, pp. 287-300. *In* Quality in Stored and Processed Vegetables and Fruit. Goodenough, P.W. and Atkin, R.K. (eds.). Academic Press, London.
182. Lund, B.M. 1982. The effect of bacteria on postharvest quality of vegetables in fruits, with particular reference to spoilage, pp. 133-148. *In* Bacteria in Plants. Rhodes-Roberts, M.E. and Skinner, F.A. (eds.). Soc. Appl. Bacteriol., Symp. Ser. No. 10.
183. Magnen, M. 1970. Container for the preservation of fruit and vegetables. U.S. Patent No. 3,507,667.
184. Martens, M. and Baardseth, P. 1987. Sensory quality, pp. 427-454. *In* Postharvest Physiology of Vegetables. Weichmann, J. (ed.). Marcel Dekker, Inc. New York.
185. Marmé, D., Marré, E. and Hertel, R. 1982. Plasmalemma and Tonoplast: Their Functions in the Plant Cell. Elsevier Biomedical Press, Amsterdam.
186. Matile, P. 1978. Biochemistry and function of vacuoles. Ann. Rev. Plant Physiol. 29: 193-213.
187. Mattoo, A.R. and Anderson, J.D. 1984. Wound-induced increase in 1-amino-cyclopropane-1-carboxylic synthase activity: Regulatory aspects and membrane association of the enzyme, pp. 139-147. *In* Ethylene: Biochemical, Physiological and Applied Aspects. Fuchs, Y. and Chalutz, E. (eds.). Martinus Nijhoff/Dr. W. Junk Publishers, The Hague.
188. Mayer, A.M. and Harel, E. 1981. Polyphenol oxidases in fruits - changes during ripening, pp. 159-180. *In* Recent Advances in the Biochemistry of Fruits and Vegetables. Friend, J. and Rhodes, M.J.C. (eds.). Academic Press, London.
189. Mazliak, P. 1983. Plant membrane lipids. Changes and alterations during aging and senescence, pp. 123-140. *In* Postharvest Physiology and Crop Preservation. M. Lieberman (ed.). Plenum Press, New York.
190. McLachlan, A. and Stark, R. 1985. Modified Atmosphere Packaging of Selected Prepared Vegetables. Technical Memorandum No. 412. Campden Food Preservation Research Association, Campden, UK.
191. Mead, J.F. 1976. Formation of free radicals in membranes *in vitro* and *vivo*, pp. 58-68. *In* Free Radicals in Biology. Pryor, W. (ed.). Academic Press, New York.
192. Miller, W.R. and Smittle, D.A. 1987. Storage quality of hand- and machine-harvested rabbiteye blueberries. J. Am. Soc. Hort. Sci. 112: 487-490.
193. Mitchell, G.F. 1985. Cooling horticultural commodities, pp. 35-43. *In* Postharvest Technology of Horticultural Crops. Kader, A.A., Kasmire, R.F., Mitchell, F.G., Reid, M.S. Sommer, N.F. and Thompson, J.F. (eds.). Special Publ. 3311. University of California, Davis, CA.

194. Mohr, W.P. and Cocking, E.C. 1968. A method of preparing highly vacuolated, senescent, or damaged plant tissue for ultrastructural study. J. Ultrastruct. Res. 21: 171-181.

195. Mohr, W.P. and Stein, M. 1969. Fine structure of fruit development in tomato. Can. J. Plant Sci. 49: 549-553.

196. Moline, H.E. 1984. Postharvest Pathology of Fruits and Vegetables. Public. NE-87. University of California, Berkeley, CA.

197. Moline, H.E. 1984. Diagnosis of postharvest diseases and disorders, pp. 17-23. In Postharvest Pathology of Fruits and Vegetables. Moline, H.E. (ed.). Publ. NE-87. University of California, Davis, CA.

198. Monning, A. 1983. Studies on the reaction of Krebs cycle enzymes from apple (cv. Cox Orange) to increased levels of CO_2. Acta Hort. 138: 113-118.

199. Morris, L.L. 1982. Chilling injury of horticultural crops: An overview. Hort. Sci. 17: 161-162.

200. Morris, J.R. 1983. Influence of harvesting methods on quality of small fruits and grapes. Hort. Sci. 18: 412-417.

201. Myer, R.A. 1985. Modified atmosphere package and process. U.S. Patent No. 4,515,266.

202. Myers, R.A. 1989. Packaging considerations for minimally processed fruits and vegetables. Food Technol. 43(2): 129-131.

203. Nobel, P.S. 1970. Plant Cell Physiology. W.H. Freeman and Co., San Francisco, CA.

204. O'Beirne, D. 1988. Modified atmosphere packaging of ready-to-use potato strips and apple slices. Ir. J. Food Sci. Technol. 12: 94-95.

205. O'Beirne, D. 1990. Modified atmosphere packaging of fruits and vegetables, pp. 183-199. In Chilled Foods. The State of the Art. Gormley, T.R. (ed.). Elsevier Applied Science, New York.

206. O'Beirne, D. and Ballantyne, A. 1987. Some effects of modified atmosphere packaging and vacuum packaging in combination with antioxidants on quality and storage-life of chilled potato strips. Int. J. Food Sci. Technol. 22: 515-523.

207. O'Brien, M., Cargill, B.F. and Fridley, R.B. 1983. Principles and Practices for Harvesting and Handling Fruits and Nuts. AVI Publ. Co., Westport, CT.

208. Ogawa, J.M. and Manji, B.T. 1984. Control of postharvest diseases by chemical and physical means, pp. 55-66. In Postharvest Pathology of Fruits and Vegetables. Moline, H.E. (ed.). Publ. NE-87. University of California, Berkeley, CA.

209. Omarkhayyam, R. 1986. Free radicals and senescence, p. 116. In Processes and Control of Plant Senescence. Leshem, Y.Y., Halevy, A.H. and Frenkel, C. (eds.). Elsevier Press, New York.

210. Paine, F.A. 1987. Modern Processing, Packaging and Distribution Systems of Food. Blackie, Glasgow.

211. Paliyath, G., Poovaiah, B.W., Munske, G.R. and Magnuson, J.A. 1984. Membrane fluidity in senescencing apples: Effects of temperature and calcium. Plant Cell. Physiol. 25: 1083-1087.

212. Parkin, K.L., Marangoni, A., Jackman, R.L., Yada, R.Y. and Stanley, D.W. 1989. Chilling injury. A review of possible mechanisms. J. Food Bioch. 13: 127-153.

213. Pattee, H.E. 1985.   Evaluation of Quality of Fruits and Vegetables. AVI Publ. Co., Westport, CT.

214. Pearl, R.C. 1990.   Trends in consumption and processing of fruits and vegetables in the United States.   Food Technol. 44(2): 102-104.

215. Peleg, K. 1985.   Produce Handling, Packaging and Distribution. AVI Publ. Co., Westport, CT.

216. Pesis, E. Fuchs, Y. and Lauberman, G. 1978.   Cellulase activity and softening in avocado.   Plant Physiol. 61: 416-419.

217. Phan, C.T., Pantastico, E.B., Ogata, K. and Chachin, K. 1975. Respiration and respiratory climacteric. *In* Postharvest Physiology, Handling and Utilization of Tropical and Subtropical Fruits and Vegetables.   Pantastico, E.B. (ed.).   AVI Publ. Co., Westport, CT.

218. Pichia, D.H. 1986.   Postharvest fruit conditioning for reduced chilling injury in watermelons.   Hort. Sci. 21: 1407-1409.

219. Platt-Aloia, K.A. and Thomson, W.W. 1981.   Ultrastructure of the mesocarp of mature avocado fruit and changes associated with ripening.   Ann. Bot. 48: 451-465.

220. Poovaiah, B.W. 1979.   Role of calcium in ripening and senescence.   Comm. Soil Sci. and Plant Anal. 10: 83-88.

221. Poovaiah, B.W. 1986.   Role of calcium in prolonging storage life of fruits and vegetables.   Food Technol. 40(5): 86-89.

222. Ponting, J.D., Jackson, R. and Watters, A. 1972.   Refrigerated apple slices: Preservative effects of ascorbic acid, calcium and sulfites.   J. Food Sci. 37: 434-436.

223. Powrie, W.D., Wu, C.R.H. and Skura, B.J. 1990.   Preservation of cut and segmented fresh fruit pieces.   U.S. Patent No. 4,895,729.

224. Pressey, R. 1977.   Enzymes involved in fruit softening, pp. 172-191. *In* Enzymes in Food and Beverage Processing.   Ory, R.L. and St. Angelo, A.J. (eds.).   Am. Chem. Soc. Symp. Ser. 47. American Chemical Society, Washington, D.C.

225. Price, N.C. and Stevens, L. 1989.   Fundamentals of Enzymology. Oxford University Press, Oxford.

226. Priestley, R.J. 1979.   Effects of Heating on Foodstuffs.   Applied Science Publ., London.

227. Prince, T.A. 1989.   Modified atmosphere packaging of horticultural commodities. *In* Controlled/Modified Atmosphere/Vacuum Packaging of Foods.   Brody, A.L. (ed.).   Food and Nutrition Press, Trumbull, CT.

228. Prussia, S.E. and Woodroof, J.G. 1986.   Harvesting, handling and holding fruit, pp. 25-97. *In* Commercial Fruit Processing. Woodroof, J.G. and Luh, B.S. (eds.).   AVI Publ. Co., Westport, CT.

229. Putman, J.J. 1989.   Food consumption, prices and expenditures, 1966-87.   Stat. Bull. No. 73.   Econ. Res. Service, U.S. Dept. of Agric., Washington, D.C.

230. Ranson, S.L., Walker, D.A. and Clarke, I.D. 1957.   The inhibition of succinic oxidase by high CO_2 concentrations.   Bioch. J. 66: 57.

231. Ranson, S.L., Walker, D.A. and Clarke, I.D. 1960.   Effects of carbon dioxide on mitochondrial enzymes from Ricinus.   Bioch. J. 76: 216-221.

232. Reeve, R.M. 1953.   Histological investigation of texture of apples II.   Structure and intercellular spaces.   Food Res. 18: 604-617.

233. Reid, M.S. 1985. Product maturation and maturity indices, pp. 8-11. *In* Postharvest Technology of Horticultural Crops. Kader, A.A., Kasmire, R.F., Mitchell, F.G., Reid, M.S., Sommer, N.F. and Thompson, J.F. (eds.). Special Publ. 3311. University of California, Davis, CA.

234. Reid, M.S. 1985. Ethylene in postharvest technology, pp. 68-74. *In* Postharvest Technology of Horticultural Crops. Kader, A.A., Kasmire, R.F., Mitchell, F.G., Reid, M.S., Sommer, N.F. and Thompson, J.F. Special Publ. 3311. University of California, Davis, CA.

235. Rhodes, M.J.C., Wooltorton, L.S.C. and Hill, A.C. 1981. Changes in phenolic metabolism in fruit and vegetable tissues under stress, pp. 191-220. *In* Recent Advances in the Biochemistry of Fruits and Vegetables. Friend, J. and Rhodes, M.J.C. (eds.). Academic Press, London.

236. Robbins, J., Sjulin, T.M. and Patterson, M. 1989. Postharvest storage characteristics and respiration rates of five cultivars of red raspberry. Hort. Sci. 24: 980-982.

237. Roberts, J.A. and Tucker, G.A. 1985. Ethylene and Plant Development. Butterworth, London.

238. Robinson, J.E., Browne, K.M. and Burton, W.G. 1975. Storage characteristics of some vegetables and soft fruits. Ann. Appl. Biol. 81: 399-408.

239. Roepken, K.E. 1988. Consumer trends in the 1980s and implications for the dairy industry. Food Technol. 42(1): 123-125.

240. Rolle, R.S. and Chism, G.W. 1987. Physiological consequences of minimally processed fruits and vegetables. J. Food Qual. 10: 157-176.

241. Romig, W.R. and Orton, T.J. 1989. Applications of biotechnology to the improvement of quality of fruits and vegetables, pp. 381-393. *In* Quality Factors of Fruits and Vegetables. Jen, J.J. (ed.). ACS Symposium Series No. 405. American Chemical Society, Washington, DC.

242. Ronk, R.J., Carson, K.L. and Thompson, P. 1989. Processing, packaging and regulation of minimally processed fruits and vegetables. Food Technol. 43(2): 136-139.

243. Rosen, J.C. and Kader, A.A. 1989. Postharvest physiology and quality maintenance of sliced pear and strawberry fruits. J. Food Sci. 54: 656-659.

244. Rushing, N.B. and Senn, V.J. 1962. Effect of preservatives and storage temperature on the shelf-life of chilled citrus salads. Food Technol. 16(2): 77-79.

245. Ryall, A.L. and Lipton, W.J. 1972. Handling, Transportation and Storage of Fruits and Vegetables. Vol. 1. Vegetables and Melons. AVI Publ. Co., Westport, CT.

246. Ryall, A.L. and Pentzer, W.T. 1974. Handling, Transportation and Storage of Fruits and Vegetables. Vol. 2. Fruits and Tree Nuts. AVI Publ. Co., Westport, CT.

247. Salisbury, F.B. and Ross, C.W. 1985. Plant Physiology. Third Ed. Wadsworth Publ. Co. Belmont, CA.

248. Santerre, C.R., Leach, T.F. and Cash, J.N. 1991. Bisulfite alternatives in processing abrasion-peeled Russet Burbank potatoes. J. Food Sci. 56: 257-259.

249. Sapers, G.M. and Hicks, K.B. 1989. Inhibition of enzymic browning in fruits and vegetables, pp. 29-43. *In* Quality Factors of Fruits and Vegetables. Jen, J.J. (ed.). ACS Symp. Series 405. American Chemical Society, Washington, DC.

250. Sapers, G.M., Garzarella, L. and Pilizota, V. 1990. Application of browning inhibitors to cut apple and potato by vacuum and pressure infiltration. J. Food Sci. 55: 1049-1053.

251. Schallenberger, R.S. and Birch, G.C. 1975. Sugar Chemistry. AVI Publ. Co., Westport, CT.

252. Sfakiotakis, E.M. and Dilley, D.R. 1974. Induction of ethylene production in Bosc pears by postharvest cold stress. Hort. Sci. 9: 336-338.

253. Shear, C.B. and Faust, M. 1975. Preharvest nutrition and postharvest physiology of apples, pp. 35-42. *In* Postharvest Biology and Handling of Fruits and Vegetables. Haard, N.F. and Salunkhe, D.K. (eds.). AVI Publ. Co., Westport, CT.

254. Shewfelt, R.L. 1986. Postharvest treatment for extending the shelf life of fruits and vegetables. Food Technol. 40(5): 70-80.

255. Shewfelt, R.L. 1987. Quality of minimally processed fruits and vegetables. J. Food Qual. 10: 143-156.

256. Shewfelt, R.L. 1990. Quality of fruits and vegetables. Food Technol. 44(6): 99-106.

257. Shewfelt, R.L. and Erickson, M.C. 1991. Role of lipid peroxidation in the mechanism of membrane-associated disorders in edible plant tissue. Trends in Food Sci. Technol. 2: 152-154.

258. Shipway, M.R. and Bramlage, W.J. 1973. Effects of carbon dioxide on activity of apple mitochondria. Plant Physiol. 51: 1095-1098.

259. Sinclair, W.B. and Eny, D.M. 1946. The organic acids of grapefruit juice. Plant Physiol. 21: 140-147.

260. Singh, B., Wang, D.J., Salunkhe, D.K. and Rahman, A. 1972. Controlled Atmosphere Storage of Lettuce. 2. Effects on biochemical composition of the leaves. J. Food Sci. 37: 52-55.

261. Singh, B., Yang, C.C., Salunkhe, D.K. and Rahman, A.P. 1972. Controlled atmosphere storage of lettuce. 1. Effects on quality and respiration rate of lettuce heads. J. Food Sci. 37: 48-51.

262. Siriphanich, J. and Kader, A.A. 1986. Changes in cytoplasmic and vacuolar pH in harvested lettuce tissue as influenced by CO_2. J. Am. Soc. Hort. Sci. 111: 73-77.

263. Smith, W.H. 1957. Accumulation of ethyl alcohol and acetaldehyde in black currents kept in high concentrations of carbon dioxide. Nature 178: 876.

264. Smith, W.L. and Worthington, J.T. 1965. Reduction of postharvest decay of strawberries with chemical and heat treatments. Plant Dis. Rep. 49: 619-623.

265. Smith, S., Geeson, J. and Stow, J. 1987. Production of modified atmospheres in deciduous fruits by the use of films and coatings. Hort. Sci. 22: 772-776.

266. Smock, R.M. 1979. Controlled atmosphere storage of fruits. Hort. Rev. 1: 301-336.

267. Sofos, J.N. 1989. Sorbate Food Preservatives. CRC Press, Boca Raton, FL.

268. Somer, N.F. 1985.   Strategies for control of postharvest diseases of selected commodities, pp. 83-99.   *In* Postharvest Technology of Horticultural Crops.   Kader, A.A., Kasmire, R.F., Mitchell, F.G. Reid, M.S., Somer, N.F. and Thompson, J.F. (eds.).   University of California, Davis, CA.

269. Splittstoesser, D.F. 1987.   Fruits and fruit products, pp. 101-128. *In* Food and Beverage Mycology. Second Ed. Beuchat, L.R. (ed.). AVI Van Nostrand Reinhold, New York.

270. Spotts, R.A. 1984.   Environmental modification for control of postharvest decay, pp. 67-72.   *In* Postharvest Pathology of Fruits and Vegetables. Moline, H.E. (ed.).   Publ. NE-87. University of California, Berkeley, CA.

271. Sterling, C. 1975.   Anatomy of toughness in plant tissue, pp. 43-54. *In* Postharvest Biology and Handling of Fruits and Vegetables.   Haard, N.F. and Salunkhe, D.K. (eds.).   AVI Publ. Co. Inc., Westport, CT.

272. Stelzig, D.A. 1984.   Physiology and pathology of fruits and vegetables, pp. 36-41.   *In*   Postharvest Pathology of Fruits and Vegetables.   Moline, H.E. (ed.).   Publ. NE-87.   University of California, Berkeley, CA.

273. Stewart, J.K. and Wells, J.M. 1970.   Heat and fungicide treatments to control decay of cantaloupes. J. Am. Soc. Hort. Sci. 95: 226-229.

274. Stryer, L. 1988.   Biochemistry. Third Ed. W.H. Freeman and Co., New York.

275. Studer, H.E. 1983.   Influence of mechanical harvesting on the quality of fruit vegetables. Hort. Sci. 18: 417-421.

276. Swanson, B.G. and Bonorden, W.R. 1989.   Chemistry and safety of acidified vegetables, pp. 216-223.   *In* Quality Factors of Fruits and Vegetables. Jen, J.J. (ed.). ACS Symp. Series 405. American Chemical Society, Washington, D.C.

277. Thai, C.N., Prussia, S.E., Shewfelt, R.L. and Davis, J.W. 1986. Latent damage simulation and detection for horticultural products. Am. Soc. Agr. Engrs. Tech. Paper 86-6552.

278. Toledo, R., Steinberg, M.P. and Nelson, A.I. 1969.   Heat of respiration of fresh produce as affected by controlled atmosphere.   J. Food Sci. 34: 261-264.

279. Tomkins, R.G. 1962.   The conditions produced in film packages by fresh fruits and vegetables and the effect of these conditions on storage life. J. Appl. Bacteriol. 25: 290-307.

280. Tucker, M.L., Christoffersen, R.E., Woll, L. and Laties, G.G. 1985. Induction of cellulase by ethylene in avocado fruit, pp. 163-171. *In* Ethylene and Plant Development.   Roberts, J.A. and Tucker, G.A. (eds.). Butterworths, Oxford.

281. Ulrich, R. 1970.   Organic acids, pp. 89-118. *In* The Biochemistry of Fruits and Their Products. Vol. 1. Hulme, A.C. (ed.). Academic Press, New York.

282. Uritani, I. 1978.   Temperature stress in edible plant tissues after harvest, pp. 136-160.   *In* Postharvest Biology and Biotechnology. Hultin, H.O.   and Milner, M (eds.).   Food and Nutrition Press, Westport, CT.

283. USDA. 1982.   Composition of Foods: Fruits and Fruit Juices. Agriculture Handbook 8-9.   U.S. Dept. of Agriculture, Washington, D.C.

284. USDA. 1984. Composition of Foods: Vegetables and Vegetable Products. Agriculture Handbook 8-11. U.S. Dept. of Agriculture, Washington, D.C.

285. USHHS. 1988. The Surgeon General's Report on Nutrition and Health. U.S. Dept. of Health and Human Services. U.S. Govt. Print Office. Washington, D.C.

286. Wade, N.L. 1979. Physiology of cool-storage disorders of fruit and vegetables, pp. 81-96. *In* Low-temperature Stress in Crop Plant. Lyons, J.M., Graham, D. and Raison, J.K. (eds.). Academic Press, New York.

287. Wagner, G.J. 1982. Compartmentation in plant cells: the role of the vacuole, pp. 1-45. *In* Cellular and Subcellular Organization Plant Metabolism. Creasy, L.L. and Hrazdina, G. (eds.). Plenum Press, New York.

288. Wang, C.Y. 1982. Physiological and biochemical responses of plants to chilling stress. Hort. Sci. 17: 173-186.

289. Wang, C.Y. 1990. Physiological and biochemical effects of controlled atmosphere on fruits and vegetables, pp. 197-223. *In* Food Preservation by Modified Atmospheres. Calderon, M. and Barkai-Golan, R. (eds.). CRC Press, Boca Raton, FL.

290. Wang, C.Y. and Adams, D.O. 1980. Ethylene production by chilled cucumbers (*Cucumis sativus,* L). Plant Physiol. 66: 841-843.

291. Wang, C.Y. and Baker, J.E. 1979. Effects of two free radical scavengers and intermittent warming on chill injury and polar lipid composition of cucumber and sweet pepper fruits. Plant Physiol 20: 243-251.

292. Wang, S.S., Haard, N.F. and Di Marco, G.R. 1971. Chlorophyll degradation during controlled-atmosphere storage of asparagus. J. Food Sci. 36: 657-661.

293. Wankier, B.N., Salunkhe, D.K. and Campbell, W.F. 1970. Effect of controlled atmosphere storage on biochemical changes in apricot and peach fruits. J. Am. Soc. Hort. 95: 604-609.

294. Wasserman, B.P. 1990. Expectation and role of biotechnology in improving fruit and vegetable quality. Food Technol. 44(2): 68-70.

295. Watada, A.E., Abe, K. and Yamuchi, N. 1990. Physiological activities of partially processed fruits and vegetables. Food Technol. 44(5): 82-85.

296. Watada, A.E., Herner, R.C., Kader, A.A., Romani, R.J. and Staby, G.L. 1984. Terminology for the description of developmental stages of horticultural crops. Hort. Sci. 19: 20-21.

297. Wei, C., Cook, D.L. and Kirk, J.R. 1985. Use of chlorine compounds in the food industry. Food Technol. 39(1): 107-115.

298. Weichmann, J. 1987. Postharvest Physiology of Vegetables. Marcel Dekker, Inc., New York.

299. Weier, T.E. and Stocking, C.R. 1949. Histological changes induced in fruits and vegetables by processing, pp. 297-342. *In* Advances in Food Research. Vol. 2. Mrak, E.M. and Stewart, G.F. (eds.). Academic Press, New York.

300. Williams, A.A. 1981. Relating sensory aspects to quality, pp. 17-33. *In* Quality in Stored and Processed Vegetables and Fruit. Goodenough, P.W. and Atkin, R.K. (eds.). Academic Press, London.

301. Wills, R.B.H., McGlasson, W.B., Graham, D., Lee, T.H. and Hall, E.G. 1989. Postharvest: An Introduction to the Physiology and Handling of Fruit and Vegetables. Van Nostrand Reinhold, New York.

302. Worthington, J.T. and Smith, W.L. 1965. Postharvest decay control of red raspberries. Plant Dis. Rep. 45: 783-786.

303. Yang, S.F. 1981. Biosynthesis of ethylene and its regulation, pp. 89-106. *In* Recent Advances in the Biochemistry of Fruits and Vegetables. Friend, J. and Rhodes, M.J.C. (eds.). Academic Press, London.

304. Yang, S.F. 1985. Biosynthesis and action of ethylene. Hort. Sci. 20: 41-45.

305. Yang, S.F. and Hoffman, N.E. 1984. Ethylene biosynthesis and its regulation in higher plants. Ann. Rev. Plant Physiol. 35: 155-189.

306. Yang, S.F. and Pratt, H.K. 1978. The physiology of ethylene in wounded plant tissue, pp. 595-622. *In* Biochemistry of Wounded Plant Tissues. Kahl, G. (ed.). Walter de Gruyter and Co., Berlin.

307. Yang, S.F., Liu, Y., Su, L., Peiser, G.O., Hoffman, N.E. and McKeon, T. 1985. Metabolism of 1-aminocyclopropane-1-carboxylic acid. *In* Ethylene and Plant Development. Roberts, J.A. and Tucker, G.A. (eds.). Butterworths, London.

308. Yu, Y.B. and Yang, S.F. 1980. Biosynthesis of wound ethylene. Plant Physiol. 66: 281-285.

309. Zagory, D. and Kader, A.A. 1988. Modified atmosphere packaging of fresh produce. Food Technol. 42(9): 70-77.

310. Zagory, D. and Kader, A.A. 1989. Quality maintenance in fresh fruits and vegetables by controlled atmospheres, pp. 174-188. *In* Quality Factors of Fruits and Vegetables. Jen, J.J. (ed.). ACS Symp. Series 405. American Chemical Society, Washington, D.C.

311. Zapsalis, C. and Beck, R.A. 1985. Food Chemistry and Nutritional Biochemistry. John Wiley and Sons, New York.

312. Zind, T. 1988. Fresh Trends 1988. The Packer Focus, Vance Publ. Corp., Lincolnshire, IL.

313. Zind, T. 1989. Fresh trends '90 - A profile of fresh produce consumers. Packer Focus, Vance Publ. Corp., Lincolnshire, IL.

Chapter 8

MODIFIED ATMOSPHERE PACKAGING OF MISCELLANEOUS PRODUCTS

M.G. Fierheller
Food Processing Development Center, Alberta Agriculture

8.1 Convenience foods (minimally processed)

Convenience foods is as ambiguous a title as the chapter title of "Miscellaneous Products". For the purpose of this discussion convenience foods include refrigerated ready to consume meals and entrees, prepared salads, sandwiches, pizza, fresh pasta, soups and sauces. These products have received some form of heat treatment (minimally processed), are for the most part low acid, are marketed refrigerated (between -2°C and +4°C) but not frozen and require little preparation before consumption.

Recent marketing trends show increased sales of convenience foods. These trends are influenced by changing demographics. The ones commonly cited are increasing double-income families with limited time for home preparation, an ageing population with more disposable income, and the large market penetration of microwave ovens. The consumer perceives refrigerated as "fresh" and frozen or canned products as "processed". Manufacturers believe that consumers are looking for convenience, i.e. the ease and speed of preparation before consumption. They want quality in taste and freshness and the products to be nutritionally sound (low fat, salt, calories and cholesterol). Preservatives have a negative connotation and are to be avoided. The consumer wants variety, products that are similar to restaurant quality; products that are difficult to prepare at home from scratch. The popularity of ethnic flavoured entrees (Italian, Chinese, Mexican, Indian) is evident in North American and Western European markets [3, 9, 27, 34].

The popularity of the in-store supermarket 'deli' is increasing rapidly. This market is growing in the United States at an annual rate of 14.3% with total sales prediction of $28.1 billion by 1992 [27]. Another large retail market is convenience store food takeout. This market is growing at an annual rate of 17.3% to reach $6.2 billion in 1992. A third market for minimally processed refrigerated foods is the institutional market. Many restaurant chains, fast food outlets and hospitals have set up central commissaries to prepare soups, sauces, salads and main menu items.

Refrigerated minimally processed or chilled foods have a limited shelf life. In most situations an extended shelf life is required for effective marketing. The market size and location dictates the shelf life required. For example Marks and Spencers can access the UK market with a 3-d shelf life [27]. U.S. processors require 3 to 4 weeks [11, 27]. Canadian sandwich manufactures require 5 weeks' shelf life for national distribution [14]. In general, European processors can supply a market with chilled foods within a smaller radius from production facilities than can the North American processor. They therefore require a shorter shelf life.

Extended shelf life can be achieved through heat treatment (pasteurization, cooking); vacuum or modified-atmosphere (MA) packaging; where applicable pH and water activity reduction and, of course, good temperature control. A partial cook (pasteurization) or full cook for chilled entrees and ready-to-eat meals is a necessity for convenience. The consumer's only interest in preparation is to re-heat the product as quickly as possible, most often using a microwave oven. The heat treatment for cooking is usually sufficient to reduce the viable microbial population to very low levels and to denature enzymes but not to destroy bacterial spores.

8.1.1 Modified atmosphere packaging of minimally processed foods

There are several approaches to heat treatment and packaging. One method is to cook, chill and package. This approach is commonly used for processed meats. This method will assure a certain level of recontamination of microorganisms after cooking. The traditional method for packaging processed meats has been vacuum packaging. Vacuum packaging has provided an adequate shelf life (4 to 6 weeks). Vacuum packaging is not suitable for products that will suffer from the hypobaric pressures. One example is thin sliced processed meats that are difficult to separate if vacuum packaged. These products are packaged in a pillow pouch; the terms pillow pouch, gas-flushed and modified-atmosphere packaging (MAP) being synonymous.

The primary gas used for flushing is carbon dioxide (CO_2). Chapter 2 goes into detail on the effects of CO_2 on the growth of foodborne microorganisms and so is only superficially covered here.

CO_2 is very effective for controlling mould growth at temperatures less than 15°C. Chilled foods in air show signs of mould growth within 2 to 3 weeks. CO_2 levels greater than 20% control mould growth under good refrigeration temperatures (4°C). Yeasts are not easily controlled using CO_2 but in most cases they are not a factor limiting the shelf life of cooked foods. CO_2 will also effectively control the growth of spoilage bacteria especially gram-negative aerobic bacteria such as *Pseudomonas* spp. Lactic acid-producing bacteria such as *Lactobacillus* spp. are not easily controlled with CO_2 and will predominate [12].

Gas supply companies, equipment manufacturers and trade magazines give varying reports and recommendations on the optimum gas mixture to use. This information is sketchy and often conflicting. There are many statements similar to "the film used, the gas formulation and the process are trade secrets that took large amounts of money to put right" [16] or CO_2 concentrations greater than 20% to 30% cause sour odours and flavours, discoloration or weeping [2, 25].

What is the optimum CO_2 level? In general, the shelf life of cooked chilled foods increases with increasing CO_2 levels. A minimum CO_2 level of 20% is required for total inhibition of mould growth. This is also sufficient to shift the microflora to become dominated by lactic acid bacteria. CO_2 levels greater than 20% increase the shelf life of some cooked products but with others it has little effect. Too high a CO_2 concentration can result in package collapse and an unacceptable appearance.

In the example (Figure 8.1) of parfried French fries, a similar growth rate of lactic bacteria was shown at all three levels of CO_2. The cooking process reduced the initial bacterial load to low levels but the counts reach 10^6 and greater after 2 to 3 weeks, independent of CO_2 concentration.

Figure 8.2 shows similar results for precooked perogies (Ukranian product made of extruded wheat dough filled with cooked potato paste and cheese). This figure is a one-dimensional plot of Response Surface Methodology predictions of the number of days for lactic acid bacteria to reach one million colony forming units per gram (10^6 CFU/g). The differences resulting from the three levels of CO_2 are not significant but the storage temperature does have a significant effect. The time required for spoilage at 4°C was 21 to 25 d and at 0°C it was 55 to 60 d. The extra 4 to 5 d of storage could be attributed to the increased CO_2 level.

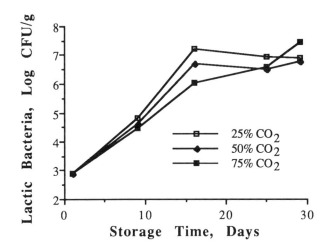

Figure. 8.1.   Parfried French fries packaged in 25, 50 and 75% CO_2 and stored at 4°C.

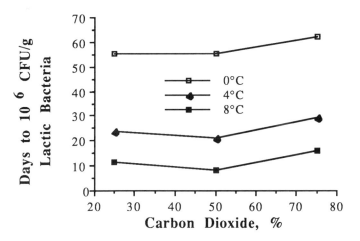

Figure 8.2.   Precooked perogies packaged in 25, 50 and 75% CO_2 and stored at 0, 4 and 8°C.

The addition of sauces to foods affects the shelf life. The growth rate of lactic acid bacteria on cooked pasta (linguini) with and without sauces is illustrated in Figure 8.3. One sauce was a low-acid, cream-based Alfredo sauce (pH 6.6) and the other was a tomato-based meat sauce (pH 4.5). The addition of the sauces reduced the shelf life from approximately 6 weeks for plain cooked pasta, to 4 weeks for the pasta with meat sauce and less than 3 weeks for the pasta with Alfredo sauce. The optimal CO_2 levels for sandwiches has been reported as 50% to 70% [14]. For Finnish ready-to-eat foods the optimal levels are reported to be 50% to 60% [2]. A review by Farber shows tables of gas combinations and shelf life of products being marketed in Canada [12]. The optimum gas mixture may at best only increase the shelf life by 10% over a gas mixture that is less than optimal. Should excessive CO_2 absorption at 60% or 70% create an unacceptable package appearance the level could be reduced, for example to 40%, with little loss of shelf life. CO_2 levels greater than 60% have no increased benefit [10].

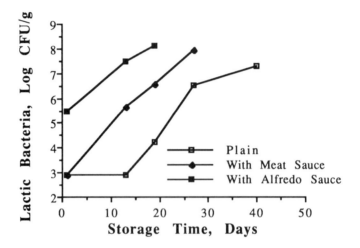

Figure 8.3.   Cooked linguini packaged in 50% CO_2 and stored at 4°C.

Chilled foods are being packaged successfully in MA on thermoforming and vertical or horizontal forming fill and seal equipment. A major difference between the two types of equipment is the level of O_2 remaining in the package. Thermoforming equipment is capable of achieving a residual O_2 of less than 0.5%. Form, fill and seal equipment can achieve a residual O_2 of 2% or greater depending on the rate of gas flushing.

The level of residual O_2 does not affect the growth rate of lactic acid bacteria and, as long as CO_2 levels are greater than 20%, moulds will not be a problem. High residual O_2 can be a factor in the shelf life if the product is susceptible to rancidity. The rate of growth of lactic acid bacteria was not influenced by the level of CO_2 or the level of residual O_2 in a study with prepared sandwiches but the shelf life of hamburger sandwiches was reduced from 5 weeks to 2 weeks in the presence of 5% to 10% O_2 [26].

8.1.2   Hot fill and pasteurized products

Hot fill for high-acid foods (pH <4.6) is a well-established process for shelf-stable products. Hot fill for low-acid foods is a recent development and presents new concerns should the product be abused after processing. This process is being promoted by the Cryovac division of WR Grace and Groen in the United States [15]. It is particularly useful for soups, sauces and other pumpable products. The products are filled into flexible film while still at pasteurizing temperatures, vacuum sealed and rapidly cooled. The products can be reheated by immersing the unopened package in hot water or pouring into steam trays, stock pots or kettles. This system is being used by central commissaries for bulk packaging and distribution to hospitals, restaurants and cafeterias.

The second approach to pasteurization of low-acid foods is to vacuum package before thermal processing. This process is referred to as "sous vide". It is described in detail elsewhere [5, 18]. Cooking can be accomplished using steam cabinets or hot water immersion for batch processes or continuously on equipment similar to Alfastar, Sweden [35] and TW Kutter, USA [19]. Both of these companies utilize microw?·-e energy for heating. The key step in both processes is temperature control. The time and temperature criteria is tailored for an optimum cook but it is usually sufficient to destroy all vegetative cells of microorganisms. Rapid cooling is employed to terminate the cooking and to move the temperature quickly through the optimum microbial growth zone (10°C to 50°C).

If the thermal process is optimally performed the product is cooked sufficiently and is likely be devoid of a vegetative microflora. There is very little information on microbial development and shelf life of "sous vide" or hot fill processed products. It is doubtful the spoilage organisms that grow under vacuum (lactic acid bacteria and yeasts) could survive the cooking process. Both methods of pasteurization eliminate post-processing contamination (assuming the packaging remains intact). The refrigerated shelf life would not be terminated because of the usual spoilage microorganisms.

The claimed shelf life of these products varies considerably. It is based more on the time required to market the product than on changes due to microbial growth. The maximum shelf life in France for vacuum-pasteurized chilled foods is 6 to 10 d [18], Culinary Brands in California claim 21 d [31] and Sonora in Canada claim 45 d for hot fill soups [15].

8.1.3 Health concerns

MAP and vacuum packaging of minimally processed foods is a relatively new technology. With it comes the unquantified risk from temperature abuse. There have been a number of research papers in the last ten years that have demonstrated a risk of growth of pathogenic bacteria exists under modified atmospheres and abusive storage conditions. One paper demonstrated nitrogen-flushed hamburger sandwiches inoculated with *Clostridium botulinum* and stored at room temperature for four days were organoleptically acceptable but toxic [23]. Results were similar in another experiment using similar products and conditions but inoculating the sample with *Staphylococcus aureus* [4].

The presence of aerobic spoilage bacteria (*Pseudomonas*) on pathogenic growth has been studied. Hintlian and Hotchkiss [20] demonstrated that a gas mixture of 75% CO_2, 10% O_2 and the balance nitrogen was optimal for inhibition of *P. fragi, C. perfringens* and *S. aureus* at 12.8°C in inoculation studies on cooked beef. In another study, *P. fragi* and *Listeria   monocytogenes* were inoculated on cooked chicken loaf [22]. High CO_2 (50% and 80%) inhibited the growth of *P. fragi* but was ineffective at preventing the growth of *L. monocytogenes*.

CO_2 effectively controls *Pseudomonas* growth but not the lactics. In many cases a lactic microflora predominates and reaches levels of 10^6 CFU/g in 2 to 3 weeks. The development of a healthy lactic population is a positive attribute. These bacteria will eventually terminate shelf life but their numbers increase to 10^7 to 10^8 CFU/g before there is a detectable change to the odour and flavour (souring). These high levels of lactic acid bacteria may also inhibit the growth of pathogenic bacteria should temperature abuse occur. In a study by Nielsen and Zeuthen [28], *Lactobacilli* and pathogenic bacteria were inoculated into vacuum-packaged processed meats. Their results showed that at 5°C and 8°C the lactics effectively controlled the growth of all pathogenic bacteria. At 12°C, *C. perfingens, B. cereus* and *S. enteritidis* growth was restricted but *S. aureus* and *Y. enterocolitica* grew rapidly.

A survey of fresh (wet) MAP pasta in Eastern Canada found three out of the five manufacturers' products had levels of 10^2 to 10^4 CFU/g of *S. aureus* in 20% of their products [29]. Many of the smaller Canadian pasta processors do not use pasteurization. A study evaluating the growth of lactics and *S. aureus* with temperature abuse gave the results shown in Fig. 8.4 [13]. In this case the pasta was not intentionally inoculated with contaminating microflora. The counts reported are the log numbers of bacteria after 2 weeks. Initial counts were approximately 10^3 to 10^4 for lactics and $>10^2$ for *S. aureus*. *S. aureus* competed favourably with the lactics reaching the highest numbers ($>10^6$) between 10°C and 15°C. Growth of *S. aureus* was somewhat inhibited at temperatures higher than 15°C because of strong lactic growth.

The probability that some level of post-thermal processing contamination occurs before MAP is high but there is no guarantee that this is always the case. Some commercial products have been observed to develop no spoilage microflora. Hot fill or pasteurizing of packaged product is not likely to result in post-processing contamination. On the positive side this would be beneficial to the shelf life of the product and reduce the risk of growth of some pathogenic bacteria such as *S. aureus* and *Salmonella*. However, if a margin of safety relies on the survival of spoilage organisms (lactics) then lack of post-processing contamination could be detrimental. There is no doubt that the risk of outgrowth of spores of *Clostridium* is real if given the chance (temperature abuse). To date there have been no reported cases of foodborne illness linked to the consumption of products prepared by these packaging systems [12].

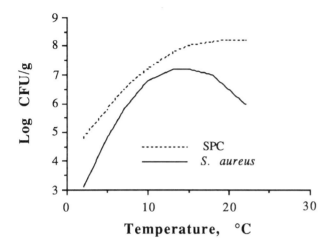

Figure 8.4.   Effect of storage temperature on microbial counts of fresh pasta containing 32% moisture, packaged in 60% CO_2 and stored for 2 weeks at various temperatures.

The majority of MAP, hot fill and sous vide-type products are low acid and high moisture. The only barrier to pathogenic out-growth is temperature control. As a result a popular topic at chilled foods and MAP conferences is the hazard analysis and critical control point (HACCP) programs for these products. Temperature control is the backbone of most HACCP programs. There are several papers published on this subject [6, 17].

8.2   Dairy products

One of the first food products to be gas flushed was retail packaged natural cheese [36]. Horizontal form and seal equipment (primarily Hayssen in North America) was adapted for CO_2 flushing. Adsorption of the high CO_2 head space created the desired vacuum package. Thermoforming vacuum packaging is being used success-fully.

Cultured dairy products by nature are reduced in pH owing to the growth of the desired bacteria starter culture. Cheeses are very susceptible to mould growth. With the exception of a few types of cheese (blue, Camembert and Roquefort cheese) moulds are an un-desirable spoilage organism. A tight-fitting package produced by CO_2

flushing or vacuum packaging limits the available O_2 and head space area required for mould growth.

Shredded cheese is in growing demand for food service [36] (i.e. mozzarella for pizza operations) and some retail packaging. This product is very susceptible to mould growth and requires MAP. Pillow packaging is also necessary to maintain the integrity of the product. A CO_2 level of 30% (balance N_2) is sufficient to control mould growth.

Some CO_2 flushing of the head space of yogurt and sour cream is taking place in Europe. Most North American dairies are relying on an adequate level of sanitation to control mould growth. Nitrogen flushing has been proposed for bulk storage of raw milk [33]. The method would be similar to N_2 sparging of bulk oil storage tanks to reduce the level of dissolved O_2. N_2 inhibited the growth rate of *Pseudomonas fluorescens* and the production of proteinase.

8.3   Intermediate   moisture   foods

Water activity (a_w) reduction will add a hurdle or barrier to microbial growth. Most spoilage and pathogenic bacteria grow most rapidly in the a_w range of 0.98 to 0.995 [26]. Reducing the a_w increases the length of the lag phase. There are few bacteria species that can grow below a_w 0.90. An a_w of less than 0.70 is necessary to prevent the growth of xerophilic yeasts and moulds [30]. Most food products are not palatable below a_w 0.70 without reconstitution.

CO_2 will effectively control mould growth in the range of a_w 0.70 to 0.90 at ambient temperatures. As an example, beef jerky, a popular North American snack food, has a moisture content between 30% and 40% (wet basis). Figure 8.5 shows a typical relationship between moisture content and a_w for beef jerky. The range for commercial jerky snacks is a_w 0.80 to 0.90. The jerky remains more pliable and organoleptically acceptable than if it is dried to an a_w less than 0.80. Many mould species are quite capable of growing in this range. High CO_2 (50%) or more commonly vacuum packaging is used to prevent growth at ambient temperatures.

The effectiveness of CO_2 at a_w levels greater than 0.90 is dependent on temperature, CO_2 concentration, pH and the use of antimicrobial agents. Most products that fall in this category are baked goods which are covered in Chapter 4. It is worthwhile to note that many countries have a_w regulations for shelf-stable food products. The Canadian regulations for "Low-Acid Foods Packaged in Hermetically Sealed Containers" use a cut-off a_w of 0.85 and pH of 4.6 for nonrefrigerated products [8]. The regulations do not apply to alcoholic beverages nor do they appear to apply to baked goods. Agriculture Canada, Veterinary Inspection Directorate will allow fermented meat products (pH <5.4) with an increased a_w of 0.90 to be packaged in a hermetically sealed package for ambient temperature storage.

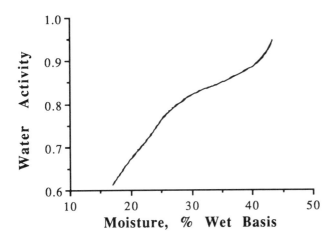

Figure 8.5   Sorption isotherm of beef jerky.

8.4   Dehydrated   foods

Below a_w 0.70 microbial growth is not a factor but deterioration can result from oxidation of lipids, vitamins (C and thiamin) and
pigments such as chlorophyll and carotene. Other changes such as
moisture gain and enzymatic and nonenzymatic browning can not
be controlled using modified atmospheres.

Oxygen, of course, is required for oxidative rancidity. Very
low levels of O_2 (<0.1%) were required to inhibit rancidity of
freeze-dried shrimp [32]. An O_2 level of 3.5% gave a 45-d shelf life
compared to less than 20 d with 7% O_2 and less than 1 d with 21% O_2.
Reduction of oxygen levels by nitrogen flushing or vacuum packaging has been used commercially for freeze-dried products, ground
coffee, roasted nuts, powdered whole milk, and dehydrated potato
flakes.

Snack foods would be the largest category of dried foods
(potato chips, corn chips, extruded snacks, roasted salted nuts and
confectionaries). Most are low-priced products and can not bear the
increased packaging costs associated with MAP. Many are deep fat
fried (high oil content) such as potato chips (North American) or
crisps (UK). In most cases a shelf life of 3 months is sufficient to successfully market these products. This length of shelf life can be attained through other means (than MAP). Proper maintenance of the
frying oil such as temperature control during frying and storage,
properly designed equipment and rapid oil turnover (product uptake) will result in finished products with low free fatty acid levels
[7]. Different oils have different stabilities and the selection of the
frying oil will influence shelf life. The use of antioxidants such as
butylated hydroxyanisole (BHA) and butylated hydroxytoluene (BHT)
are commonly used in fats and oils and the packaging liners of

breakfast cereals. Although consumer resistance to antioxidants is increasing they are more cost-effective than MAP.

Temperature has a marked influence on the rate of oxidation. It is important not only during frying but also during storage of the finished product. The Q_{10}, the rate change associated with a $10°C$ temperature change, can be 1.3 to 3 depending on the fatty acid profile [24].

Adverse effects of light (uv wavelength) induced rancidity and texture losses resulting from moisture changes can be controlled through packaging film selection. Foil and metallized laminates are commonly used for snack foods. Although they offer a good moisture and gas barrier the packages are not commonly nitrogen flushed.

If a shelf-life longer than 2 to 6 months is required, dehydrated foods will require protection from oxygen. Gas flushing or vacuum packaging in high gas barrier containers is necessary for ambient storage. The form fill and seal equipment commonly used for snack foods can be adapted for gas flushing. This equipment does not result in the total elimination of O_2. Residual O_2 levels between 2% and 3% are typical. This level is sufficient to oxidize the flavour compounds in coffee [24]. Equipment that pulls a vacuum then gas flushes can produce residual O_2 levels less than 0.5%. The use of O_2 scavengers can reduce the residual O_2 to one part per million [17].

The usual gas for flushing dehydrated foods is nitrogen. It is inert with a low fat and moisture solubility. CO_2 has been proposed for the packaging of nuts [21]. Adsorption of the CO_2 by the nuts creates a vacuum. The adsorption phenomenon is similar to gas adsorption by charcoal and can be used on a variety of grains, oil seeds, legumes, rice and corn. The shelf-life extension for pecans was from 2 to 4 weeks under ambient atmospheres to 27 weeks for CO_2-flushed product.

8.5 Conclusion

The big growth in future MAP utilization will be fresh, refrigerated partial- and full-cooked entrees ("convenience foods"). Processors need MAP or vacuum packaging with hot fill or pasteurization technology to achieve the shelf life required to meet consumer demands for these value-added, preservative-free food products.

I do not foresee new developments or major changes in the MAP technology. There will still be conflicting information on the optimal gas mixtures and packaging barrier properties required for various products, but generally the technology has been well researched. Advances will be made in the control of critical points from the initial process to distribution and retailing. The primary control is temperature. Process control of cooking and cooling time/temperature is required to assure a product with low microbial load and free of viable pathogenic organisms. Excellent temperature control ($0°\pm1°C$) is required to "guarantee" the product is free of pathogenic outgrowth. This may require new designs in retail display cases. It will require sound HACCP and quality assurance programs. Improvements in the present time/temperature indicators are required if they are to be a reliable tool. Better education of the consumers on proper handling techniques of refrigerated, perishable food products would be a positive step. This would not be a difficult task if part of the school curriculum could be set aside for this important life skill. After all, the processor and retailer can only assume partial liability for product abuse.

MAP will see limited use with intermediate moisture and dried food products. The largest category of food that can benefit from MAP is snack foods, which are high-volume, low-priced products. Some of the up-scaled high-protein, low-fat snack foods may be able to absorb the increased cost of MAP. There will be a niche market for MAP dairy products such as shredded cheese and products containing cheese such as Oscar Mayer's "Lunchables" (a sliced cheese, processed meat, cracker combo package). However, as a category MAP of dairy products will see limited use.

References:

1. Abe, L. and Kondoh, Y. 1989. Oxygen absorbers. *I n* Controlled/Modified Atmosphere/Vacuum Packaging. Brody, A. (ed.). Food & Nutrition Press, Inc., CT.
2. Ahvenainen, R. 1990. MAP of cooked cured meats and ready meals. Proceedings of International Conference on Modified Atmosphere Packaging, Oct. 15-17, 1990, Stratford upon Avon, England, published by the Campden Food and Drink Research Association, Chipping Campden, U.K.
3. Baker, J. 1990. The third wave of convenience foods. Proceedings of the Chilled Foods Association Annual Conference, Nov. 4-6, 1990, Orlando, FL, published by Chilled Foods Association, Atlanta, GA.
4. Bennett, R.W. and Amos, W.T. 1982. *Staphylococcus aureus* Growth and Toxin Production in Nitrogen-Packed Sandwiches. J. Food Prot. 45: 157-161.
5. Bristol, P. 1989. Sous Vide: gourmet meals in a vacuum package. Food in Canada 49(6): 15-18.
6. Bryan, F.L. 1990. Application of HACCP to ready-to-eat chilled foods. Food Tecnol. 44(7): 70-77.
7. Burdon, T.A.1983. Rancidity in snack foods, pp. 131-139. *I n* Rancidity in Foods. Allen, J.C. and Hamilton, R.J. (ed.). Applied Science Publishers, London and New York.
8. Canada: the Foods and Drug Act and Regulations 1989. Divison 27 Low acids foods packaged in hermetically sealed containers. Health and Welfare Canada, Ottawa, p. 73J.
9. Cristol, R.E. 1990. An overview of technical, regulatory and marketing issues of chilled, prepared foods. Proceedings of the Fifth Internatonal Conference on Controlled/Modified Atmosphere/Vacuum Packaging, January 17-19, 1990, San Jose, CA, published by Schotland Business Research, Inc., Princeton, NJ., pp. 185-196.
10. Daniels, J.A., Krishnamurthi, R. and Rizvi, S.S. 1984. A review of effects of carbon dioxide in microbial growth and food quality. J Food Prot. 48: 532-537.
11. El-Hag, N. 1989. The refrigerated food industry: current status and developing trends. Food Tech. 43(3): 96-98.
12. Farber, J.M. 1991. Microbiological aspects of modified-atmosphere packaging technology - a review. J. Food Prot. 54: 58-70.
13. Fierheller, M. 1989. Microwaveable modified atmosphere packaged sandwiches for convenience stores. Proceedings of Pack Alimentaire '89, June 13, 1989, Chicago, IL., published by Innovative Expositions, Inc., Princeton, NJ.
14. Fierheller, M. 1990. Commercialization of MAP: the Alberta scene. Proceedings of International Conference on Modified Atmosphere Packaging, Oct. 15-17, 1990, Stratford upon Avon, England, published by the Campden Food and Drink Research Association, Chipping Campden, U.K.
15. Food in Canada 1985. Albondigas soup anyone? call Sonora. Food in Canada 44(6): 18-21.
16. Food in Canada 1985. Natural Fry Inc - new process, bright future. Food in Canada 44(8): 67-68.

17. Food Technology Overview 1990. Hazard analysis and critical control point (HACCP) system and food safety. Food Technol. 44(5): 156-180.

18. Gorlich, M.P. 1987. Vacuum packaging and cooking of haute cuisine meals for prolonged refrigeration distribution. Proceedings of the Third Internatonal Conference on Controlled /Modified Atmosphere/Vacuum Packaging, September 16-18, 1987. Itasca, IL., published by Schotland Business Research, Inc., Princeton, NJ.

19. Harlfinger, L. 1990. Microwave pasteurizaton to prolong refrigerated shelf-life. Proceedings of the Fifth International Conference on Controlled/Modified Atmosphere/Vacuum Packaging, January 17-19, 1990, San Jose, CA, published by Schotland Business Research, Inc., Princeton, NJ., pp. 185-196.

20. Hintlian, C.B. and Hotchkiss, J.H. 1987. Comparative growth of spoilage and pathogenic organisms in modified atmosphere-packaged cooked beef. J. Food Prot. 50: 218-223.

21. Holaday, J.L., Pearson, J.L. and Slay, W.O. 1979. A new packaging method for peanuts and pecans. J. Food Sci. 44: 1530-1533.

22. Ingham, S.C., Escude, J. and McCown, P. 1990. Comparative growth rates of *Listeria monocytogenes* and *Pseudomonas fragi* on cooked chicken loaf stored under air and two modified atmospheres. J. Food Prot. 53: 289-291.

23. Kautter, D.A., Lynt, R.K., Lilly, T. Jr. and Solomon, H.M. 1981. Evaluation of the botulism hazard from nitrogen-packed sandwiches. J. Food Prot. 44: 59-61.

24. Labuza, T.P. 1982. Shelf-Life Dating of Foods. Food & Nutrition Press Inc., Westport, CT.

25. Leeson, R. 1987. The use of gaseous mixtures in controlled and modified atmosphere packaging. Food Technol. New Zealand, 22(6): 24-25.

26. McMullen, L.and Stiles, M.E. 1989. Storage life on selected meat sandwiches at 4°C in modified gas atmospheres. J. Food Prot. 52: 792-798.

27. Morris, C.E. 1989. Convenience foods: the second revolution. Food Eng. 61(9): 69-82.

28. Nielsen, H.-J. and Zeuthen, P. 1985. Influence of lactic acid bacteria and the overall flora on development of pathogenic bacteria in vacuum-packed, cooked emulsion-style sausage. J. Food Prot. 48: 28-34.

29. Park, C.E., Szabo, R. and Jean, A. 1988. A survey of wet pasta packaged under a $CO_2:N_2$ (20:80) mixture for staphylococci and their enterotoxins. Can. Inst. Food. Sci. Technol. J. 21: 109-111.

30. Pitt, J.I. 1974. Xerophilic fungi and the spoilage of foods of plant origin, pp. 273-307. *In* Water Relations of Foods. Duckworth, R.B. (ed.). Academic Press, London.

31. Przybyla, A.E. 1989. State-of-the-art sous vide plant under construction in California. Food Eng. 61(6): 103-106

32. Simon, I.B., Labuza, T.P. and Karel, M. 1971. Computer-aided predictions of food storage stability: Oxidative deterioration of a shrimp product. J. Food Sci. 36: 280-286.

33. Skura, B.J., Craig, C. and McKellar, R.C. 1986. Effect of disruption of an N_2-overlay on growth and proteinase producton in milk by

Pseudomonas fluorescens. Can. Inst. Food Sci. Technol. J. 19: 104-106.

34. Smith, L. 1990. The right product, at the right time, in the right place. Proceedings of the Chilled Foods Association Annual Conference, Nov. 4-6, 1990, Orlando, FL, published by Chilled Foods Association, Atlanta, GA.

35. Uhnbom, P. 1990. Alfastar ready meal system. Proceedings of the Fifth International Conference on Controlled/Modified Atmosphere/Vacuum Packaging, January 7-19, 1990, San Jose, CA, published by Schotland Business Research, Inc., Princeton, NJ., pp. 135-139.

36. Vichos, T. 1989. Update on trends in cheese packaging. Proceedings of Pack Alimentaire '89, June 13, 1989, Chicago, IL., published by Innovative Expositions, Inc., Princeton, NJ.

Chapter 9

FURTHER RESEARCH IN MODIFIED ATMOSPHERE PACKAGING

B. Ooraikul
Department of Food Science, University of Alberta

As research into and industrial applications of modified atmosphere packaging (MAP) evolved many constraints as well as opportunities related to this technology have become apparent. The constraints are largely the results of incomplete understanding of scientific principles underlying this system of food preservation, and the lack of technical know-how to apply the scientific principles to the best advantage. The effects of MAs on the microflora, physical, chemical and organoleptic properties of various products need to be thoroughly studied. This will form the necessary scientific base for this fledgling technology. Continued technical developments on packaging equipment, packaging films, and the methods by which the headspace atmosphere may be modified are essential. Substantial investment in research time and money is required to ensure survival and continued success of MAP system. It must be remembered that MAP is but one of the technologies available to food processors in providing high-quality, near-fresh products to increasingly sophisticated and demanding consumers. Other processes, such as aseptic packaging, microwave pasteurization or sterilization, and irradiation, particularly the last when it becomes accepted by consumers, are the major competitors. However, when fully understood and properly applied MAP should have several economic and quality advantages over its competitors.

There are many aspects of MAP that require immediate as well as long-term studies. The following are topics that should be considered for further investigation.

1. Shelf-life extension and safety
   a. Promotion of the growth of certain microorganisms on products
   b. Use of preservatives in combination with MAP
2. Packaging materials
   a. Ongoing development of barrier films
   b. Use of edible films as intermediate barriers
3. Headspace atmospheres
   a. Gas combinations and modifications
   b. Modelling of MAP regimes for produce
4. Effect of MAP on organoleptic quality of products
5. Quality assurance for MAP operations

9.1 Shelf life extension and safety

Safety is one of the cornerstones of an efficient food supply. Without guarantees of safety for consumer health a food product is useless. This implies that the product must be free from known microbial, chemical or physical hazards.

Assurance of microbial safety is particularly important with MAP products, as has been discussed in previous chapters. This is because of the susceptibility of some products to growth and toxin production of some pathogenic organisms, especially spore-forming anaerobes such as *Clostridium botulinum*. The major problem stems from the fact that gases used in MAP, especially CO_2, retard the spoilage microorganisms, thus minimizing competition for clostridia, which are not affected by CO_2. These conditions may allow pathogens to grow and produce toxin while the food remains organoleptically acceptable. The worst fate that could happen to MAP technology would be an outbreak of botulism from an apparently perfect MAP product.

Hintlian and Hotchkiss [30] recommended the use of the ratio of counts of spoilage organisms to counts of pathogens under certain MAP atmospheres as an indicator of the relative safety of MAP products. Higher ratios represent lower risk as spoilage organisms outgrow the pathogens, resulting in the product being spoiled before toxin production. Their studies on cooked roast beef have led to the suggestion that a combination of 75% CO_2 with more than 2% O_2 in the headspace atmosphere, and low-temperature storage, is a safeguard against clostridial hazards. However, the level of O_2 in the headspace and/or the storage temperature in the food chain cannot be guaranteed without expensive and rigorous quality control and monitoring programs. Other foolproof safety measures must be devised before MAP technology is universally accepted.

Research is needed to elucidate the effect of CO_2 and other gases and chemicals such as CO, SO_2, ethanol, ethylene oxide and propylene oxide on pathogens, especially clostridia. Mechanisms through which these gases affect the organisms must be well understood so that their applications in MAP is effective. The following are some of the measures which deserve further study to extend shelf life of MAP products and to assure their safety.

9.1.1 Promotion of growth of certain microorganisms on product

Use of other microorganisms and preservatives to prevent the outgrowth of clostridia and other pathogens has been actively investigated. Storage studies of vacuum-packaged and MAP meat products have revealed that lactobacilli predominate after a few weeks of storage [4, 17, 18, 37, 63, 71]. This has been used to advantage in extending safe shelf life of some MAP products. For example, in crumpets with an initial plate count of 8×10^3 CFU/g, packaged with CO_2 and N_2 (3:2) and stored at 25°C for 21 d, lactic acid bacteria (LAB) increased from undetectable levels to almost 100% of the total plate count of 10^8 CFU/g [63]. This rapid growth of LAB was accompanied by the depletion of residual O_2 and an increase in CO_2 content in the headspace. Two major problems have been encountered in the ambient storage of MAP crumpets, i.e. development of mould spots and/or swelling of packages due to excessive CO_2 production. The former was due to the existence of residual O_2 in the headspace during the first 5-7 d of storage, and the latter was due to the vigorous growth of LAB and other CO_2-producing organisms.

Experience had shown that if the growth of LAB and other common bacteria such as *Bacillus licheniformis* was slow in the first 5 d of storage so that >0.5% O_2 remained in the headspace, mould spots would develop (see Chapter 4). On the other hand, if bacteria grew rapidly during the first 5 d and remained quite vigorous thereafter mould spots would not develop but the package would swell. Therefore, if balanced growth of LAB and bacilli could be achieved, so that their growth could be drastically curtailed after 5-7 d, both mould spots and swelling problems would be avoided. Indeed, this was accomplished in MAP crumpets by a slight modification of the product recipe and an addition of a small quantity of δ-gluconolactone to reduce product pH and a_w. More than 2 months' mould-free shelf life at ambient temperature, without swelling, has been achieved for MAP crumpets.

Growth of LAB is also promoted on meats and meat products under MAP. LAB have been known to outgrow pathogens under MAP conditions. Certain strains of LAB are now being studied for their effectiveness in controlling growth of contaminating pathogens on MAP meats and meat products [2]. LAB produce a variety of metabolites such as lactic acid, H_2O_2 and antibiotic-like bactericidal proteins called bacteriocins [20, 39]. These substances, especially bacteriocins produced by *Pediococcus acidilactici* and *P. pentosaceus*, are active against a wide spectrum of bacteria [33]. However, to be suitable for application on MAP meats the bacteria must be able to grow and produce bacteriocins at low storage temperatures, at the natural pH of meat of 5.5 or lower, and should not produce an excessive amount of lactic acid to cause souring [57].

The use of harmless organisms to combat harmful organisms on foods is an interesting concept that deserves extensive study. Many naturally occurring microorganisms found on foods have the ability not only to outgrow pathogens but also to produce metabolites which are potentially harmful to them. Through selection processes these organisms may be identified, and with genetic engineering methods their beneficial abilities as "biological preservatives" may be enhanced. Indeed, some controversial chemical preservatives may be replaced with these microorganisms.

9.1.2   Use of preservatives in combination with MAP

Addition of certain preservatives, especially antimicrobial compounds, to MAP products may provide an additional safety measure while assisting in the extension of shelf life. Delvocid and parabens, for example, have been shown to be effective against yeasts in cherry cream cheese cake, while sorbic acid or sorbates help prevent the development of mould spots on crumpets (Chapter 4).

The various types of antimicrobials have been reviewed by Dziezak [23], Beuchat and Golden [9] and Wagner and Moberg [68]. These compounds may be naturally occurring, in animals, plants or microorganisms, or they may be synthetic. For example, lysozyme, which is present in milk and eggs, is inhibitory to *Listeria monocytogenes*, *Campylobacter jejuni*, *Salmonella typhimurium*, *Bacillus cereus* and *Clostridium botulinum* [34]. A combination of citric and ascorbic acids can inhibit growth and toxin production of *C. botulinum* [53]. Other organic acids such as succinic, malic, tartaric, benzoic, lactic and propionic acids are inhibitory to varying degrees to some bacteria, moulds or yeasts. Some medium-chain-length fatty acids such as lauric, myristic and palmitic acids are effective inhibitors of bacteria while capric and lauric acids are more effective against yeasts [35]. Compounds from some herbs, e.g. thymol from thyme and oregano, cinnamic aldehyde from cinnamon and eugenol from cloves, are known to have wide spectra of antimicrobial activity [9]. Some phenolic compounds such as anthocyanins are inhibitory to *Escherichia coli*, *Staphylococcus aureus* and *Lactobacillus casei* [55]. Caffeine is effective against several mycotoxigenic moulds [14, 15, 47].

Common chemical preservatives include benzoic acid and sodium benzoate, alkyl esters of *p*-hydroxybenzoic acid (parabens), sorbates, propionates, sulphites, acetates and diacetates, nitrite and nitrate, ethylene and propylene oxides, diethyl pyrocarbonate and alcohols such as ethanol [16]. Benzoic acid and sodium benzoate are most active against yeasts and bacteria between pH 2.3 and 4.0. Parabens are effective against moulds and yeasts over a much wider range of pH (3-9). Sorbic acid and sorbates have broad spectrum activity up to pH 6.5 against moulds and yeasts and some against bacteria. Propionates are more active against moulds than benzoates, but have no activity against yeasts and little activity against bacteria. Sulphites are probably the most universal food preservatives with broad spectrum activity against moulds, yeasts and bacteria over a wide range of pH. Unfortunately, owing to the sensitivity of some groups of people, e.g. asthmatics, the use of these preservatives has been greatly curtailed [66]. Acetic acid and its salts are generally more effective against yeasts and bacteria than moulds, especially at low pH. Nitrite and nitrate are commonly used in cured meats for proper development of colour and to inhibit outgrowth of bacterial spores, notably spores of *C. botulinum*. Like sulphites their use has been curtailed due to the possible formation of carcinogenic nitrosamines. Ethylene and propylene oxides are normally used as gaseous sterilants for dry products such as spices and flours. They kill all microorganisms, the former being more effective. However, the presence of residues, especially ethylene glycol, in some products poses a safety problem. Propylene oxide is about one-third as toxic as ethylene oxide. It is generally recognized as safe, therefore it is more acceptable as a food additive. Diethyl pyrocarbonate has broad spectrum antimicrobial activity, especially against yeasts. It has no toxicity or residue problems. Ethanol has been used largely as a sterilant for equipment. Recently it has been used to control microbial growth and extend shelf life of some bakery products (see Chapter 4).

Antibiotics which have been used as food preservatives include chlortetracycline, oxytetracycline, nisin, pimaricin (natamycin) and nystatin [16]. They exhibit selective antimicrobial activity, and a few are broad spectrum inhibitors. They are generally more expensive than other preservatives, and because of their static action it has been suggested that they may be used as adjuncts to other methods of preservation.

As the demand for foods with fresh or near-fresh quality grows there has been a decrease in the use of traditional heat processes and in the severity of heat treatment. While extended shelf life and microbiological stability and safety are desired, there is increasing concern over excessive use of chemical preservatives. Faced with these challenges food scientists and technologists must become more creative in developing new processing and preservation techniques, or balancing the use of the existing techniques.

Microbiological stability and safety of food products is the consequence of preservative factors acting in combination, often at levels at which singly they would not be inhibitory [56]. Therefore, use of preservatives at moderate concentrations in combination with mild heat treatment and/or other preservation methods may be one of the keys to provide minimally processed foods. In this connection, research into the use of naturally occurring preservatives should be accelerated as adding "natural" substances to the food would be much more acceptable than adding "synthetic" chemicals.

A comprehensive research program should be developed to evaluate the suitability and effectiveness of various preservatives, or combinations of preservatives, together with MAP, to provide high quality and safe products with extended shelf life. However, experiments such as these would be very complicated because many factors come into play in determining the growth of various microorganisms and other changes on the product. Appropriate tools in experimental designs, e.g. Response Surface Methodology (RSM) [11, 29], can be used to screen the factors and focus only on important ones while keeping the number of experiments to a manageable level. The technique involves factorial experimental designs and multiple regression analyses. Generally, first- and second-order polynomials are generated to define the levels of the most significant factors required to optimize the desired response. Commercial computer software is available to simplify the application of this statistical technique, handling of data and generation of graphical response surfaces which enable the prediction of responses (e.g. microbial growth) under various conditions. Smith *et al.* [64] used RSM to predict optimum a_w, headspace CO_2 concentration and storage temperature for desired mould-free shelf life of crumpets. Roberts [56] proposed the concept of "predictive microbiology" in which the growth response of microorganisms of concern could be modelled with respect to the main controlling factors of temperature, pH and a_w. Using this database, the effects of other factors on growth could be evaluated. This concept has been used to develop mathematical models to predict the probability of toxin production in a model system as a function of certain preservatives, heat treatment and storage temperature. Obviously, much more work is needed to generate data for various models involving preservatives and MAP.

Another possible method of combining preservatives with MAP is by dipping or soaking products in marinades prior to the application of MAP. This method is particularly suitable for meats and seafoods. Marinades which consist of vinegar, salt, sugar, soy sauce and spices should have good antimicrobial properties. Some other preservatives may also be added to the marinades. In addition to improved safety and shelf life, marinated products would also have an improved organoleptic quality. These products should be popular for home use and for the foodservice industry, as they are ready for use, e.g. barbecuing, roasting, stewing, etc., having superior flavour and taste due to the long period of marination during storage.

9.2 Packaging materials

Important new developments of packaging films generally concern physical integrity, barrier properties, adaptability of some physical properties to changing headspace atmosphere and storage conditions, and microwavability. The following are some of the areas that should be considered for further research and development.

9.2.1 Ongoing development of barrier films

One of the major costs in MAP operation is the cost of packaging materials. Aboagye *et al.* [1] estimated that packaging film accounted for 95% of packaging cost of MAP crumpets. This cost increases with the increase in the thickness of the film and its barrier against gases and water vapour. Many products require high-barrier films that provide packaging integrity to prevent product damage during the evacuation cycle. Therefore, to make MAP technology more competitive the film cost must be reduced, but not at the expense of the desired properties.

A high-barrier film of thinner gauge, strengthened with paperboard, polystyrene, aluminum foil or other low-cost materials may considerably reduce packaging cost. A recent development in Japan pioneered the use of electron beam technology to coat polyethylene terephthalate (PET), polyester, nylon and polypropylene films with silicon dioxide to provide superior strength, and improved O_2 and moisture barriers [3]. A combination of aluminium or tin plate with flexible films may also be considered when both strength and barrier are required.

Developments of "smart films" and microwavable films have been discussed in Chapter 3. Smart films would be much more difficult to develop, but if such films are available many problems encountered in the application of MAP to respiring produce, meats and seafoods would be solved. Another possible development in barrier films is the incorporation of atmosphere modifiers to the films. Thus, coating the inner surface of the films with "Ageless" or "Ethicap", for example, would greatly improve the effectiveness of MAP for some products. This form of application of these materials would be more acceptable to consumers than the inclusion of small sachets in the package as is currently the case.

9.2.2 Use of edible films as intermediate barriers

Coating of food with edible materials to preserve its quality and extend shelf life has been in practice for centuries [38]. The most common practice is coating of fruits and vegetables with wax to retard respiration, thus slowing down dehydration and senescence. Hot-melt paraffin waxes and carnauba oil-in-water emulsions have been used effectively with citrus fruits, apples, tomatoes, cucumbers, egg plants, etc. Preservation of meats and other products has also been attempted by coating with gelatin films.

Interest in edible films has intensified over the past thirty years owing to the increasing popularity of fresh, near-fresh, frozen and fabricated foods. The most important function of an edible film is resistance to migration of moisture. Multicomponent foods may become spoiled or organoleptically unacceptable if a critical a_w is not maintained during storage [43]. For example, pie crust may become soggy and gummy if there is excessive moisture transmission from the filling [44]. The level of a_w is also important for microbial, chemical and enzymatic activities [24, 58, 67]. Other important functions of edible films are gas- and solute-barrier properties. One of the most desirable benefits of coating foods with edible films is the fact that they may be used as vehicles for incorporating additives such as antioxidants, antimicrobial agents, colours, nutrients, etc. [38].

There are three major types of edible films, viz. polysaccharide, protein and lipid [38]. In addition, composite films made from mixtures of any or all of the three basic films are also possible. Polysaccharide films may be made from hydrocolloids such as alginate, pectin, carrageenan, starch, dextrins and cellulose derivatives such as methylcellulose, carboxymethylcellulose, hydroxypropyl methylcellulose or hydroxypropylcellulose. The films may be formed by casting aqueous solutions of the materials on the surface of foods, followed by solvent evaporation. Alginate and pectin films may be strengthened and barrier characteristics modified by cross-linking them with divalent cations such as Mg^{2+} or Ca^{2+}. Protein films may be made from polypeptides such as collagen, gelatin, casein, serum albumin, ovalbumin, wheat gluten and zein, also by solvent evaporation or cross-linking with ionized Ca. Zein, the alcohol-soluble protein from corn gluten, has been reported to form relatively good water-resistance films [27]. Lipid films may be made from acetylated monoglycerides, natural waxes and surfactants. These materials may be applied to the surface of food in either molten form, solution in appropriate solvent or oil-in-water emulsion. Generally, lipid films are good moisture barriers.

Application of edible films as intermediate barriers would complement MAP in preservation of quality and extending shelf life of foods. Fresh meat, fish, fruits and vegetables could all benefit from coating with films such as alginate, gelatin or acetylated monoglyceride. Drip loss in meat and fish would be greatly reduced, while microbial growth would be retarded if appropriate preservatives were added to the films. Fruits and vegetables coated with lipid films would be able to maintain freshness longer as there would be less moisture loss to headspace atmosphere. Fungicides added to the films would reduce mould and yeast problems common to fruits and vegetables. However, the coating must not be so impervious as to induce anaerobic respiration causing physiological disorders. The barrier characteristics of the intermediate films must be taken into account in determining the optimum gas atmosphere for these products.

Bakery products may also benefit from coating with edible films prior to MAP. For example, polysaccharide or protein films may be placed between layers of products such as layer cakes or cream cheese cakes with fruit toppings to prevent excessive migration of moisture and solutes between layers. Crispiness or flakiness of fruit pies may also be preserved if a layer of edible film is formed between the crust and the filling. In addition, mould and yeast problems would be reduced if appropriate additives were incorporated into these films.

With an edible film acting as the first barrier, freshness of the product is likely better preserved and surface microbial growth retarded by preservatives added to the film. Consequently, spoilage of the product due to an incorrect gas mixture, or changes in the headspace atmosphere or failure of the barrier films should be reduced.

9.3 Headspace atmospheres

9.3.1 Gas combinations and modifications

Effects of various gases on microbial growth, physicochemical and organoleptic properties of food products have not been exhaustively studied. Traditional gases used in MAP have been CO_2, N_2 and O_2. Other gases such as CO, SO_2, ethylene oxide and propylene oxide may be equally or more effective in controlling microbial growth in some products. CO is effective in reducing ageing processes and inhibiting some yeast and mould growth on some fruits and vegetables, possibly through its ability to interact with cytochromes, thereby blocking oxidative decay processes [70]. At 5-10% CO, with less than 5% O_2, the gas is an effective fungistat which can be used on commodities that do not tolerate high CO_2 levels [36]. CO combines with myoglobin to form a bright red pigment, carboxymyoglobin, which has a very similar colour to oxymyoglobin. Therefore, addition of CO at low concentrations together with CO_2 will maintain the desirable colour on MAP red meat [70]. Kramer *et al.* [42] found CO effective in inhibiting enzymatic browning in cut apples and sliced potatoes. However, possible health hazards to the operators may prevent this gas from being used in MAP. Kramer *et al.* [42] also found SO_2 effective against enzymatic browning and microbial growth on cut fruits and vegetables. However, its corrosivity and residual acid-sulphurous odour and taste make it unsuitable to be used as a sole gas. Ooraikul (unpublished results) used SO_2 to effectively control microbial growth on waffles and extend their ambient shelf life to more than four weeks. However, owing to its extremely high water solubility the gas becomes totally absorbed by the product in a short period of time, drawing the packaging film tightly around the product as though it was vacuum packaged. Nevertheless, SO_2 may be used as a purging gas to sanitize the products and reduce enzymatic activities on produce prior to the introduction of other gases. It may also be used at relatively low concentrations to increase effectiveness of other gases in MAP.

As mentioned earlier, ethylene and propylene oxides are normally used as gaseous sterilants for some products. Kramer *et al.* [42] used ethylene oxide in combination with CO and CO_2 to prolong shelf life of diced potatoes and sliced apples. The products were microbiologically stable and retained satisfactory appearance and odour at ambient temperatures for up to 11 months. However, they found that ethylene oxide at concentrations of 1×10^{-3} M or higher in potatoes showed mutagenic effects when the product was fed to lung cells. Propylene oxide is one-third as toxic as ethylene oxide and its reaction product, propylene glycol, is generally recognized as safe. It may be used to replace ethylene oxide without causing any hazard to health. In any case, these gases may be used for purging or in combination with other harmless gases for MAP.

The use of atmosphere modifiers such as O_2 scavengers, CO_2 absorbers or generators, or ethanol generators has been discussed in Chapters 3 and 4. More research should be done in this area. Absorbers, scavengers and generators for various gases which are more acceptable to consumers, cheaper and more easily applied should be developed, and some should have the capability of being automatically activated whenever needed.

9.3.2 Modelling of MAP regimes for produce

Respiring produce poses a unique challenge in application of MAP. The headspace atmosphere of the produce is continually modified through O_2 uptake and CO_2 evolution, and at some critical O_2 concentration respiration becomes anaerobic. This results in increased production of CO_2 and accumulation of ethanol, acetaldehyde and other compounds which are toxic to plant cells and cause development of off-flavours. Therefore, balancing of CO_2 and O_2 levels in the headspace of MAP produce is essential to avoid anaerobiosis while providing extension of shelf life. This has been variously achieved through selection of specific film permeability and/or use of gas scavengers, absorbers or generators. To provide a more reliable shelf life and product quality, mathematical modelling of MAP regimes for various types of produce must be developed based on the knowledge of the interaction between respiration and package permeability. The model should be able to predict the equilibrium conditions inside the package and how long it will take to achieve that equilibrium, i.e. when the gas flux due to respiration equals the gas flux due to permeation. Several models have been attempted but none have been comprehensive and general enough to take all the salient variables into account [36, 72].

An efficient model should take into account the effects of changing O_2 and CO_2 concentrations on respiration, the possibility of the respiration quotient being other than 1, the permeability of the film to O_2 and CO_2, the effect of temperature on permeability, the surface area and headspace of the package, the resistance of the produce to diffusion of various gases, and the optimum atmosphere for the extension of shelf life and retention of quality of produce. If gas scavengers, absorbers or generators are used, their effect should also be taken into account. The development of such a model would require a comprehensive database on respiration characteristics, permeability of the film at various temperatures, diffusivity of the produce toward various gases, rates of gas absorption and generation by the absorbers, generators, etc. This information is generally still lacking at the present time. However, if a general model can be developed for all produce, MAP could be a viable method for the shelf-life extension of fresh fruits and vegetables.

The principles of modelling are generally based on the exchange of three major gases, O_2, CO_2 and ethylene, between the commodities and the headspace and between the headspace and the atmosphere outside the package. CO_2 and ethylene are products of respiration and biosynthesis which are controlled by such factors as the type of produce, O_2 and CO_2 levels in the headspace and storage temperature. These gases have to diffuse through the tissue of the commodity, in which they are produced, to the headspace, while O_2 in the headspace is being consumed by the tissue. At the same time the three gases permeate in or out of the package at rates which are controlled by such factors as the film permeability, surface area and thickness, pressure or concentration gradient across the film, temperature and humidity. There are certain concentrations of these gases in the headspace for each commodity where its respiration and biosynthesis rates are minimized, thus delaying the ageing process, without causing injuries to the produce. This atmospheric condition may be created actively by providing the desired gas combination in the headspace during packaging, or passively by packaging the commodity in air and allowing the gas exchange to take place until steady state is reached and, depending on film permeability, the desired headspace atmosphere is achieved [36]. In modelling the packaging system for a commodity the desired headspace gas combination must be known so that the length of time to reach steady state and/or a suitable type of film to be used that would allow the maintenance of the headspace gas combination can be determined. Fick's Law of Diffusion may be used to predict the rates at which the gases diffuse through the commodities or permeate through the packaging films. In a simplified form the Law may be written for gas diffusion through the commodity as follows [36]:

$$J_{gas} = \frac{A.\Delta C_{gas}}{R}$$

where J = total flux of gas ($cm^3.sec^{-1}$); A = surface area of the barrier (cm^2); ΔC = concentration gradient across the barrier; and R = resistance to diffusion of gas ($sec.cm^{-1}$)

Using the steady-state approach the resistance, R, of plant tissues to diffusion of the gases may be estimated based on the production or consumption rates of the gases by the tissue and the concentration gradient of the gases across the tissue. Thus:

$$R = \frac{concentration\ gradient}{production\ (consumption)\ rate}$$

For permeation of the gases through the packaging film, a slightly modified version of the Law may be written as follows:

$$J_{gas} = \frac{KA}{x}(P_a - P)$$

where K = film permeability to gas $(cm^3.ml/sec.m^2.atm)$; P_a = atmospheric partial pressure of gas (atm); P = partial pressure of gas inside the package (atm); and x = film thickness (mil)

Difficulties in modelling result from problems experienced in defining the factors involved in the above equations in such a way that all the variables listed above are taken into account [72] so that the predicted and observed results are not significantly different. One of a few efficient models to date was developed for peaches by Deily and Rizvi [21]. They calculated the film permeabilities necessary to achieve a predetermined headspace atmosphere based on peach respiration values. The respiration rates and headspace atmospheres were then verified experimentally while the quality attributes of the MAP and control peaches were also evaluated during storage. They concluded that peaches respond well to MAP. A great deal more research is required for the whole range of fruits and vegetables.

9.4 Effects of MAP on organoleptic quality of products

Much research has been done on the effects of vacuum packaging and MAP on organoleptic quality of the products, most of which has been on meat and meat products. With red meats one of the most important sensory qualities is colour. Hood and Riordan [32] reported that the ratio of sales of discoloured beef to bright red beef was 1:2 when 20% metmyoglobin was present in the former. However, Lynch *et al.* [50] found that if properly informed the consumers are as likely to accept the purple red vacuum-packaged ground beef as the bright cherry red product.

Other important sensory attributes in meat and meat products are flavour and texture. Seideman *et al.* [59] stored roast beef packaged in combinations of CO_2, O_2 and N_2 at 1-3°C for up to 35 d and found that increasing O_2 concentration resulted in an increase in off-odour, surface discolouration and a reduction in overall appearance and palatability. Similar results were obtained with turkey breast packaged in a similar manner [52]. The sensory quality of CO_2-packaged fresh poultry stored at 4°C remained acceptable for 8-16 d compared with 4-5 d for the control, and the usual pale colour was restored when the package was opened [10]. Hall *et al.* [28] reported no significant difference in sensory scores between pork loins packaged in 20% CO_2 and 80% N_2, 40% CO_2 and 60% N_2 or vacuum packaged and stored at 1-3°C for up to 28 d, though MAP appeared more advantageous for product appearance. Pork packaged under CO_2 and stored at -1.5°C was little affected in eating qualities for about 18 weeks, but prolonged storage resulted in loss of colour due to loss of myoglobin in the exudate [26]. Lee *et al.* [46], on the other hand, reported that vacuum- or N_2-packaged pork could not be kept at >0°C for more than 14 d without obvious microbial, physicochemical and sensory changes. Lee *et al.* [45] reported that an N_2 atmosphere lowered the incidence of greening and exudate loss in veal chucks stored at 0-7°C. Simard *et al.* [60] obtained an improved shelf life of beef by packaging in an N_2 atmosphere. This retarded discoloration and minimized exudate losses. Sliced, cooked roast beef packaged in 75% CO_2, 15% N_2 and 10% O_2 and stored at 4.4°C deteriorated within 7 d [31]. Korkeala *et al.* [41] found lactobacilli at 10^7 CFU/g or more to be the main microorganisms responsible for spoilage of vacuum-packaged, cooked, ring sausages stored at 8°C. The product was judged acceptable up to 29 d at 8°C, 43 d at 4°C and 55 d at 2°C [40]. McMullen and Stiles [51] reported that commercially produced processed meat and roast beef sandwiches packaged in 50% CO_2 and 50% air and stored at 4°C were judged acceptable up to 35 d. They found that cooked hamburgers, similarly packaged and stored, were acceptable up to 14 d, and up to 35 d if packaged without O_2.

Not as much work has been reported on organoleptic quality of other MAP products. On fish, fresh trout packaged in a CO_2-enriched atmosphere was judged only marginally acceptable after 25 d storage, but cooked trout similarly packaged and stored was acceptable [7]. Scallops packaged in 100% CO_2 and stored at 4°C had a shelf life of 22 d, i.e. 12 d longer than in air, based on texture, flavour and overall acceptability scores [12]. Addition of potassium sorbate and polyphosphate to morwong fillets and packaging them in 100% CO_2 was more effective in maintaining microbial and sensory qualities of the product compared with vacuum-packaged or frozen samples [65]. On fruits and vegetables, Ballantyne [5] reported shelf-life extension, based on colour change and development of off-flavours and odours, of Brussels sprouts, shredded lettuce and whole and sliced mushrooms packaged in polymer films which allowed equilibrium gas concentrations of 2-4% O_2 and 3-6% CO_2 while stored at 5°C. For broccoli, similarly packaged and stored at 10°C, favourable shelf life extension was achieved. Ballantyne *et al.* [6] also reported that shredded lettuce packaged in 35-μm low-density polyethylene (LDPE) with the equilibrium atmosphere of 1-3% O_2 and 5-6% CO_2, whether achieved naturally or through gas flushing, could be kept for 14 d at 5°C, which almost doubled the shelf life of controls. Changes in firmness and skin colour of Bramleys and Cox's apples could be retarded for 2-4 weeks with MA retail packaging in 30–μm LDPE and stored at 15°C [25]. However, MAP was more effective in the immediately pre-climacteric and early climacteric apples than in those harvested 4 d before the onset of the climacteric or late-picked apples with much higher rates of respiration [62]. Potter [54] reported proprietary studies which revealed that MAP peaches retained a fresh texture and flavour for one year and sliced apples for 5 months. Unfortunately, the headspace atmospheres and storage temperatures used for this study were not given in the report.

The effects of MAP on organoleptic quality of bakery products have not been extensively studied. While the technology has been quite widely used with meats and produce, the bakery industry has been slow in accepting MAP [48]. This may be due to the fact that the traditional bakery industry consists largely of small, regionally based bakers with limited production and distribution. Therefore, the need for shelf-life extension is insufficient to justify the high costs

of an MAP operation. Nevertheless, with improved efficiency and reduced costs the technology should be attractive to medium and large
bakery operations, especially for high-value specialty products. This
would accelerate research on various aspects of MAP with respect to
bakery products.

Sensory studies of bakery goods can be quite complex. This is
because there is an extremely wide range of products with varying
composition, colours, flavours, textures, etc. Simple products such as
crumpets, bread, rolls and waffles are quite similar in their chemical
and physical properties and, therefore, they share similar sensory
problems, which are related mainly to "staling". On the other hand,
complex products such as pies, cakes, especially the multicomponent, multi-layer-type products, e.g. fruit pies, layer cakes,
fruit-filled dairy-based cakes, etc., have a wide variety of colour,
flavour, aroma, texture and taste, even within one product. There
may be a migration of components from one part of the product to
the other during storage. For example, moisture migrates from the
moist fruit filling to the drier crust in apple pies, making it soft and
less flaky; or red colour from cherry topping diffuses into the white
layers of cherry cream cheese cake. These make organoleptic study
of these products much more complicated.

Commercial MAP crumpets, with a possible ambient shelf life
of 3 months [48], become increasingly crumbly after a few weeks.
The colour also fades slightly after extended storage. The textural
change in crumpets is not unlike that in the staling of bread where
the crumb becomes firm, losing its elasticity and juiciness, and it
crumbles easily. The staling of bread is a phenomenon of starch ret-
rogradation, which involves the increased realignment of starch
molecules from an amorphous to a semi-crystalline form [8].
However, unlike bread texture which has protein as a continuous
phase interspersed with a starch gel matrix, crumpet structure is
devoid of a continuous protein network but is made up of tightly
packed deformed starch granules with protein masses dispersed
throughout [69]. This difference in structural make-up makes the
crumbliness problem more severe in crumpets than in bread, especially when stored at low temperatures. Fortunately, if physical
damage to the product is avoided during storage and shipment,
crumpets regain their original textural quality almost totally when
they are toasted, which is the usual preparation method for serving.
As such, organoleptic changes have not been a major problem in
commercial MAP crumpets.

Brümmer [13] purged sliced bread with a mixture of 90% CO_2
and 10% N_2 in pillow-type pouches and stored it for 14 d. He found
the crumb structure, chewing quality, flavour and taste of the
product to be superior to those similarly stored but packaged in deep-
drawn trays where the packages were first evacuated before backflushing with the same gas mixture. He attributed the inferior
quality of the latter to the vacuum treatment through which some
moisture and flavours were lost making the product taste dry and
stale. Apple pies purged with the same gas mixture deteriorated in

aroma, taste and appearance after two weeks [13]. Yeast or mould activity in the fruit layer of the pie was suspected to be the cause. These problems were not experienced when preservatives were used. The same researcher reported changes in the texture and flavour of MAP cheese cake after 5-6 d and of Danish pastry after 8 d.

On the other hand, Doerry [22] reported that neither CO_2 nor N_2 had any effect on the rate of staling of sliced bread, pound cake or sponge cake after 7 d storage at 25°C. Bread stored in N_2 developed off-odours faster than in air, but when stored in CO_2 it developed less off-odour and after a longer time. Stale taste appeared to be accentuated by CO_2 and N_2 atmospheres, and this was attributed to anaerobic microbial activity in bread. Pound and sponge cakes which were microbiologically stable in all three atmospheres showed no appreciable difference in their moisture content or flavour during the 7 d storage period.

Obviously, different products react differently to MAP and their storage conditions. Research on organoleptic properties of MAP products must be vigorously pursued to ensure acceptability of the products at the end of the extended storage period. It is possible that fresh meats will be more tender and juicier with MAP and judicious use of food additives. Crispness and colour of fresh fruits and vegetables may be better preserved with different headspace atmospheres or storage temperatures. Textural integrity of various components or layers of some bakery products may be preserved by modifying the recipes or processing methods, or by the application of edible films or coatings on components or layers susceptible to changes. Sensory evaluation must be used in conjunction with microbial analysis to determine appropriate pull dates for MAP products to ensure their safety and organoleptic acceptability.

9.5 Quality assurance for MAP operations

The processing operations of MAP require precision in the control of the many factors involved. The safety and quality of the products depend on these factors. Therefore, quality assurance is an extremely important part of the operations. Quality assurance integrates all aspects of the product and its production. It takes into consideration product composition, specifications, processing, packaging, storage, distribution, microbiological safety, equipment sanitation and pest and rodent control. It is a corporate program, the failure of which may seriously influence the economic well-being of the corporation, its competitiveness and even its survival.

Safety and quality of MAP products depend on initial product characteristics which include composition, pH, a_w, microbial quality and organoleptic properties. They further depend on the packaging operations, the success of which is influenced by properties of the packaging material, effectiveness in providing the desired headspace atmosphere and package integrity. They also depend on conditions of storage, handling and distribution of the products, particularly temperature control during storage and distribution.

The quality assurance program must ensure that everything is done right the first time.

Quality assurance or quality control programs in MAP operations require research on identification and removal or modification of steps or factors in the operations that may cause product failure. "Hazard Analysis Critical Control Point" (HACCP) is a system that could meet these demands. The system was introduced in 1971 at the National Conference on Food Protection and it was used by the Pillsbury Company for the U.S. Space Program in 1972 [19]. It was designed specifically for the control of microbiological hazards in foods. However, with some modifications it may also be applied to the control of other quality parameters such as nutritional values or sensory qualities.

Though HACCP has not been widely used for perishable foods its principles lend themselves well to quality assurance in MAP programs. Application of the HACCP to a food processing operation should involve the following steps [19]:

1. Describe the product and how it will be used by the consumer.
2. Prepare a manufacturing, storage and distribution flow diagram of the product.
3. Conduct risk analysis for ingredients, product, packaging and storage; reduce the risks by making changes to the design; and incorporate these changes into the processing, packaging and storage schemes.
4. Select critical control points, designate their locations on the flow diagram, describe them in detail and establish monitoring procedures for these points.
5. Implement HACCP in routine activities.

Based on Corlett [19], integration of HACCP into the design of safe MAP products should include: a. identification of risk factors in the initial product design; b. formulation of the product to reduce risks by employing back-up microbiological barriers such as low pH, low a_w, competitive microbial flora, preservatives, thermal processing and partial reliance on MA; c. selection of appropriate packaging to produce, store and distribute safe products; d. education of consumers. The last point is particularly important with MAP products because they are the "new generation" products which are prone to consumer abuse, especially with respect to handling and storage.

Research should involve developing effective ways in which HACCP can be integrated into the MAP operations, by developing monitoring and analytical procedures for every critical control point. Appropriate statistical methods should be used, for example in establishing sampling plans, in process control and in evaluation and interpretation of the results of the analyses. Suitable equipment or instruments should be developed for monitoring and control of important parameters such as the permeability and integrity of packaging films under various atmospheric conditions, accuracy of gas mixture, headspace gas atmosphere, microbial activity and storage temperature and humidity.

Some instruments have been developed and are in commercial use, or they are being tested for use, for example, time-temperature indicators and oxygen detection tablets. The I-POINT time-temperature monitor (I-POINT, Malmo, Sweden) has been used successfully to monitor temperature fluctuation during the storage of frozen hamburger [61]. The indicator is based on enzymatic hydrolysis of the lipid substrate, which causes a change in pH of the solution. The pH change is monitored by a pH indicator, causing a gradual colour change. This change is irreversible and it is accelerated by an increase in temperature. Some time-temperature indicators have already been used with MAP products. For example General Foods Culinova products have been labelled with a scannable time-temperature indicator on the bottom of each package [49]. The indicators on all packages are checked each time a delivery is made to determine whether they have been subjected to an unacceptably high storage temperature. Packages that fail the test are rejected. "Freshness indicators" which can be "read" by consumers at home are also available. As "pull date" is normally determined under a reasonable storage and handling condition, the use of freshness indicators together with the date-coding will further reduce the danger caused by possible consumer abuse of the product.

A colour-changing O_2 detection tablet is being tested for use with MAP and vacuum-packaged products [49]. The tablet is pink in colour when the O_2 concentration inside the package is 0.1% or less and it turns blue at O_2 concentrations of 0.5% or higher. This O_2 detection tablet would be very valuable for products with mould problems which require an O_2 level <0.5%.

More research into various aspects of quality assurance of MAP foods is needed. Beside the aspects of safety and fresh quality, not much has been done to determine the effect of various MAP regimes on the nutritional value of the products. Residual activity of some anaerobes, for example, may cause changes in certain nutrients in the product, or possible reaction between CO_2 or other gases in the headspace and some components of the food may also alter its nutritional properties.

A perfect MAP product with absolute safety and natural freshness which retains all of the original nutrients and organoleptic qualities may never be possible, but with continuous improvement through research and development, MAP technology can become one of the most successful innovations in food preservation of this century.

References

1. Aboagye, N.Y., Ooraikul, B., Lawrence, R. and Jackson, E.D. 1986. Energy costs in modified atmosphere packaging and freezing processes as applied to a baked product. Proceedings of the Fourth International Congress of Engineering and Food, Edmonton, AB. Food Engineering and Process Applications Vol. 2 Unit Operations. Ed.: Le Maguer, M and Jelen, P. Elsevier Appl. Sci. Publishers, London. pp. 417-425.
2. Ahn, C. and Stiles, M.E. 1990. Antibacterial activity of lactic acid bacteria isolated from vacuum-packaged meats. J. Appl. Bacteriol. 69: 302-310.
3. Ashton, R. 1989. What was new at Pack Expo. Packaging 34(2): 86-89.
4. Bailey, J.S., Reagan, J.O., Carpenter, J.A., Schuler, G.A. and Thomson, J.E. 1979. Types of bacteria and shelf-life of evacuated carbon dioxide-injected and ice-packed broilers. J. Food Prot. 42: 218-221.
5. Ballantyne, A. 1987. Modified atmosphere packaging of selected prepared vegetables. Technical Memorandum, Campden Food Preservation Research Association No. 464. 68 pp.
6. Ballantyne, A., Stark, R. and Selman, J.D. 1988. Modified atmosphere packaging of shredded lettuce. Int. J. Food Sci. Technol. 23: 267-274.
7. Barnett, H.J., Conrad, J.W. and Nelson, R.W. 1987. Use of laminated high and low density polyethylene flexible packaging to store trout (*Salmo gairdneri*) in a modified atmosphere. J. Food Prot. 50: 645-651.
8. Belitz, H.-D. and Grosch, W. 1986. Food Chemistry. Translated by D. Hadziyev. Springer Verlag, Berlin.
9. Beuchat, L.R. and Golden, D.A. 1989. Antimicrobials occurring naturally in foods. Food Technol. 43(1): 134-142.
10. Bohnsack, U., Knippel, G. and Höpke, H.U. 1987. Effects of CO_2 atmosphere on keeping quality of fresh poultry. Fleischwirtschaft 67: 1131-1136.
11. Box, G.E.P., Hunter, W.G. and Hunter, J.S. 1978. Statistics for Experimenters. An Introduction to Design Data Analysis & Model Building. John Wiley and Sons, New York.
12. Bremner, H.A. and Statham, J.A. 1987. Packaging in CO_2 extends shelf-life of scallops. Food Technol. Aust. 39: 177-179.
13. Brümmer, J.-M. 1986. West German experience with controlled carbon dioxide packaging for bakery products. CAP '86. Proceedings of the Second International Conference and Exhibition on Controlled Atmosphere Packaging. Schotland Business Research, Inc., Princeton, NJ., pp. 359-380.
14. Buchanan, R.L., Harry, M.A. and Gealt, M.A. 1983. Caffeine inhibition of sterigmatocystin, citrinin and patulin production. J. Food Sci. 48: 1226-1228.
15. Buchanan, R.L., Tice, G. and Marino, D. 1981. Caffeine inhibition of ochratoxin A production. J. Food Sci. 47: 319-321.
16. Chichester, D.F. and Tanner, F.W. 1975. Antimicrobial food additives. *In* Handbook of Food Additives, Second Edition, Furia, T.E. (ed.), pp. 115-184. CRC Press, Cleveland, OH.

17. Christopher, F.M., Seideman, S.C., Carpenter, Z.L., Smith, G.C. and Vanderzant, C. 1979. Microbiology of beef packaged in various gas atmospheres. J. Food Prot. 42: 240-244.

18. Christopher, F.M., Smith, G.C., Dill, C.W., Carpenter, Z.L. and Vanderzant, C. 1980. Effect of CO_2-N_2 atmospheres on the microbial flora of pork. J. Food Prot. 43: 268-271.

19. Corlett, D.A. Jr. 1989. Refrigerated foods and use of Hazard Analysis and Critical Control Point principles. Food Technol. 43(2): 91-94.

20. Daeschel, M.A. 1989. Antimicrobial substances from lactic acid bacteria for use as food preservatives. Food Technol. 43(1): 164-167.

21. Deily, K.R. and Rizvi, S.S. 1981. Optimization of parameters for packaging of fresh peaches in polymeric films. J. Food Proc. Eng. 5: 23-41.

22. Doerry, W.T. 1987. Packaging bakery foods in controlled atmospheres. CAP '87. Proceedings of the Third International Conference on Controlled/Modified Atmosphere/Vacuum Packaging. Schotland Business Research, Inc. Princeton, NJ., pp. 203-233.

23. Dziezak, J.D. 1986. Preservatives: antimicrobial agents, a means toward product stability. Food Technol. 40(9): 104-111.

24. Eichner, K. 1986. The influence of water content and water activity on chemical changes in foods of low-moisture content under packaging aspects. *In* Food Packaging and Preservation: Theory and Practice, Mathlouthi, M. (ed.), p. 93, Elsevier Applied Science Publishers, London, England.

25. Geeson, J.D., Smith, S.M., Everson, H.P., Genge, P.M. and Browne, K.M. 1987. Responses of CA-stored Bramley's seedling and Cox's Orange Pippin apples to modified atmosphere retail packaging. Int. J. Food Sci. Technol. 22: 659-668.

26. Gill, C.O. and Harrison, J.C.L. 1989. The storage life of chilled pork packaged under carbon dioxide. Meat Sci. 26: 313-324.

27. Guilbert, S. 1986. Technology and application of edible protective films. *In* Food Packaging and Preservation: Theory and Practice, Mathlouthi, M. (ed.), p. 371, Elsevier Applied Science Publishers, London, England.

28. Hall, L.C., Smith, G.C., Dill, C.W., Carpenter, Z.L. and Vanderzant, C. 1980. Physical and sensory characteristics of pork loins stored in vacuum or modified atmosphere packages. J. Food Prot. 43: 272-276.

29. Henika, R.G. 1982. Use of Response Surface Methodology in sensory evaluation. Food Technol. 36(11): 96-101.

30. Hintlian, C.B. and Hotchkiss, J.H. 1986. The safety of modified atmosphere packaging: A review. Food Technol. 40(12): 70-76.

31. Hintlian, C.B. and Hotchkiss, J.H. 1987. Microbiological and sensory evaluation of cooked roast beef packaged in a modified atmosphere. J. Food Proc. Preserv. 11: 171-179.

32. Hood, D.E. and Riordan, E.B. 1973. Discoloration in pre-packaged beef: measurement by reflectance spectrophotometry and shopper discrimination. J. Food Technol. 8: 333-343.

33. Hoover, D.G., Walsh, P.M., Kolaetis, K.M. and Daly, M.M. 1988. A bacteriocin produced by *Pediococcus* species associated with a 5.5-megadalton plasmid. J. Food Prot. 51: 29-31.
34. Hughey, V.L. and Johnson, E.A. 1987. Antimicrobial activity of lysozyme against bacteria involved in food spoilage and food-borne disease. Appl. Environ. Microbiol. 53: 2165-2170.
35. Kabara, J.J. 1983. Medium-chain fatty acids and esters. *In* Antimicrobials in Foods, Branen, A.L. and Davidson, P.M. (eds.), p. 109. Marcel Dekker, Inc. New York.
36. Kader, A.A., Zagory, D. and Kerbel, E.L. 1989. Modified atmosphere packaging of fruits and vegetables. CRC Crit. Rev. Food Sci. Nutr. 28: 1-30.
37. Kempton, A.G. and Bobier, S.R. 1970. Bacterial growth in refrigerated vacuum-packed luncheon meats. Can J. Microbiol. 16: 287-297.
38. Kester, J.J. and Fennema, O.R. 1986. Edible films and coatings: a review. Food Technol. 40(12): 47-59.
39. Klaenhammer, T.R. 1988. Bacteriocins of lactic acid bacteria. Biochimie 70: 337-349.
40. Korkeala, H., Alanko, T., Mäkelä, P. and Lindroth, S. 1989. Shelf-life of vacuum-packed cooked ring sausages at different chill temperatures. Int. J. Food Microbiol. 9: 237-247.
41. Korkeala, H., Lindroth, S., Ahvenainen, R. and Alanko, T. 1987. Interrelationship between microbial numbers and other parameters in the spoilage of vacuum-packed cooked ring sausages. Int. J. Food Microbiol. 5: 311-321.
42. Kramer, A., Solomos, T. Wheaton, F., Puri, A., Sirivichaya, S., Lotem, Y., Fowke, M. and Ehrman, L. 1980. A gas-exchange process for extending the shelf life of raw foods. Food Technol. 34(7): 65-74.
43. Labuza, T.P. 1982. Moisture gain and loss in packaged foods. Food Technol. 36(4): 92-97.
44. Labuza, T.P. 1984. Moisture Sorption: Practical Aspects of Isotherm Measurement and Use. American Association of Cereal Chemists, St. Paul, MN.
45. Lee, B.H., Simard, R.E., Laleye, L.C. and Holley, R.A. 1983. Microflora, sensory and exudate changes of vacuum- or nitrogen-packed veal chucks under different storage conditions. J. Food Sci. 48: 1537-1542, 1563.
46. Lee, B.H., Simard, R.E., Laleye, L.C. and Holley, R.A. 1985. Effects of temperature and storage duration on the microflora, physicochemical and sensory changes of vacuum- or nitrogen-packed pork. Meat Sci. 13: 99-112.
47. Lenovich, L.M. 1981. Effect of caffeine on aflatoxin production on cocoa beans. J. Food Sci. 46: 655, 657.
48. Lingle, R. 1988. CAP for U.S. bakery products: To be or not to be? Prepared Foods 157(3): 91-93.
49. Lingle, R. 1990. Safeguarding product integrity. Prepared Foods. 159(3): 162-164.
50. Lynch, N.M., Kastner, C.L. and Kropf, D.H. 1986. Consumer acceptance of vacuum packaged ground beef as influenced by product color and educational materials. J. Food Sci. 51: 253-255, 272.

51. McMullen, L. and Stiles, M.E. 1989. Storage life of selected meat sandwiches at 4°C in modified gas atmospheres. J. Food Prot. 52: 792-798.
52. Mead, G.C., Griffiths, N.M., Jones, J.M., Grey, T.C. and Adams, B.W. 1983. Effect of gas-packaging on the keeping quality of turkey breast fillets stored at 1°C. Lebensm. Wíss. u. Technol. 16: 142-146.
53. Notermans, S., Dufrenne, J. and Keybets, M.J.H. 1985. Use of preservatives to delay toxin formation by *Clostridium botulinum* (type B, strain Okra) in vacuum-packed, cooked potatoes. J. Food Prot. 48: 851-855.
54. Potter, C. 1986. UBC team boost fresh product shelf life. Food in Canada 46(8): 34-36.
55. Powers, J.J., Somaatmadja, D., Pratt, D.E. and Hamdy, M.K. 1960. Anthocyanins. II. Action of anthocyanin pigments and related compounds on the growth of certain microorganisms. Food Technol. 14(12): 626-632.
56. Roberts, T.A. 1989. Combinations of antimicrobials and processing methods. Food Technol. 43(1): 156-163.
57. Schoebitz, R.P. 1988. Production of inhibitory substances by lactic acid bacteria in ground beef. M.Sc. thesis, University of Alberta, Edmonton, AB.
58. Schwimmer, S. 1980. Influence of water activity on enzyme reactivity and stability. Food Technol. 34(5): 64-74, 82.
59. Seideman, S.C., Carpenter, Z.L., Smith, G.C., Dill, C.W. and Vanderzant, C. 1979. Physical and sensory characteristics of beef packaged in modified gas atmospheres. J. Food Prot. 42: 233-239.
60. Simard, R.E., Lee, B.H., Laleye, C.L. and Holley, R. 1985. Effects of temperature and storage time on the microflora, sensory and exudate changes of vacuum- or nitrogen-packed beef. Can. Inst. Food Sci. Technol. J. 18: 126-132.
61. Singh, R.P. and Wells, J.H. 1985. Use of time-temperature indicators to monitor quality of frozen hamburger. Food Technol. 39(12): 42-50.
62. Smith, S.M., Geeson, J.D. and Genge, P.M. 1988. The effect of harvest date on the responses of Discovery apples to modified atmosphere retail packaging. Int. J. Food Sci. Technol. 23: 81-90.
63. Smith, J.P., Jackson, E.D. and Ooraikul, B. 1983. Microbiological studies on gas-packaged crumpets. J. Food Prot. 46: 279-284.
64. Smith, J.P., Khanizadeh, S., van de Voort, F.R., Hardin, R., Ooraikul, B. and Jackson, E.D. 1988. Use of response surface methodology in shelf life extension studies of a bakery product. Food Microbiol. 5: 163-176.
65. Statham, J.A., Bremner, H.A. and Quarmby, A.R. 1985. Storage of morwong (*Nemadactylus macropterus* Bloch and Schneider) in combinations of polyphosphate, potassium sorbate, and carbon dioxide at 4°C. J. Food Sci. 50: 1580-1584, 1587.
66. Taylor, S.L., Higley, N.A. and Bush, R.K. 1986. Sulfites in foods: uses, analytical methods, residues, fate, exposure assessment, metabolism, toxicity, and hypersensitivity. *In* Advances in Food Research, Vol. 30, pp. 1-76. Chichester, C.O., Mrak, E.M. and Schweigert, B.S. (eds.). Academic Press, Inc. Orlando, FL.
67. Troller, J.A. 1980. Influences of water activity on microorganisms in foods. Food Technol. 34(5): 76-80, 82-83.

68. Wagner, M.K. and Moberg, L.J. 1989. Present and future use of traditional antimicrobials. Food Technol. 43(1): 143-147, 155.
69. Weatherall, A.E. 1986. Studies on starch in crumpets. Ph.D. thesis. Department of Food Science, University of Alberta, Edmonton, AB, Canada.
70. Wolfe, S.K. 1980. Use of CO- and CO_2-enriched atmospheres for meat, fish, and produce. Food Technol. 34(3): 55-58, 63.
71. Young, H., Youngs, A. and Light, N. 1987. The effects of packaging on the growth of naturally occurring microflora in cooked, chilled foods used in the catering industry. Food Microbiol. 4: 317-327.
72. Zagory, D. and Kader, A.A. 1988. Modified atmosphere packaging of fresh produce. Food Technol. 42(9): 70-77.

INDEX